T0237172

Engineering Materials

This series provides topical information on innovative, structural and functional materials and composites with applications in optical, electrical, mechanical, civil, aeronautical, medical, bio- and nano-engineering. The individual volumes are complete, comprehensive monographs covering the structure, properties, manufacturing process and applications of these materials. This multidisciplinary series is devoted to professionals, students and all those interested in the latest developments in the Materials Science field.

More information about this series at http://www.springer.com/series/4288

Manuel Laso · Nieves Jimeno

Representation Surfaces for Physical Properties of Materials

A Visual Approach to Understanding Anisotropic Materials

 Springer

Manuel Laso
Universidad Politécnica de Madrid ETSI
Industriales
Madrid, Spain

Nieves Jimeno
Universidad Politécnica de Madrid ETSI
Industriales
Madrid, Spain

ISSN 1612-1317   ISSN 1868-1212   (electronic)
Engineering Materials
ISBN 978-3-030-40872-5   ISBN 978-3-030-40870-1   (eBook)
https://doi.org/10.1007/978-3-030-40870-1

© Springer Nature Switzerland AG 2020
This work is subject to copyright. All rights are reserved by the Publisher, whether the whole or part
of the material is concerned, specifically the rights of translation, reprinting, reuse of illustrations,
recitation, broadcasting, reproduction on microfilms or in any other physical way, and transmission
or information storage and retrieval, electronic adaptation, computer software, or by similar or dissimilar
methodology now known or hereafter developed.
The use of general descriptive names, registered names, trademarks, service marks, etc. in this
publication does not imply, even in the absence of a specific statement, that such names are exempt from
the relevant protective laws and regulations and therefore free for general use.
The publisher, the authors and the editors are safe to assume that the advice and information in this
book are believed to be true and accurate at the date of publication. Neither the publisher nor the
authors or the editors give a warranty, expressed or implied, with respect to the material contained
herein or for any errors or omissions that may have been made. The publisher remains neutral with regard
to jurisdictional claims in published maps and institutional affiliations.

This Springer imprint is published by the registered company Springer Nature Switzerland AG
The registered company address is: Gewerbestrasse 11, 6330 Cham, Switzerland

To Socorro, Blanca and Sol, in chronological order of appearance.

Preface

This book is about the visualization of anisotropy. Representation surfaces (RS from now on) are helpful in developing an intuitive understanding of the directional dependence of physical magnitudes and properties of materials. RSs are a graphical generalization to tensors of higher order of the familiar "arrow" used for vectors.

The book is meant to be a self-contained introduction to the topic of anisotropy. But the book is also somewhat more ambitious than its title suggests. We have taken RSs as a pretext to introduce the reader to a few essential concepts of crystallography and to Cartesian tensors.

The intended readership of the book includes advanced (7–8th semester) undergraduate students in materials science, mechanical and electrical engineering, inorganic or physical chemistry, and geology.

The text requires no special previous knowledge of either crystallography or Cartesian tensors. Two chapters devoted to them should help the novice get acquainted with all necessary background. Readers familiar with these subjects may freely skip the corresponding part, but even without previous knowledge of crystallography or Cartesian tensors the entire material of the book can be covered in a one-semester, 15-week course.

Assuming no background in crystallography or Cartesian tensors, approximately 2 and 3 weeks should be devoted to Chaps. 2 and 3, respectively. The concept of projection and the RS itself are introduced in Chap. 4 (approx. 2 weeks).

It is especially important for the student to acquire a fundamental understanding of and sufficient proficiency in the material of Chaps. 3 and 4 before moving on to the second half of the book.

The remaining five chapters are organized around properties of increasing rank and can easily be covered in 8 weeks, giving priority to RSs for second and fourth rank properties (2 weeks at least to each) because of their more numerous applications. In a pinch, the more specialized section on polycrystalline averages in Chap. 8 can be omitted.

Although specific properties or physical magnitudes have been used for illustration and to make ideas more specific, the concept of RS is a general one. For this reason, and in order to keep the length of the book suitable for a one-semester course, we have omitted detailed explanations of the physical phenomena responsible for structure-property relationship. We have likewise excluded from the scope of the book more specialized topics like magnetic groups, the use of the representation quadric in optics, visualization of non-homogeneous tensorial fields in medical imaging, etc.

All RSs appearing in the figures were drawn using ParaView. Numerical input to ParaView was generated with simple f90 code.

In many aspects of notation, crystallographic conventions, and naming of variables, we have closely followed Walter Borchard-Ott's, *Kristallographie: Eine Einführung für Naturwissenschaftler*, and *Dynamics of Polymeric Liquids* by Bird et al. Needless to say, all errors, omissions, unclear explanations, etc. are the sole responsibility of the authors.

We are deeply indebted to many individuals who directly or indirectly made a decisive contribution to the completion of this book. Chiefly among them, Profs. Hans Christian Öttinger, Ueli W. Suter (ETH Zürich), Marco Picasso and Michel Rappaz (EPF Lausanne), Kurt Kremer (Max-Planck Institut für Polymerforschung), Greg Rutledge (MIT), Patrik Schmuki (Friedrich-Alexander Universität Erlangen), Ignacio Romero, Katerina Foteinopoulou, Nikos Karayiannis, Juan Luis Prieto and David Portillo (UPM), Paco Chinesta (ENSAM ParisTech), and Elías Cueto (Universidad de Zaragoza), and Prof. Ravi Jagadeeshan. We would also like to thank a few generations of students who served as experimental audience. Last, but by no means least, to JF and Dr. BH, whose help was decisive at the start and at the end of the book.

We would like to express our huge debt of gratitude to the late Profs. John Nye (University of Bristol), Bob Newnham (Pennsylvania State University), Walter Borchardt-Ott (Westfälische Wilhelms-Universität Münster) for their authoritative and unfailing support.

Tres Cantos, Langkawi and Jávea   Manuel Laso
January 2020   Nieves Jimeno

Contents

Chapter 1
Introduction

A representation surface (RS from now on) is a simple graphical tool that helps understand how a physical magnitude depends on direction. Without worrying for the time being about the meaning of the two objects shown in the figure below, a RS conveys the information contained in the diagram on the left by means of the shaded surface on the right hand side. Most people find it easy to determine what the symmetries of the RS are, or in which directions it is elongated or shortened.

Although the diagram on the left below these lines is the basis of the RS on the right, the former lends itself rather less well than the RS to visual understanding (Fig. 1.1).

Representation surfaces have been around for quite a long time. Early use of RSs goes back to the visualization of complex fluid flow fields [1, 2], where the denominations "Reynolds glyph" and "polar plot" were often employed for representation surfaces.

RSs belong to a group of visualization techniques that find widespread use in graphic analysis of complex data fields, together with Lamé ellipsoids, Dupin indicatrix, Haber glyphs, HWY tensor glyphs, implicit ellipse, exponentially scaled ellipse glyphs, superquadrics, tensor splats, hyper-streamline tubes and a long etc. [3–9].

A very large body of literature on visualization methods for tensor fields has been published over the last two decades. The comprehensive works by Weickert and Hagen [10] and Westin et al. [11] are excellent overviews of the field.

Most of these references are devoted to visualization of inhomogeneous fields, chiefly second rank tensor fields. Emphasis is put on the suitability of large sets of a given glyph type for the extraction of overall features of the field.

In the area of materials science, Nye's [2] and Newnham's [12] works make use of *single* RSs to display orientational dependence of properties. This book follows along the same lines and focuses on the visualization of properties (matter tensors) for anisotropic materials and not on visualization of non-homogeneous fields. RSs are equally applicable to tensorial magnitudes that are not material properties (field tensors).

© Springer Nature Switzerland AG 2020
M. Laso and N. Jimeno, *Representation Surfaces for Physical Properties of Materials*, Engineering Materials, https://doi.org/10.1007/978-3-030-40870-1_1

Fig. 1.1 The str(s) of a trigonal material and its representation surface

Many of these magnitudes from physics and materials science can conveniently be represented by Cartesian tensors and, through appropriate condensation, matrices. The symmetry of a material or of an experimental setup determine relationships among the components of these magnitudes, e.g. making some of them cancel, making others equal, etc. These relationships among components have an effect on the shape of the RS of the physical magnitude, and hence on its directional or orientational dependence. The books by Nye [2], Newnham [12] and Tinder [13] are canonical references on the topic.

In order to turn the previous statements into quantitative results it is necessary to acquire some knowledge of symmetry as usually taught in crystallography, and also to familiarize oneself with the manipulation of Cartesian tensors.

Chapters 2 and 3 deal with these two subjects at the level required to follow the rest of the book. In Chap. 2 the reader gets familiar with geometric elements of symmetry, with crystallographic and limit classes and learns how to assign a given material to a class.

Chapter 3 starts by presenting the basic definitions and manipulation rules of Cartesian tensors. A sufficient degree of proficiency in handling Cartesian tensors should be acquired before progressing to the next chapter. To this end, a number of exercises have been interspersed throughout Chap. 3. In spite of the somewhat derogatory term "index gymnastics" sometimes applied to the non-coordinate free approach, it is strongly adviced to work through them in detail in order to achieve fluency in subindex manipulation.

At first sight Chaps. 2 and 3 seem almost entirely unrelated: while Chap. 2 is mostly visual and requires minimal, if any, mathematical background, Chap. 3 deals with the manipulation of Cartesian tensors and their representation as matrices, and contains a few figures only. Yet both come together in Sect. 3.6, where the effect of symmetry on property components is explained.

Chapter 4 introduces the idea of RS and applies the skills acquired in Chap. 3 to a selected few physical magnitudes of particular importance, such as electric resistivity, stress and strain, etc. whose basic physical or geometrical meaning should be well understood at an early stage.

The construction of the RS of a given physical magnitude is solely dependent on its tensorial rank, on its symmetry or skew-symmetry and on the particular crystallographic or limit class to which the geometry of the material or of the field tensor belongs.

For this reason, the remaining chapters are devoted to properties of increasing tensorial rank and to skew symmetric properties. Instead of using a generic property of the appropriate rank in each chapter, calculations and illustrations have been carried out for specific properties, chosen because of their importance or probable familiarity to potential readers.

As an example, piezoelectricity and, to a lesser extent, the Pöckels coefficient have been chosen as representative properties in Chap. 7; elastic compliance and stiffness in Chap. 8. It goes without saying that the treatment will be identical for any other property of the same rank and index symmetry.

In the next section we have included for reference a short alphabetical list of properties and physical magnitudes together with their symbols and units in the SI system.

1.1 Variables and Symbols

See Table 1.1.

1.2 Constitutive Equations

The reader will very likely encounter many magnitudes of the previous section in so called *constitutive equations* or CEs, i.e. in cause-effect relationships among them [14, 15].

CEs are not universal principles and they tend to carry names, most often that of the person credited with their invention.[1] A particularly important type among CEs is the one that relates gradients and fluxes through a linear relationship with proportionality constant called *phenomenological coefficient*. They have especial relevance in the field of transport phenomena [16].

A useful classification of CEs, properties and phenomenological coefficients is based on the reversibility or irreversibility of the physical process involved. Although

---

[1]This is a fine labeling arrangement for CEs. They do not represent fundamental laws imposed by Nature, but are more or less plausible quantitative statements that seem to passably or accurately reproduce the behavior of a material. Agreement with experiment is the key to the naming and survival of successful ones.

Table 1.1 Variables and symbols

Symbol	Name	Unit
A	Area	m^2
$\underline{B}$	Magnetic induction (magnetic flux density)	T ($V s/m^2$, Wb/m^2)
$\underline{B}$	Relative dielectric impermeability	–
$\underline{\underline{c}}$	Elastic stiffness	Pa
$\underline{\underline{d}}$	Piezoelectric modulus	C/N
C_i	Concentration of species i	kg/m^3, mol/m^3, átomos/m^3
$\underline{D}$	Electric displacement	C/m^2
$\underline{\underline{D}}$	Mass diffusivity	m^2/s
E	Young's modulus	Pa
$\underline{E}$	Electric field	N/C (V/m)
$\underline{F}$	Force	N
G	Shear modulus	Pa
$\underline{H}$	Magnetic field	A/m
I	Electric current	A
$\underline{J}$	Electric current density	A/m^2
$\underline{J_i}$	Mass flux of species i	$kg/m^2 s$, $mol/m^2 s$, átomos/$m^2 s$
$\underline{k}$	Thermal conductivity	W/m K
$\underline{K}$	Dielectric constant	–
$\underline{\underline{K}}$	Kerr coefficient	C^2/N^2 (m^2/V^2)
$\underline{\underline{M}}$	Electrostriction coefficient	C^2/N^2 (m^2/V^2)
p	Pressure	Pa
$\underline{p}$	Pyroelectric modulus	$C/m^2 K$
$\underline{P}$	Electric polarization	C/m^2
$\underline{\underline{p}}^{opt}$	Elasto-optic coefficient	–
$\underline{q}$	Heat flux	W/m^2
$\dot{Q}$	Specific power dissipation	W/m^3
R	Electric resistance	ω
$\underline{r}$	Thermal resistance	m K/W
$\underline{\underline{r}}$	Pöckels coefficient	m/V
$\underline{R}$, $\underline{\underline{R}}$	Hall or galvanomagnetic coefficient	$V m^2/A^2$
S	Volumetric entropy density	$J/K m^3$
$\underline{\underline{s}}$	Elastic compliance	Pa^{-1}
T	Absolute temperature	K
$\underline{u}$	Displacement	m

(continued)

Table 1.1 (continued)

Symbol	Name	Unit
U	Internal energy per unit mass	J/kg
V	Volume; electric potential	m^3; V
$\underline{v}$	Velocity	m/s
$\underline{\underline{\alpha}}$	Thermal expansion coefficient	1/K
$\underline{\underline{\epsilon}}$	Strain (displacement gradient)	–
η	Viscosity	Pa s
$\underline{\underline{\gamma}}^i$	Thomson coefficient of species i	V/K
$\underline{\underline{\dot{\gamma}}}$	Velocity gradient	s^{-1}
$\underline{\underline{\kappa}}$	Dielectric permittivity	F/m (C/V m)
λ	Lamé constant	Pa
μ	Lamé constant	Pa
ν	Poisson's ratio	–
$\underline{\underline{\mu}}$	Magnetic permeability	H/M (V s/A m)
$\underline{\underline{\underline{\pi}}}$	Coefficient of piezoresistivity	1/Pa
$\underline{\underline{\underline{\pi}}}^{opt}$	Piezo-optic coefficient	Pa^{-1}
$\underline{\underline{\Pi}}^i$	Peltier coefficient of species i	V
ρ	Mass density	kg/m^3
$\underline{\underline{\rho}}$	Electric resistivity	ω m
$\underline{\underline{\underline{\rho}}}^{mag}$	Magnetoresistivity coefficient	$V m^3/A^3$
σ_{pol}	Polarization surface charge	C/m^2
$\underline{\underline{\sigma}}$	Electric conductivity	S/m
$\underline{\underline{\Sigma}}^i$	Seebeck coefficient of species i	V/K
$\underline{\underline{\tau}}$	Stress	Pa
$\underline{\omega}$	Angular velocity	rad/s

reversibility does not play a role in the construction and interpretation of the RS, the reader is encouraged to consult [17–19] for more information on this important topic.

We have also included a list of several usual CEs from different fields of materials science for the reader's convenience. It is likely that only a few of them will be of interest to any particular reader. We have taken especial care to avoid collisions among symbols for variables. As a consequence, some symbols have been changed where confusion would have been unavoidable (Tables 1.2 and 1.3).

Table 1.2 Constitutive equations

Pöckels effect	$\Delta \underline{\underline{B}} = \underline{\underline{r}} \cdot \underline{E}$
Kerr effect	$\Delta \underline{\underline{B}} = \underline{\underline{K}} : \underline{E}\,\underline{E}$
Piezo-optic effect	$\Delta \underline{\underline{B}} = \underline{\underline{\pi}}^{opt} : \underline{\underline{\tau}}$
Elasto-optic effect	$\Delta \underline{\underline{B}} = \underline{\underline{p}}^{opt} : \underline{\underline{\epsilon}}$
Thermal expansion	$\underline{\underline{\epsilon}} = \underline{\underline{\alpha}} \Delta T$
Piezocaloric effect	$\Delta S = \underline{\underline{\alpha}} : \underline{\underline{\tau}}$
Direct piezoelectricity	$\underline{P} = \underline{\underline{d}} : \underline{\underline{\tau}}$
Inverse piezoelectricity	$\underline{\underline{\epsilon}} = \underline{E} \cdot \underline{\underline{d}}$
Pyroelectricity	$\underline{P} = \underline{p} \Delta T$
Electrocaloric effect	$\Delta S = \underline{p} \cdot \underline{E}$
Electrostriction	$\underline{\underline{\epsilon}} = \underline{\underline{M}} : \underline{E}\,\underline{E}$
Piezoresistivity	$\Delta \underline{\underline{\rho}} = \underline{\underline{\pi}} : \underline{\underline{\tau}}$
Hall effect	$\underline{E} = \underline{\underline{R}} : \underline{H}\,\underline{J}$
Magnetoresistivity	$\underline{\underline{\rho}} \equiv \underline{\underline{\rho}}^{mag} : \underline{H}\,\underline{H}$
Seebeck effect	$\underline{\nabla} V = \underline{\underline{\Sigma}}^{A} \cdot \underline{\nabla} T$
Peltier effect	$\underline{q} = \underline{\underline{\Pi}}^{A} \cdot \underline{J}$
Thomson effect	$\dot{Q} = \underline{\underline{\gamma}}^{A} : (\underline{\nabla} T)\underline{J}$
Volumetric electric energy density stored in dielectric	$U_{\mathrm{el}} = \frac{1}{2}\underline{D} \cdot \underline{E}$
If the material is linear	$U_{\mathrm{el}} = \frac{1}{2}\underline{D} \cdot \underline{E} = \frac{1}{2}\underline{\underline{\kappa}} : \underline{E}\,\underline{E}$
Volumetric elastic energy density stored in an elastic material	$U_{\mathrm{elast}} = \frac{1}{2}\underline{\underline{\tau}} : \underline{\underline{\epsilon}}$
If the material is linear (Hookean)	$U_{\mathrm{elast}} = \frac{1}{2}\underline{\underline{s}} \vdots \underline{\underline{\tau}}\,\underline{\underline{\tau}} = \frac{1}{2}\underline{\underline{c}} \vdots \underline{\underline{\epsilon}}\,\underline{\underline{\epsilon}}$
Power dissipation in a conductor per unit volume	$\dot{Q} = \underline{E} \cdot \underline{J}$
If the conductor is linear (Ohmic)	$\dot{Q} = \underline{E} \cdot \underline{J} = \underline{\underline{\rho}} \cdot \underline{J} \cdot \underline{J} = \underline{\underline{\rho}} : \underline{J}\,\underline{J}$
Power dissipation in a viscous fluid per unit volume	$\dot{Q} = \underline{\underline{\tau}} : (\underline{\nabla}\,\underline{v})^{T}$
If the fluid is linear (Newtonian)	$\dot{Q} = -\eta \underline{\underline{\dot{\gamma}}} : (\underline{\nabla}\,\underline{v})^{T}$
If the fluid is generalized Newtonian	$\dot{Q} = -\eta\,(|\underline{\underline{\dot{\gamma}}}|)\underline{\underline{\dot{\gamma}}} : (\underline{\nabla}\,\underline{v})^{T}$

Table 1.3 Flux-gradient constitutive equations

Flux = −phenomenological coefficient × gradient	
$\underline{J} = -\underline{\underline{\sigma}} \cdot \underline{\nabla} V$	Inverse Ohm
$\underline{q} = -\underline{\underline{k}} \cdot \underline{\nabla} T$	Fourier
$\underline{J}_A = -\underline{\underline{D}} \cdot \underline{\nabla} C_A$	Fick
$\underline{\underline{\tau}} = -\eta \left[\underline{\nabla} \underline{v} + (\underline{\nabla} \underline{v})^T \right] = -\eta \dot{\underline{\underline{\gamma}}}$	Newton
$\underline{\underline{\tau}} = -\eta \left(|\dot{\underline{\underline{\gamma}}}| \right) \dot{\underline{\underline{\gamma}}}$	Generalized Newtonian fluid
$\underline{\underline{\tau}} = \underline{\underline{c}} : \frac{1}{2} \left[\underline{\nabla} \underline{u} + (\underline{\nabla} \underline{u})^T \right] = \underline{\underline{c}} : \underline{\underline{\epsilon}}$	Hooke
−Phenomenological coefficient^{-1} × flux = gradient	
$-\underline{\underline{\sigma}}^{-1} \cdot \underline{J} = -\underline{\underline{\rho}} \cdot \underline{J} = \underline{\nabla} V$	Ohm
$-\underline{\underline{k}}^{-1} \cdot \underline{q} = -\underline{\underline{r}} \cdot \underline{q} = \underline{\nabla} T$	
$-\underline{\underline{D}}^{-1} \cdot \underline{J}_A = \underline{\nabla} C_A$	
$-\eta^{-1} \underline{\underline{\tau}} = \dot{\underline{\underline{\gamma}}}$	
$\underline{\underline{c}}^{-1} : \underline{\underline{\tau}} = \underline{\underline{s}} : \underline{\underline{\tau}} = \underline{\underline{\epsilon}}$	

References

1. Hinze, J.O.: Turbulence. McGraw-Hill, New York (1975)
2. Nye, J.F.: Physical Properties of Crystals. Oxford University Press, Oxford (1985)
3. Westin, C.-F., Maier, S.E., Mamata, H., Nabavi, A., Jolesz, F.A., Kikinis, R.: Processing and visualization for diffusion tensor MRI. Med. Image Anal. **6**(2), 93–108 (2002)
4. Barr, A.H.: Superquadrics and angle-preserving transformations. IEEE Comput. Graph. Appl. **1**(1), 11–23 (1981)
5. Haber, R.B.: Visualization techniques for engineering mechanics. Comput. Syst. Eng. **1**(1), 37–50 (1990)
6. Hotz, I., Feng, L., Hagen, H., Hamann, B., Joy, K., Jeremic, B.: Physically based methods for tensor field visualization. In: Proceedings of the Conference on Visualization'04, pp. 123–130. IEEE Computer Society (2004)
7. Jankun-Kelly, T.J., Mehta, K.: Superellipsoid-based, real symmetric traceless tensor glyphs motivated by nematic liquid crystal alignment visualization. IEEE Trans. Vis. Comput. Graph. **12**(5), 1197–1204 (2006)
8. Kindlmann, G.: Superquadric tensor glyphs. In: Proceedings of the Sixth Joint Eurographics-IEEE TCVG Conference on Visualization, pp. 147–154. Eurographics Association (2004)
9. Schultz, T., Kindlmann, G.L.: Superquadric glyphs for symmetric second-order tensors. IEEE Trans. Vis. Comput. Graph. **16**(6), 1595–1604 (2010)
10. Weickert, J., Hagen, H.: Visualization and Processing of Tensor Fields. Springer Science & Business Media, Berlin (2005)
11. Westin, C.F., Vilanova, A., Burgeth, B.: Visualization and Processing of Tensors and Higher Order Descriptors for Multi-Valued Data. Springer, Berlin (2014)
12. Newnham, R.E.: Properties of Materials: Anisotropy, Symmetry, Structure. Oxford University Press, Oxford (2005)
13. Tinder, R.F.: Tensor Properties of Solids: Phenomenological Development of the Tensor Properties of Crystals. Morgan & Claypool Publishers, San Rafael (2008)
14. Truesdell, C., Noll, W.: The Non-Linear Field Theories of Mechanics. Springer, Berlin (2004)

15. Smith, R.R.: An Introduction to Continuum Mechanics: After Truesdell and Noll. Springer Science & Business Media, Berlin (2013)
16. Bird, B., Stewart, W.E., Lightfoot, E.N.: Transport Phenomena, 2nd edn. Wiley, New York (2002)
17. Öttinger, H.C.: Beyond Equilibrium Thermodynamics. Wiley, New York (2005)
18. Prigogine, I.: Introduction to Thermodynamics of Irreversible Processes. Interscience Publishers, New York (1961)
19. De Groot, R., Mazur, P.: Non-Equilibrium Thermodynamics. North-Holland, Amsterdam (2013)

Chapter 2
Geometric Symmetry

This is a mostly visual and almost equation-free chapter which is usually appealing to readers with good spatial intuition. Its first half is a minimal but self-contained presentation of crystallographic concepts needed in the rest of the book. The second half is devoted to the determination of the crystallographic or limit class (alias for point group) to which a material belongs, based on its geometric elements of symmetry.

Since point group determination is the primary goal of this chapter, a great deal of rich and beautiful crystallographic results has had to be omitted or simplified, but without compromising correctness or consistency of the material presented. No mention is made of the group theoretical foundation of crystallography, space groups, non-symmorphic symmetry elements, diffraction, and a very long list of crystallographic essentials. The books by Borchardt-Ott [1] and Giacovazzo [2] are excellent general texts.

The ability to identify geometric symmetry elements in a material will also prove useful when dealing with the symmetry of material properties and of their representation surfaces.

2.1 Homogeneity and Isotropy

These two concepts are generally well understood, yet it is useful to start by making their meaning more precise. Their definition involves the concept of *invariance*.

In loose terms we will say that either the structure of a material or a property P of a material is invariant under a geometrical transformation g if it is not possible to detect any difference between P and $g(P)$. The transformation g may be geometrical, temporal or of other kinds, but this book is concerned with geometrical

© Springer Nature Switzerland AG 2020
M. Laso and N. Jimeno, *Representation Surfaces for Physical Properties of Materials*, Engineering Materials, https://doi.org/10.1007/978-3-030-40870-1_2

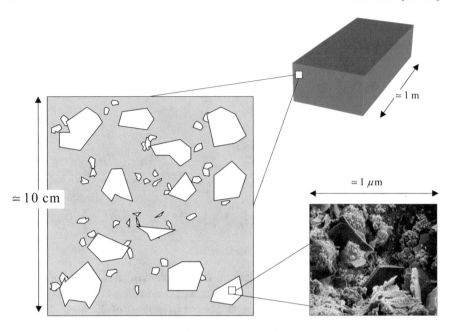

Fig. 2.1 Concrete is a composite material for which the natural scale of use/observation is of the order of 10^{-1} m. Below that scale a hierarchy of increasingly finer components (sand, gravel, cement) and geometrical details can be observed. If properly mixed and poured, concrete is nevertheless homogeneous and isotropic at its macroscopic scale of use

transformations only.[1] More specifically, with those associated with the geometric elements of symmetry described in Sect. 2.3. The term "symmetric" will often be used as a synonym of "invariant": a material or a property invariant under a transformation g is symmetric with respect to g.

When judging isotropy and invariance it is necessary to correctly identify the *scale of observation* or scale of use of a material: homogeneity and isotropy are to be judged at a length scale commensurate with or "reasonable" for the application.

The choice of the correct scale of observation is a matter of the specific application, of common sense and of experience (Fig. 2.1). Needless to say, the term "observe" does not necessarily mean visual observation: differences in surface roughness of a material may lead to differences in friction coefficient which are "observed" by means of mechanical experiments.

The same material may have to be considered at different scales, depending on the application: the scales at which to judge the homogeneity of steel for use in a razor blade or in armor plating are very different. If we take an extreme point of view, all materials are strictly inhomogeneous and anisotropic at the atomic scale, but it would be contrary to common sense to choose this atomic scale to judge these characteristics.

---

[1]Magnetic properties and magnetic point groups will not be dealt with.

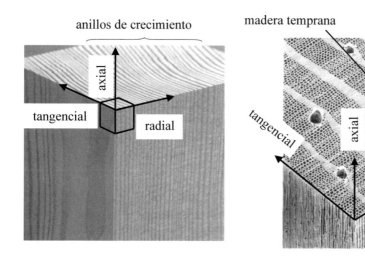

Fig. 2.2 The structure of wood. Prominent growth rings in some types of wood are a consequence of the different rates of growth in spring and fall. Most properties of wood are markedly different in the radial, tangential and axial directions

With few exceptions, homogeneity and isotropy are to be judged in the bulk of the material, *irrespective* of the shape and boundaries of a concrete specimen. Abstracting the shape of an specimen is not always a simple task.

Wood is a particularly clear example. Many wood types display easily observable growth rings created by different rates of growth in spring and fall (Fig. 2.2, left). Although it is tempting to conclude that wood belongs to the cylindrical limit class ∞/mm because of the approximately cylindrical shape of a wood log, an inspection of its structure (Fig. 2.2, right) proves that it is an orthorhombic material that belongs to the crystallographic class *mmm*.

The structure of wood is responsible for most of its properties, including mechanical properties. The external, approximately cylindrical shape of the log has nothing to do with the internal structure of wood.

With these previous notions in mind, the definitions of homogeneity and isotropy in Euclidean space are simple:

H: a material is *homogeneous* if it is invariant under translation, i.e. under the action of any element of the translation group T.

I: a material is *isotropic* if it is invariant under rotation, i.e. under the action of any member of the group of 3×3 orthogonal matrices $O(3)$.[2]

when considered at the scale of observation.

In practical terms, H means that the measurement of a given property will yield the same result when performed at any point within the material. I means that all measurements of a property will give the same result, no matter in which direction

---

[2]See Sect. 2.3.3.

the measurement is carried out at a given point in the material. For isotropy to make sense, the property to be measured has to be *directional*, i.e. be a tensor of rank ≥ 1.

As a final remark, a material may simultaneously be homogeneous/isotropic for some properties, but not for others. Materials belonging to the five cubic classes are the clearest example: they are isotropic for all second rank properties, but anisotropic for fourth rank properties. Strictly speaking, one should specify the rank of the property when referring to the homogeneity/isotropy of a material.

2.2   Lattice + Base = Crystal

In spite of the title of this section, and of the repeated use of the word "crystal" throughout this chapter, the geometric concepts it deals with apply to any kind of material with a regular, repetitive structure that fills, covers, or tessellates space without leaving gaps or producing overlaps. The material structure is not necessarily regular at the atomic scale. Its regularity lies at the scale of observation defined in the previous section.

As we will see, the geometric symmetry of many materials (wood, laminates, fiber composites, muscle, oriented polycrystalline materials, etc.) can be analyzed and described by the same methods used for "standard" crystals.

For our purposes, the primary goal is the assignment of the structure of a material to one of the 32 *crystallographic classes* based on the symmetry of the material.

In addition to these crystallographic classes there exist another seven *Curie* or *limit classes*,[3] for a total of 39 classes, one of which ($\infty\infty m$) is the isotropic class. While there is only one way for a material to be isotropic, one could say that there are 38 possible distinct ways of its being anisotropic.

The simplest way to define a crystal is by means of a *lattice* and a *base*.

> The crystal is constructed by placing (without rotating it) a single copy of the base at all lattice sites,

a definition that is based on the concepts "lattice" and "base".

Several definitions of a lattice can be given:

- a discrete set of points that "looks" the same when observed from any one of them,
- the set of all points whose position vectors are of the form: $\underline{R} = n_1\underline{a}_1 + n_2\underline{a}_2 + n_3\underline{a}_3$ where $\underline{a}_1, \underline{a}_2, \underline{a}_3$ are any three non-coplanar *primitive vectors*, and n_1, n_2, n_3 are three integers,
- a countably infinite set of vectors, closed under the operations of vector addition and subtraction.

These definitions are equivalent. A key point in all three is that a lattice is made up of points in the mathematical sense, as specified by their coordinates in Cartesian space, and are therefore not material, not tangible. It is best to consider lattice points

---

[3]From now on, the terms "Curie class" and "limit class" will be used interchangeably.

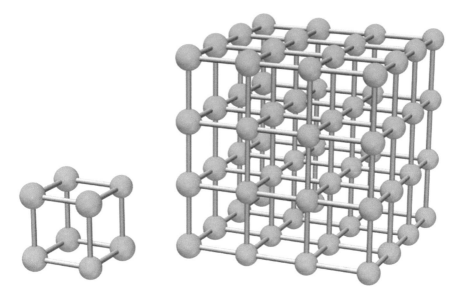

Fig. 2.3 Left: typical, and also typically misleading, representation of a cubic lattice. Right: 27 copies of the cell on the left

as position markers where something tangible (the base) is to be placed in order to build the crystal.

The usual representations of lattices in textbooks (Fig. 2.3) are notoriously, if almost unavoidably, misleading in this respect.

The left part of Fig. 2.3 is a typical representation of a cubic P lattice, a so-called *unit cell*, while the right panel is a group of 27 copies. These images convey the impression that "spheres" (presumably spherical atoms) occupy the lattice sites (wrong!), and that some kind of "bonds" join adjacent sites (also wrong). It is also tempting to conclude that the unit cell contains eight lattice sites (no surprise once again: also wrong). Upon seeing the "lattice" of Fig. 2.3 for the first time it is very easy to mistake it for a crystal.

The lattice is a countably infinite set of points,[4] nothing occupies its sites, no bonds join sites. The spheres and the lines in these figures are just visual aids.

We could remove the "bonds" to avoid confusion and just leave some kind of mark at the lattice sites, as in the left part of Fig. 2.4. On the paper the sense of three-dimensionality is now quite lost. If, faithful to the definition of lattice, we go further and remove the marks, nothing would be left, as in the deliberately empty right part of Fig. 2.4.

The use of *some* of kind of representation for the lattice is thus a necessity and also acceptable, as long as its meaning is properly understood: the "bonds" in Fig. 2.3

---

[4]Imagine it weren't; it would then be finite in extent and would therefore have a boundary (in Euclidean space). Lattice points close to or on the boundary would have a different environment from those far away from the boundary, contradicting the first definition.

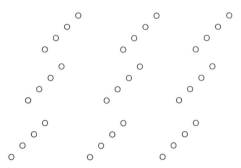

Fig. 2.4 A lattice represented as markers on sites (left); lattice with "bonds" and site markers removed; nothing tangible or visible is left (right)

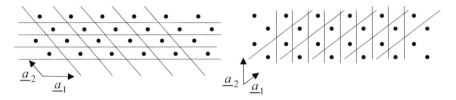

Fig. 2.5 The same 2D lattice (schematically represented by dots), two different primitive cells and their primitive base vectors

help visualize the edges of the cell, the "atoms" in the same figure are no atoms, just markers that indicate where the lattice points are located.

A few basic concepts related to the lattice:

- the *coordination number* of a lattice is the number of lattice points closest to a given one than to any other lattice point,
- a *primitive cell* is a region of space which, when translated by *all* the lattice vectors tessellates space without overlaps with other copies of itself and without leaving gaps. A primitive cell contains a single lattice point.

  Given a lattice, there are infinite ways of choosing a primitive cell. In Fig. 2.5 two different types of primitive cells are shown for the same lattice. In both cases the cell is the smallest parallelogram (parallelepiped in 3D) defined by the two sets of parallel lines (three sets of parallel planes in 3D). By translating this parallelogram (parallelepiped) by *all* lattice vectors, 2D (3D) space is covered completely without either gaps or overlaps.

- a *conventional cell* is a region of space which, when translated by *a subset* of the lattice vectors tessellates space without overlapping other copies of itself and without leaving gaps. The same lattice of Fig. 2.5 is shown in Fig. 2.6 but with non-primitive cells which are larger than the primitive cell. These larger cells cannot be translated by *all* lattice vectors, i.e. those of the form: $n_1\underline{a}_1 + n_2\underline{a}_2$, with n_1, n_2 integers, because they would overlap.

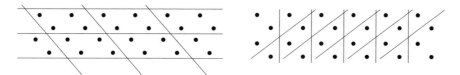

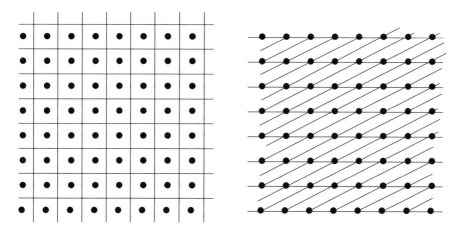

Fig. 2.6 The same 2D lattice as in Fig. 2.5 and two different non-primitive cells. The one on the left contains four lattice points, the one on the right two lattice points

Fig. 2.7 Two different cells for the same square lattice. The one on the left is as symmetric as the lattice

The cell on the left side of Fig. 2.6 can only be translated by vectors of the form: $2n_1\underline{a}_1 + 2n_2\underline{a}_2$, with n_1, n_2 integers.[5]

While primitive cells often are *less* symmetric than the lattice, the conventional cell is chosen such that its symmetry elements are the same as those of the lattice. Conventional cells can be made up of several primitive cells, its volume is then a multiple of the primitive cell volume.

In Fig. 2.7 two different primitive cells have been drawn for the 2D square lattice (represented by filled circles). The square cell on the left side has the same symmetry as the lattice; this is not the case for the parallelogram cell on the right hand side.

A *Wigner–Seitz* cell [3] (Fig. 2.8) is a cell which is simultaneously primitive and as symmetric as the lattice. Its definition is straightforward:

A Wigner–Seitz cell is the set of points (points of space, *not* lattice points) which are closer to a given lattice point than to any other lattice point.

The construction of the Wigner–Seitz cell in 2D is illustrated in Fig. 2.8: a lattice point is joined by straight segments to its immediate neighbors. These segments are

---

[5]Which is the set of allowed translations for the cell on the right side of Fig. 2.6?

Fig. 2.8 The Wigner–Seitz cell: points interior to the polygonal domain drawn with a thick line are closer to the lattice point P than to any other. Its sides bisect the segments marked with arrows

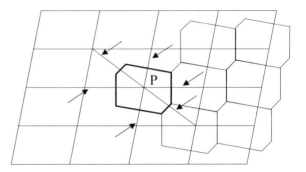

then bisected by straight lines. The convex polygonal region defined by the bisecting lines is the 2D Wigner–Seitz cell.

The construction in 3D is entirely analogous: instead of bisecting the segments by lines, they are bisected by planes. The Wigner–Seitz cell is the convex polyhedron defined by the bisecting planes.

The definition of the base is so much simpler than that of the lattice: the base is literally any object. It can be as simple as a single atom or as complicated as a whole virus or even a macroscopic object.[6]

The symmetry of the resulting crystal depends on the symmetry of both the lattice and the base: the symmetry elements of the crystal are those common to both the lattice and the base. If the base is of equal or higher symmetry than the lattice the resulting crystal has the same symmetry as the lattice, and it is called a *holoedric* crystal.

The simplest example is the crystal formed by a spherical base and any lattice: since the sphere (limit class $\infty\infty m$, see Sect. 2.4.2) has all geometric elements of symmetry, the resulting holoedric crystal has the same symmetry as the lattice.

If a very asymmetric base (e.g. belonging to crystallographic class 1, see Sect. 2.4.1) is placed on the sites of a cubic P lattice as in Fig. 2.9 (crystallographic class $m3m$), the resulting crystal is less symmetric than the lattice. In this case it has in fact the same symmetry as the base: triclinic class 1.

2.2.1 Lattices in 3D: The 14 Bravais Lattices

In 3D there are 14 different lattices, called *Bravais lattices*, grouped in 7 crystallographic systems (Figs. 2.10, 2.11, 2.12, 2.13 and 2.14). This classification is not arbitrary. We will come back to this point in Sect. 2.4.

At the risk of being repetitive: keep in mind in the following that the lattice sketches of these figures are just helpful representations of infinite sets of points in

---

[6]We are using the term "crystal" in the most general sense of a 3D-periodic structure, and not restricted to crystals made up of atoms or molecules.

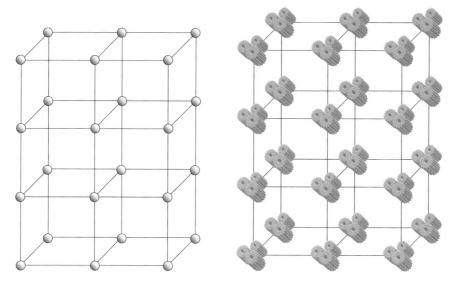

Fig. 2.9 Placing a very symmetric base (left: a sphere) on the sites of a cubic P lattice, a holoedric cubic crystal of class *m3m* results, which is as symmetric as the lattice itself. If a less symmetric base is used (right: a base formed by three different cogwheels), the resulting crystal has lower symmetry than the lattice. It needs not be cubic any more

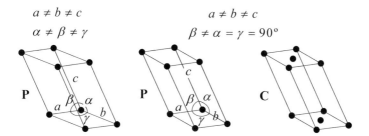

Fig. 2.10 The single triclinic lattice (left); the two monoclinic lattices

the mathematical sense: there are no "atoms" at the lattice points, no "bonds" join them, the lattice is obtained by periodically repeating the cell in the three spatial directions.

There are several types of lattice:

- the P or primitive lattice: with a single lattice point per cell,
- the I or body centered lattice: same as P but with an additional lattice point in the center of the cell,

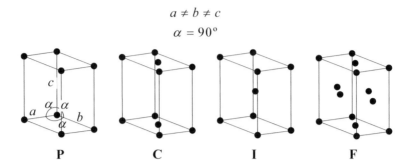

Fig. 2.11 The four orthorhombic lattices

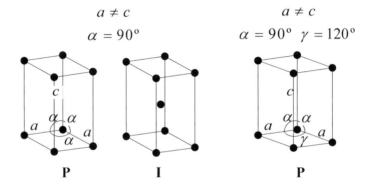

Fig. 2.12 The two tetragonal lattices (left); the hexagonal lattice

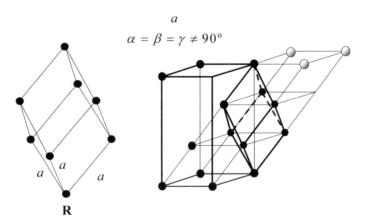

Fig. 2.13 The rhombohedral lattice (left); the trigonal lattice

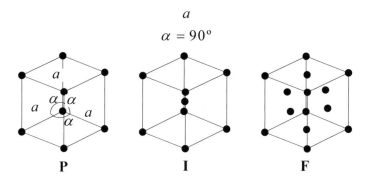

Fig. 2.14 The three cubic lattices

- the A or pinacoid centered[7] lattice: same as P but with an additional lattice point in the center of the A faces, that is, the faces that intersect the x axis. In B or C lattices, the second lattice point is in the center of the B or C faces, respectively.
- the F or face centered lattice: same as P but centered in all three pinacoids,
- the trigonal lattice, same as P but with two additional lattice points as shown in Fig. 2.13.

These lattices contain 1, 2, 2, 4 and 3 lattice points respectively. The P lattice does *not* contain 8 lattice points, even if 8 circles appear at the corners of the cell. Only one of them (*any* of the them, but just one) belongs to the cell, the others belong to the neighbor cells.

Figures 2.10, 2.11, 2.12, 2.13 and 2.14 also contain the *lattice parameters*, the lengths a, b, c of the cell edges and the angles α, β, γ between edges meeting at a lattice point.

The list of Bravais lattices in Figs. 2.10, 2.11, 2.12, 2.13 and 2.14 is exhaustive. Any 3D lattice one can imagine either is one of those 14 or can be represented by (is equivalent to) one of them. Before reading on, try to explain why the tetragonal C lattice does not appear in Fig. 2.12.

The left part of Fig. 2.15 shows one cell of this tetragonal C lattice. A top view of two such neighboring cells is shown at the right. By joining two corner and two face center lattice points a new lattice can be defined, which has the same value for the c cell parameter, a smaller base edge $a/\sqrt{2}$, and is just a P lattice. A tetragonal C lattice would be redundant.

All lattices that seem to be missing in Figs. 2.10, 2.11, 2.12, 2.13 and 2.14 are absent not because they do not exist, but because they can be reduced to at least one of those shown in these figures.

Another point worth emphasizing: the hexagonal cell of Fig. 2.12 is a prism which does *not* have a hexagon as its base, but a rhombus made up of two equilateral

---

[7]Think of a pinacoid as a pair of parallel faces.

Fig. 2.15 The tetragonal C lattice (left) is not different from a (smaller) tetragonal P lattice

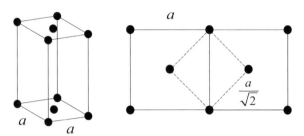

triangles. This hexagonal P cell, *if considered alone, without its infinite periodic copies*, does not have hexagonal symmetry, although the infinite lattice does.

To see the hexagonal symmetry of the hexagonal P lattice it is necessary to consider not just one cell but a larger fragment of the lattice. For this reason it is usual to use a larger, non-primitive conventional cell with a volume three times the volume of the P cell. This conventional cell *is* an hexagonal prism and, even when considered alone, does have hexagonal symmetry.

The trigonal lattice can be represented *either* by a primitive rhombohedral (R) lattice *or* by a non-primitive hexagonal lattice with three lattice points per cell. The relation between them is shown in Fig. 2.13. The R lattice is primitive, not so the trigonal lattice.[8]

2.3 Crystal Symmetry Operations and the Point Group

We will now describe a set of geometric transformations or operations to which any object, including a crystal, can be subjected. In loose terms we can say that the point group or the symmetry of a crystal will consist of the set of such transformations under which it is invariant.

There are only 32 possible different sets of transformations under which a 3D periodic structure can be invariant. These sets of geometrical transformations are called the *point groups* (because they leave at least one point of space fixed) or *crystallographic symmetry classes*.[9] A crystal will necessarily belong to one and only one of them. The main goal of this chapter is to learn how to assign the structure of a given 3D periodic material to one of these classes, or to one of the 7 non-crystallographic classes of Sect. 2.4.

The first step is to get acquainted with the so called *geometric elements of symmetry*. The transformations these geometric objects generate have the mathematical structure of a group.

---

[8]How much larger is the volume of the trigonal than the R lattice? There are two ways to answer this question without performing any geometric calculation.

[9]There are other geometric transformations which leave no point of space fixed, leading to the space group, but they are not essential for the purposes of this book.

It is important to clearly separate the "geometric elements of symmetry" from the "elements of the symmetry group":

- the *geometric elements of symmetry* are geometric entities: points, lines (axes) or planes. Each of them generates one or more geometric operations. For example, a single threefold rotation axis generates three individual operations or *actions*, by rotating space $1/3$, $2/3$ or $3/3 = 1$ turns about it.
- the *elements of the symmetry group* are the actions just described, i.e. geometric transformations, such as rotations, reflections, etc. It is these actions that have the mathematical structure of a group: the point group is formed by these actions, and *not* by the geometric elements (points, axes, planes).

Clearly, the number of elements (the *order*) of the symmetry group is always equal or larger than the number of geometric elements of symmetry.

As a matter of fact, the elements of the symmetry group can also be generated by other geometric elements of symmetry, different from the ones described below (e.g. rotation-reflection axes instead of rotation-inversion axes). Which geometric elements of symmetry we choose in order to generate the elements of the symmetry group is irrelevant. What counts is the set of elements of the symmetry group.

2.3.1 Inversion Center

An inversion center or center of symmetry, $\bar{1}$ ("minus one" o "one bar") acts as reflection point (not a plane!), i.e. a $\bar{1}$ performs the antipodal transformation $(x_1, x_2, x_3) \mapsto (-x_1, -x_2, -x_3)$ on all points of space. Only the origin is left undisturbed. A left and a right hand in the position of Fig. 2.16 are related by a $\bar{1}$. An inversion center changes the handedness or *chirality* of any object: a left hand is changed in a right hand and viceversa, a right-handed optically active molecule in its left-handed counterpart, etc.

2.3.2 Rotation Axes

Rotations about an axis leave a straight line, the axis itself, invariant. An n-fold rotation axis rotates space by integer multiples of $\frac{1}{n}$ turns, i.e. multiples of $\frac{360°}{n}$. Not all values of n are compatible with three-dimensional periodicity; the only possible values are $n = 1, 2, 3, 4, 6$. The written symbol of an n-fold rotation axis is n.

Every axis has several actions: the rotation by a fraction of $\frac{1}{n}$ turns $(\frac{360°}{n})$ can be carried out n distinct times, either clockwise or anticlockwise, but always in the same sense of rotation. Depending on n, the axes are:

Fig. 2.16 Inversion, $\bar{1}$: two hands of different handedness, but otherwise identical, are related by an inversion about a point in their middle. Every point of the left thumb is transformed into the corresponding point of the right thumb, etc.

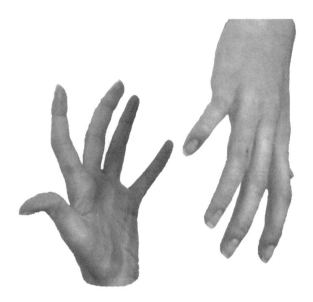

- a 1 axis rotates space by one complete turn, so it is equivalent to performing no rotation, since rotating space 360° about any axis leaves every point of space invariant. It is often also called E.[10] All objects (including the mathematical points lattices are made of) possess at the very least this geometric element of symmetry. The nth action of an n-fold rotation axis is equivalent to E.

- a twofold rotation axis, 2, symbol ◖, rotates space by ($\frac{360°}{2} = 180°$) in each of its two actions: $2^1, 2^2 = E$.

- a threefold rotation axis, 3, symbol ▲, rotates space by ($\frac{360°}{3} = 120°$) in each of its three actions: $3^1, 3^2, 3^3 = E$.

- a fourfold rotation axis, 4, symbol ◆, rotates space by ($\frac{360°}{4} = 90°$) in each of its four actions: $4^1, 4^2, 4^3, 4^4 = E$.

- a sixfold rotation axis, 6, symbol ⬢, rotates space by ($\frac{360°}{6} = 60°$) in each of its six actions: $6^1, 6^2, 6^3, 6^4, 6^5, 6^6 = E$.

These actions can be denoted by placing a superindex on the axis name; the orientation of the axis, in case it has to be made explicit, can be placed as a subindex using standard crystallographic direction indices [1] or, if the direction coincides with one of the crystallographic axes, using the name of the axis. So for example:

- $3^2_{[111]}$ is the second action (second 120° rotation) about a threefold rotation axis oriented along [111].

- $4^3_{[100]} = 4^3_x$ is the third action (third 90° rotation) about a fourfold rotation axis oriented along the [100] direction which coincides with the x crystallographic axis.

---

[10]The initial of *Einheit*, the German word for "unit".

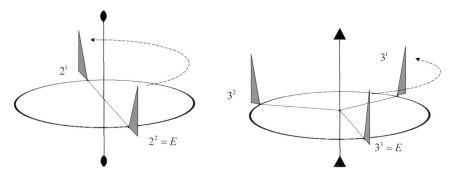

Fig. 2.17 Actions of twofold 2 and threefold 3 rotation axes

Fig. 2.18 Actions of a
fourfold rotation axis 4.
Equivalences such as
$4^2 = 2^1$, etc. are also shown

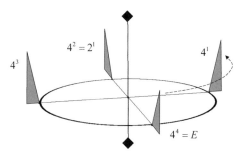

Fig. 2.19 Actions of a
sixfold rotation axis 6.
Equivalences such as
$6^4 = 3^2$, etc. are also shown

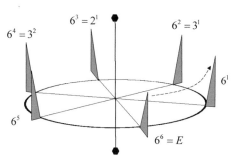

The individual actions of these axes are shown in Figs. 2.17, 2.18 and 2.19 and in Figs. 2.20 and 2.21 projected onto the paper along the axis. The dashed lines in Figs. 2.17, 2.18 and 2.19 show the first action of each type of axis.

As is obvious in Figs. 2.18 and 2.19, some axes (e.g. a 4 or a 6) may include others (a 2 or a 3). When enumerating the elements of the point group, the actions associated with these axes are counted only once: even if a sixfold rotation axis 6 automatically includes a 2 and a 3, only six actions (6^1 through $6^6 = E$) are counted, and *not* six for the sixfold axis, three for the threefold axis and another two for the twofold axis.

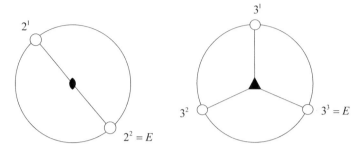

Fig. 2.20 Projection of actions of twofold and threefold rotation axes, seen along the axis, which is perpendicular to the paper. Rotation is anticlockwise

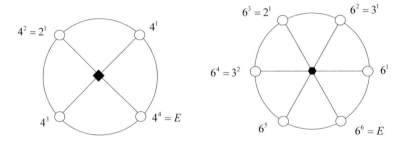

Fig. 2.21 Projection of actions of fourfold and sixfold rotation axes, seen along the axis, which is perpendicular to the paper. Rotation is anticlockwise

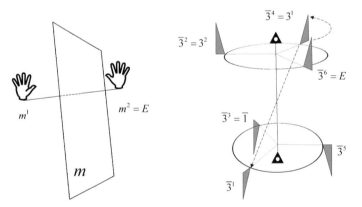

Fig. 2.22 Actions of a mirror plane m (equivalent to a $\bar{2} \perp m$) and of a threefold rotoinversion $\bar{3}$. Equivalences such as $\bar{3}^3 = \bar{1}$ are also indicated

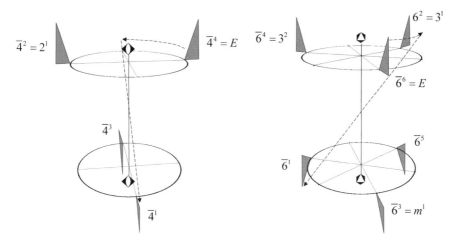

Fig. 2.23 Actions of rotoinversion axes $\bar{4}$ y $\bar{6}$. Equivalences such as $\bar{4}^2 = 2^1, \bar{6}^2 = 3^1$, etc. are also indicated

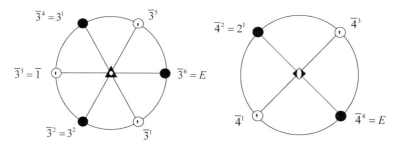

Fig. 2.24 Projection of actions of rotoinversion axes $\bar{3}$ and $\bar{4}$ along the axis, which is perpendicular to the paper. Rotation is anticlockwise. Changes of chirality are indicated by a prime

2.3.3 Rotation-Inversion Axes

A rotation-inversion or rotoinversion n-fold axis is the product of an n-fold rotation and an inversion: space is rotated by $\dfrac{1}{n}$ of a turn and then an inversion is carried out. As an example, the second action of a threefold rotoinversion axis $\bar{3}^2$ consists in rotating 120°, inverting, rotating 120° again, and inverting again (see Figs. 2.22 2.23, 2.24 and 2.25). The written symbol of an n-fold rotation axis is $\bar{n}$. Depending on n, the axes are:

- A $\bar{1}$ axis: equivalent to an inversion center.
- a twofold rotoinversion axis, $\bar{2}$, rotates space by ($\frac{360°}{2} = 180°$) and performs an inversion in each of its two actions: $\bar{2}^1, \bar{2}^2 = E$. However, this axis is equivalent to a reflection in a mirror plane perpendicular to the axis. Since it is more natural for most of us to recognize a mirror than a twofold rotoinversion axis, the name m is used instead of $\bar{2}$.

Fig. 2.25 Projection of
actions of rotoinversion axis
$\bar{6}$ along the axis, which is
perpendicular to the paper.
Rotation is anticlockwise.
Changes of chirality are
indicated by a prime

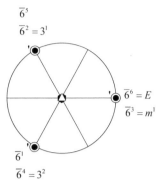

The twofold rotoinversion axis $\bar{2}$, alias mirror plane, has m as written symbol, and
as graphical symbol a *thick* line, either straight or curved.

This is an example of what has been said about two different geometric elements
of symmetry generating the same elements of the point group: the actions $\bar{2}^1$ and
m^1 carry out the same geometric transformation, although they stem from two
different geometric elements of symmetry.

- a threefold rotoinversion axis, $\bar{3}$, symbol ▲, rotates space by ($\frac{360°}{3} = 120°$) and
  performs an inversion in each of its *six* actions: $\bar{3}^1, \bar{3}^2, \bar{3}^3, \bar{3}^4, \bar{3}^5, \bar{3}^6 = E$. You
  may find surprising that a threefold inversion axis has not three, but six actions,
  as shown in Fig. 2.22. Additionally, the third action $\bar{3}^3$ is the same as an inversion
  center, i.e. a ▲ automatically includes an inversion center; the small circle in the
  symbol for a $\bar{3}$ is a reminder of this.

- a fourfold rotoinversion axis, $\bar{4}$, symbol ◇, rotates space by ($\frac{360°}{4} = 90°$) and
  performs an inversion in each of its four actions: $\bar{4}^1, \bar{4}^2, \bar{4}^3, \bar{4}^4 = E$. The action $\bar{4}^2$
  is identical to 2^1, i.e. a ◇ includes a binary; the ● inside the ◇ symbol reminds
  of this.

- a sixfold rotoinversion axis, $\bar{6}$, symbol ⬡, rotates space by ($\frac{360°}{6} = 60°$) and performs an inversion in each of its six actions: $\bar{6}^1, \bar{6}^2, \bar{6}^3, \bar{6}^4, \bar{6}^5, \bar{6}^6 = E$. Similarly,
  a sixfold rotoinversion axis includes a ▲, hence the symbol ⬡ (Fig. 2.23).

In Figs. 2.24 and 2.25 the change in chirality that takes place every time an inversion is performed has been indicated by a prime within the circle: if an action of
the axis changes the original handedness, the position to which the action carries the
original point is marked by a prime.

In the left part of Fig. 2.22 images of a left and a right hand indicate change of
chirality.

The geometric elements of symmetry are summarized in Table 2.1, as well as
equivalences such as $\bar{2} = m$, etc. It is easy to verify that:

- $\bar{6} \Leftrightarrow 3 \perp m$: a sixfold rotoinversion axis is equivalent to a threefold rotation axis
  plus a mirror perpendicular to this plane, and viceversa.

Table 2.1 Geometric elements of symmetry

	Symmetry element	Symbol in stereogram	Written name
Rotation axes	Identity	No symbol	1 or E
	Twofold	●	2
	Threefold	▲	3
	Fourfold	◆	4
	Sixfold	⬣	6
Rotation-inversion axes	Inversion center	No symbol	$\bar{1}$
	Twofold/reflection plane	Thick line	m
	Threefold	◮	$\bar{3}$
	Fourfold	◇	$\bar{4}$
	Sixfold	⬠	$\bar{6}$

- $4 + \bar{1} \Rightarrow \bar{4}$: a fourfold rotation axis plus an inversion center are equivalent to a fourfold rotoinversion axis, but not viceversa.
- $6 + \bar{1} \Rightarrow \bar{6} + \bar{3}$: a sixfold rotation axis plus an inversion center are equivalent to a sixfold rotoinversion axis *and* a threefold rotoinversion axis, but not viceversa.
- however: $3 + \bar{1} \Leftrightarrow \bar{3}$, i.e. a threefold rotoinversion axis is equivalent to a threefold rotation axis plus an inversion center, and viceversa.

A bit of nomenclature; the actions generated by the previous geometric elements of symmetry can be divided in:

(a) those that do not change chirality, e.g. all those generated by rotation axes. These are called *proper* rotations,
(b) those that change chirality, e.g. all those generated by rotoinversion axes or a mirror (alias twofold rotoinversion axis). These are the *improper* rotations.

The set of all actions in (a) and in (b) have the mathematical structure of a group. They are a discrete subgroup of the *orthogonal* group in three dimensions $O(3)$, which is the set of all 3×3 orthogonal matrices.

The proper rotations are a discrete subgroup of the *special orthogonal* group $SO(3)$ of 3×3 orthogonal matrices with determinant $+1$, whereas the improper rotations (3×3 orthogonal matrices with determinant -1) are not.[11] The sets of proper and improper rotations are the *components* of $O(3)$.

For the purposes of this book it is enough to be able to identify the geometric elements of symmetry of a material in order to be able to assign it to a class. It will not be necessary to enumerate the elements of the point symmetry group.

---

[11] The unit E is not in the set of improper rotations, because it does not change chirality; the existence of a unit E is one of the requisites for a set to have the structure of a group, hence the improper rotations do not form a group.

2.4 Crystallographic and Non-crystallographic Systems and Classes

The types of geometric symmetry we have to deal with are naturally organized in two groups: those possessing some of the elements of symmetry of the previous section, which are usually associated with (mono)crystalline materials: the *crystallographic classes*; and those that possess at least one rotation axis of infinite order: the *Curie* or *limit classes*.

2.4.1 Crystallographic Classes

The 14 Bravais lattices of Sect. 2.2.1 were grouped in seven crystallographic systems without providing an explanation. We can make up for this omission now: all lattices belonging to a system have the same point group; a given lattice can belong to only one of seven systems. Try, as a simple but useful exercise, to verify that the geometric elements of symmetry (planes, axes, inversion center) of the three cubic lattices are the same.

For the crystal, being composed of lattice plus base, there are more possibilities: these are the 32 crystallographic classes, which are again grouped in the same seven crystallographic systems. The arrangements of classes in systems is useful for the determination of the point group of a material:

- the first step is to determine the system to which the material belongs. This is done with the help of Table 2.2, which lists the *characteristic elements of symmetry* of each system. If, for example, we find that a material has a single threefold rotation axis and a plane of symmetry perpendicular to that plane, according to Table 2.2 the material is guaranteed to belong to one of the seven hexagonal classes.
- once the system has been determined, the search for the class is reduced to those in the system. This is done by identifying further geometric symmetry elements with the help of 32 *stereograms*.

Table 2.2 Characteristic symmetry elements of crystallographic systems

Triclinic	1 or $\bar{1}$
Monoclinic	⬤, or m, or both, in one direction
Orthorhombic	⬤ or m, or both, in three orthogonal directions
Tetragonal	◆ or ◇ (but three ◆ or three ◇ $\Rightarrow$ cubic)
Trigonal	▲ or △ (but one ▲ $\perp m \equiv$ ⬡ $\Rightarrow$ hexagonal)
Hexagonal	⬣ or ⬡
Cubic	Four ▲ or four △

Fig. 2.26 Stereographic projection of point P on the upper hemisphere onto P' on the equatorial plane

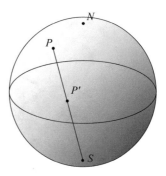

The stereogram is a very compact graphical representation of the geometric elements of symmetry of a given class. It also contains useful information on the placement of crystallographic and conventional axes with respect to the symmetry elements.

Although we will only have to read a stereogram, it is useful to understand how it is constructed. The stereogram is nothing but the stereographic projection of the geometric symmetry elements. Stereographic projection is carried out with the help of a sphere.

In stereographic projection (Fig. 2.26) a point P on the upper hemisphere is joined by a straight line with the south pole of the sphere. The intersection of this line with the plane of the equator is the projected point P'.

The stereographic projection assigns a point on the equatorial circle to each point on the upper hemisphere of the sphere. The entire upper hemisphere, including the line of the equator, is thus projected onto the equatorial circle.

The stereogram is constructed in the following steps:

- the element of symmetry (plane, axis) that is to be projected stereographically is assumed to pass through the center of the sphere,
- for each geometric element of symmetry, determine its intersection with the upper hemisphere:

  – if the element is a point (e.g. an inversion center), there is no intersection,
  – if the element is an axis, the intersection is either a point, or a pair of points in case the axis is horizontal,
  – if the element is a plane, the intersection is either half a maximum circle or the equator itself,

- all points of this intersection (one, two or infinite) are projected stereographically onto the equatorial plane,
- the symbol of the element (Table 2.1) is drawn at the projected point. Figure 2.27 illustrates how an inclined threefold axis ▲ and an inclined plane m are represented in a stereogram.

The stereogram is the result of carrying out the previous procedure for all geometric elements of symmetry of the class. Keep in mind that:

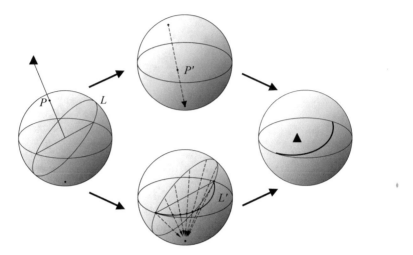

Fig. 2.27 Stereographic projection of a inclined threefold axis (upper part) and an inclined plane

- the stereogram may contain *thin* lines, in addition to the thick lines that represent planes. These thin lines are purely auxiliary,
- both the crystallographic x, y, z, (u) and the conventional sets of axes ①, ②, ③ are also represented in the stereogram,
- a horizontal axis appears as a pair of diametrically opposed symbols *not* connected by a thick line (which would represent a plane),
- depending on its inclination, a plane appears as a straight segment (vertical plane), a curved segment (inclined plane), or a circumference (horizontal plane), drawn in thick line,
- the inversion center does not appear explicitly in the stereogram. However, classes which contain an inversion center, so called *centrosymmetric* classes, are distinguished by having their name enclosed in a box (Fig. 2.28).

The stereograms and the names of the 32 crystallographic classes are presented in Figs. 2.29, 2.30, 2.31, 2.32, 2.33, 2.34 and 2.35, grouped by crystallographic systems. In each system there is a class of maximum symmetry,[12] called the *holoedric* class of the system. The point group of the other classes of a system either have half (*hemiedric* classes) or a quarter (*tetartoedric* classes) the number of elements of the holoedric class.

The lattices (considered alone, without a base) of each crystallographic system have the highest symmetry of the corresponding system: e.g. the three cubic lattices P, I, F belong to the cubic class of maximum symmetry $m3m$. A crystal formed by placing a spherical basis on any lattice is also holoedric.

---

[12]In the sense that its point group has the largest number of elements: its order is the greatest of all classes in a system.

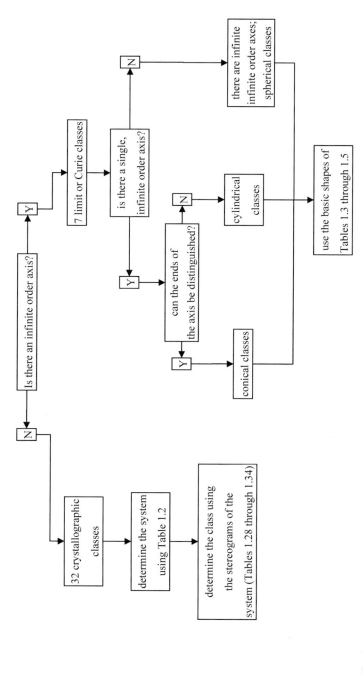

Fig. 2.28 Decision tree for the determination of crystallographic and limit classes

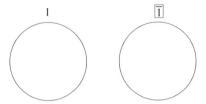

Fig. 2.29 Stereograms of the two classes of the triclinic system. The name of the centrosymmetric class is boxed

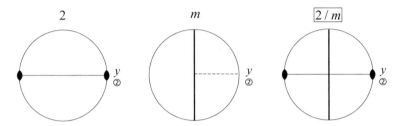

Fig. 2.30 Stereograms of the three classes of the monoclinic system. The name of the centrosymmetric class is boxed

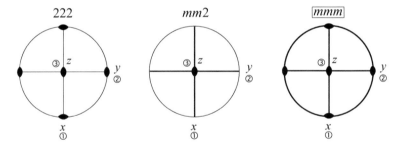

Fig. 2.31 Stereograms of the three classes of the orthorhombic system. The name of the centrosymmetric class is boxed

2.4.2 Limit or Curie Classes

A large number of materials possess non-crystallographic elements of symmetry, i.e. they are not crystals and nor do they fill up space in a periodic manner. Yet they have a high degree of order characterized by at least one so-called *infinite-fold* axis, or axis of infinite order. They belong to one of seven limit or Curie classes.

A crystal possesses a threefold axis if it is left invariant under any rotation which is a multiple of $\frac{1}{3}$ of a turn; similarly, an infinite-fold axis implies invariance under any number of rotations by $\frac{1}{\infty}$ turns, which is tantamount to saying that the structure is left invariant under a rotation by any arbitrary angle. Just consider an infinite-fold

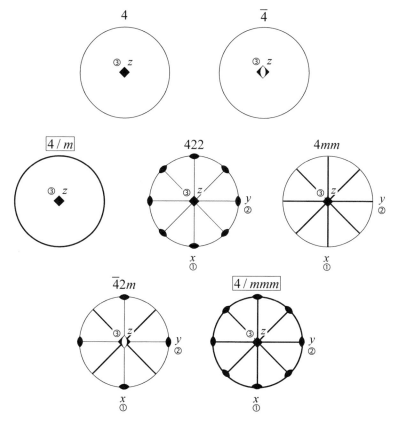

Fig. 2.32 Stereograms of the seven classes of the tetragonal system. The names of centrosymmetric classes are boxed

axis (symbol ∞ or ●) as an ordinary rotation axis of order n with $n \rightarrow \infty$, generating an infinite number of actions which form a continuous group.

An essential issue in identifying an ∞-fold axis is to judge the geometry of the material at the correct scale of use or observation. Concrete (Fig. 2.1) is clearly *not* invariant under any kind of rotation, when observed at a scale where individual particles of the components (gravel, sand) can be resolved.

At its scale of use, i.e. $O(10^{-1})$ m and upwards, as in the construction of a building or a dam, poured concrete is clearly invariant under any rotation about any axis. Nothing in its properties or in its behavior changes by rotating it by any angle, even though the positions of individual sand grains do change under the rotation. Concrete has an infinite number of ∞-fold axes and an infinite number of mirror planes m, and belongs to the limit class $\infty\infty m$, the isotropic class.

Objecting that a material does not possess an ∞-fold axis because its small-scale structure is not invariant under arbitrary rotation is not only not more precise, but is misleading as well.

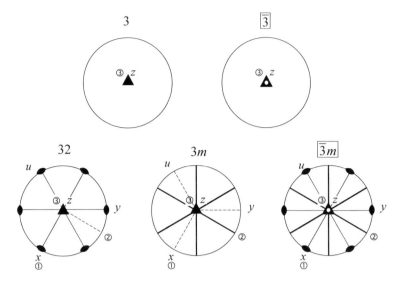

Fig. 2.33 Stereograms of the five classes of the trigonal system. The names of centrosymmetric classes are boxed

There are seven limit classes, grouped in three systems. There are no stereograms associated with these classes.[13] Instead of stereograms, objects of simple shape are used, which possess all those and only those geometric elements of symmetry of a given class. These objects are spheres, cylinders and cones, possibly "decorated".

It is essential to realize that the shape of these objects has nothing to do with the shape of the material (which should not be used to judge the symmetry of the material anyway!): concrete belongs to class ∞∞m, which is one of the spherical classes, although there is nothing "spherical" in the structure of concrete.

A simple, featureless sphere is used as a representative symbol for the ∞∞m limit class because, if judged by its shape, it has an infinite number of ∞-fold rotation and rotoinversion axes, and an infinite number of mirror planes. If the structure of a material possesses these symmetry elements, as poured concrete does, it belongs to the ∞∞m limit class.

Curie classes are grouped according to whether:

- they have an infinite number of ∞-fold rotation axes (*spherical* classes, Table 2.3),
- they have a single ∞-fold rotation axis (*cylindrical* classes, Table 2.4). The direction of this single axis is also referred to as the *preferential direction*,
- they have a single ∞-fold rotation axis (preferential direction) and, in addition, this axis has a preferential end, a "pointed" end[14] (*conical* classes, Table 2.5).

---

[13]Why? what would the stereogram of the ∞∞m class look like?

[14]The two ends of the axis can be told apart.

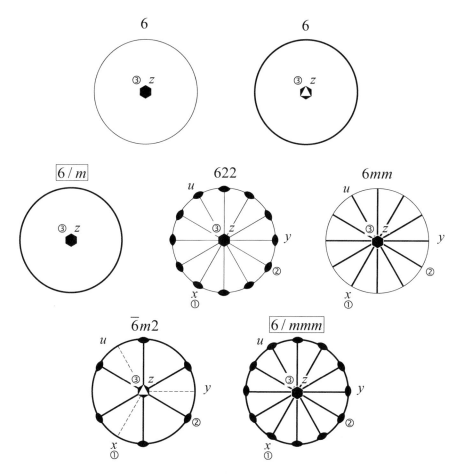

Fig. 2.34 Stereograms of the seven classes of the hexagonal system. The names of centrosymmetric classes are boxed

The limit classes ∞, $\infty\infty$ and $\infty 2$ are chiral; there are two varieties: right and left handed. Classes ∞ and ∞/m are characteristic of axial vectors and often appear in connection with magnetic materials/properties (see Sect. 9.10).

When assigning a material or structure to one of the 32 crystallographic or 7 limit classes:

- symmetry must be judged at the correct scale of use or scale of observation,
- one and the same material can be assigned to more than one class, depending on the rank of the property under consideration: a polarizing glass belongs to one of the $\infty\infty$ classes (left or right) as far as polarization is concerned; if mechanical properties are considered it belongs to $\infty\infty m$,
- it is necessary to consider the material as infinite in extent, i.e. ignore the external shape of a particular sample of object made of the material,

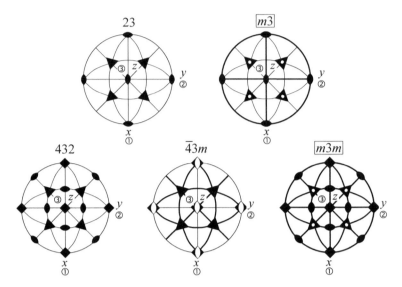

Fig. 2.35 Stereograms of the five classes of the cubic system. The names of centrosymmetric classes are boxed

- one exception to the previous rule is the *eumorphic crystal*, i.e. a crystal that is well grown, that is "perfect" in the sense that its external, macroscopic shape has the same symmetry point group as the atomistic/molecular crystal cell. The eumorphic crystal "inherits" or "reflects" the inner symmetry of the material. A eumorphic crystal will always belong to one of the 32 crystallographic classes. Most common crystalline samples are not eumorphic. Polycrystalline materials, i.e. those made up of a large number of individual crystals, very often belong to a limit class,
- the second exception are the representation surfaces themselves: when determining their symmetry, they are treated the same way as eumorphic crystals, i.e. their symmetry is judged based on their shape.

2.4.3 Standard Placement of Crystallographic and Conventional Axes

There is a widely accepted convention [1] for the placement of two sets of axes, crystallographic and conventional, with respect to the geometric elements of symmetry of the material. This convention is adhered to almost universally in crystallography but less so in engineering works.

Table 2.3 Spherical limit classes. The name of the centrosymmetric class is boxed

$\infty\infty$	Infinite ∞-fold axes	
$\boxed{\infty\infty m}$	Infinite ∞-fold axes plus infinite m	

Table 2.4 Cylindrical limit classes. The name of the centrosymmetric class is boxed

∞/m	A single ∞-fold axis plus one m perpendicular to the ∞-fold axis	
$\infty 2$	A single ∞-fold axis plus infinite 2-fold axes perpendicular to the ∞-fold axis	
$\boxed{\infty/mm}$	A single ∞-fold axis plus infinite m that contain the ∞-fold axis, plus one m perpendicular to the ∞-fold axis	

Table 2.5 Conical limit classes

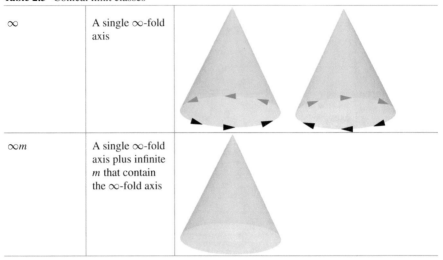

| ∞ | A single ∞-fold axis | |
| ∞m | A single ∞-fold axis plus infinite m that contain the ∞-fold axis | |

The reason for the existence of a convention is that numerical values of the components of tensor properties depend on the reference frame. Unless a common standard reference frame is used, experimental data obtained in one reference would need more or less laborious transformation (Sect. 3.3) before they could be used in a different reference frame.

The standard is described in Table 2.6 and is also included (graphically) in the stereograms of Figs. 2.29, 2.30, 2.31, 2.32, 2.33, 2.34 and 2.35.

- crystallographic axes are denoted by x, y, z, u,[15] in general they are not orthogonal, the length unit along each axis is the corresponding cell parameter, i.e. the length of the cell edge along the axis (Fig. 2.36).
- conventional axes are ①, ②, ③; they are orthogonal, the unit of length is the SI unit (m).

---

[15]The u axis is used in the trigonal and hexagonal systems only.

Table 2.6 Standard placement of crystallographic and conventional axes. Axes not appearing explicitly in the stereograms have no pre-defined, standard placement; they can be placed freely (but the placement must be specified)

Crystallographic axes		
Triclinic	$a \neq b \neq c, \alpha \neq \beta \neq \gamma$	x, y, z not specified
Monoclinic	$a \neq b \neq c, \alpha = \gamma = 90° \neq \beta$	y parallel to ●, x, z not specified
Orthorhombic	$a \neq b \neq c, \alpha = \beta = \gamma = 90°$	x, y, z parallel to the three binary axes or perpendicular to the symmetry planes
Tetragonal	$a = b \neq c, \alpha = \beta = \gamma = 90°$	z parallel to ◆ or to ◇, x, y as shown in stereogram
Trigonal	$a = b \neq c,$ $\alpha = \beta = 90°, \gamma = 120°$	z parallel to ▲ or to △, x, y as shown in stereogram
Hexagonal	$a = b \neq c,$ $\alpha = \beta = 90°, \gamma = 120°$	z parallel to ⬣ or to ⬡, x, y as shown in stereogram
Cubic	$a = b = c, \alpha = \beta = \gamma = 90°$	x, y, z parallel to the sides of the cube whose body diagonals are the four ▲ or △
Limit classes	. . . .	z parallel to an ∞-fold axis, x, y orthogonal to z and orthogonal to each other
Conventional axes		
Triclinic	. . . .	① ② ③ not specified
Monoclinic	. . . .	② parallel to y, ① ③ not specified
Orthorhombic	. . . .	① ② ③ parallel to xyz
Tetragonal	. . . .	① ② ③ parallel to xyz
Trigonal	. . . .	③ parallel to z, ① ② as shown in stereogram
Hexagonal	. . . .	③ parallel to z, ① ② as shown in stereogram
Cubic	. . . .	① ② ③ parallel a xyz
Limit classes	. . . .	③ parallel to z, both parallel to an ∞-fold axis

Fig. 2.36 Crystallographic
axes. They are, in general,
non-orthogonal, the length
unit along each axis is the
corresponding cell
parameter, i.e. the length of
the cell edge along the axis

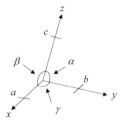

References

1. Borchardt-Ott, W.: Kristallographie. Springer, Berlin (2002)
2. Giacovazzo, C., Monaco, H.L., Artioli, G., Viterbo, D., Ferraris, G.: Fundamentals of Crystallography. Oxford University Press, Oxford (2002)
3. Ashcroft, N.W., Mermin, N.D.: Solid State Physics. Harcourt College Publishers, San Diego
   (1976)

Chapter 3
Representation of Material Properties by Means of Cartesian Tensors

The goal of this chapter is to familiarize the reader with the basic ideas and the few manipulation rules for Cartesian tensors [1–6] that will be required in the rest of the book. If you have previously been exposed to and perhaps intimidated by expressions like "covariant" and "contravariant", you are about to make a quantum leap in your study of tensor material properties: when working with Cartesian tensors the concepts of "covariant" and "contravariant" are not necessary. The manipulation rules for Cartesian tensors are but a minor extension of the familiar rules for vector and matrices. In the rest of the book the adjective "Cartesian" will often be dropped for brevity.

3.1 Definition, Rank and Components

For the purposes of this book a tensor can most usefully be understood as a physical property or geometrical magnitude that has "real", "observable", "objective" meaning, irrespective of the reference frame in which it is expressed.

It is very usual to represent vectors (rank one tensors) by an arrow. If we assume the arrow of Fig. 3.1 to be the constant electric field at a point, the field E has objective meaning: it is there for everyone to measure. There is no need to refer to its components to grasp its meaning, or to understand how a field twice as strong would act on a given charge.

The components of a tensor only appear when a frame of reference is defined (Fig. 3.1, right). Depending on the reference frame, the components of one and the same electric field will take on different values. A good part of this chapter deals with manipulating tensors using components.

This simple viewpoint deviates from more usual or more rigorous ones, such as a tensor being a linear transformation on a finite-dimensional real Euclidean space; a magnitude that transforms in a particular way when the frame of reference is rotated;

© Springer Nature Switzerland AG 2020
M. Laso and N. Jimeno, *Representation Surfaces for Physical Properties of Materials*, Engineering Materials, https://doi.org/10.1007/978-3-030-40870-1_3

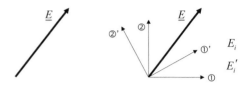

Fig. 3.1 The electric field at a point exists whether (right) or not (left) reference frames ①②, ①'②' are defined

or "quantities that transform according to unspeakable formulae" in Dodson and Poston's idiosyncratic book [7]; or a multilinear application that accepts n arguments (Sect. 3.3) [8–10].

In this book tensors come, from the physical point of view, in two varieties:

- *material* or *matter tensors*: they represent material properties, such as its density, its thermal conductivity or its elastic compliance.
- *field tensors*: they represent physical magnitudes which may exist in a material but are not material properties, such as temperature, velocity, electric field and stress.

A tensor should be handled whenever possible as a physical or geometrical object that has an autonomous, coordinate-free or reference-frame-free meaning. Yet it is often necessary to express it in a particular frame of reference. In this case it is possible and necessary to speak of its *components*, much as we are used to referring to the components of a vector. The individual components are identified by subindices: d_{ijk}.

The number of subindices (*not* the numerical value of a subindex!) necessary to specify all the components of a tensor is called the *rank* of the tensor. The rank of a tensor is hence the number of subindices it has when written in component form.

A common typographical choice to refer to tensors is to take lightface italic for scalars, boldface italic for first rank tensors, boldface Greek for second rank tensors, boldface sans serif for tensors of arbitrary rank. We will not follow that style. Instead, and for the sake of greater clarity, we will always make the rank of a tensor explicit by placing under its name a number of underlines equal to its rank. Thus:

$$\underset{\equiv}{T} \quad \text{is a third rank tensor.}$$

$$T_{ijk} \quad \text{is its } ijk\text{th component.}$$

Apart from being useful for the blackboard, the "underline" notation has the advantage of dispensing with special fonts to denote tensors of different ranks. In addition, the rank of tensors in an equation will always be in plain sight: if written in underline notation, just count the number of underlines; if written in components, count the number of subindices.

Another advantage is that the same literal symbol can stand for different physical magnitudes without confusion:

- the scalar ρ is the mass density (kg/m^3),
- the second rank tensor $\underline{\underline{\rho}}$ (Ω m), with components ρ_{ij}, is the electrical resistivity,
- the fourth rank tensor $\underline{\underline{\rho}}^{mag}$ is the coefficient of magnetoresistivity (V m^3)/A^3, with components ρ_{ijkl}^{mag}.

Each of the subindices of a tensor can take integer values that range from one to the number of dimensions d of the physical space in which we are working. In most applications, we will restrict ourselves to the usual three dimensional (3D) Euclidean space with $d = 3$. Thus i, j and k in the previous expression will take the numerical values $i = 1, 2, 3$, $j = 1, 2, 3$, $k = 1, 2, 3$. The tensor $\underline{\underline{T}}$ will have 27 components:

$$T_{111}, T_{112}, \ldots, T_{211}, T_{212}, \ldots, T_{311}, T_{312}, \ldots, T_{332}, T_{333}$$

In general, a tensor of rank n defined in a space of d dimensions will have d^n components. A tensor of rank 0 will have $d^n = 3^0$, i.e. just one component, and we will call it a *scalar*. A fourth rank tensor will have $d^n = 3^4 = 81$ components in 3D, and so on.

An important point to keep in mind: rank one tensors (like $\vec{p}$ with $d^n = 3^1 = 3$ components) and second rank tensors (like $\underline{\underline{\rho}}$ with $d^n = 3^2 = 9$ components) look temptingly similar to vectors and matrices; their components can actually be presented as such. Furthermore, even higher rank tensors can be written in matrix form by means of index condensation (Sect. 3.5).

But in this book the distinction between tensors, and vectors and matrices is essential. Tensors will behave in a specific way when the reference frame is rotated, while vectors and matrices will have no pre-defined transformation rules and will have a precise meaning in specific frames of references only (Sect. 2.4.3).

To make the distinction clear in the text, vectors will be denoted by an arrow placed over the name: $\vec{v}$ will be a vector. Matrices (not necessarily square) will be denoted by a tilde placed under the name of the variable: $\underline{\underline{\rho}}$ is a second rank tensor, $\underset{\sim}{\mu}$ is a matrix; their components will be ρ_i and μ_{ij} respectively.

That being said, and just for the sake of brevity, we will also refer to first rank tensors as "vectors", but only when no confusion can possibly arise.

It is often useful to visualize the components of a tensor as occupying the entries of an n-dimensional table. Viewed in this fashion, a zero rank tensor is just a number, a rank one tensor a one dimensional table (a list) of length d, a second rank tensor a square table (a chess board) of size $d \times d$, a third rank tensor a cubic table of size $d \times d \times d$, and a tensor of rank n a hypercube of size $\underbrace{d \times d \times \ldots \times d}_{n \text{ times}}$ (see Fig. 3.2).

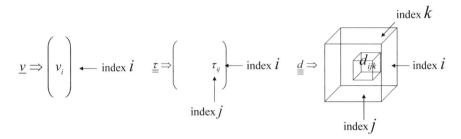

Fig. 3.2 First, second and third rank tensors viewed as a list, a checkerboard and a cube of components. The situation is similar for higher rank tensors, but cannot easily be drawn on paper

3.2 Basic Operations with Tensors

We first construct tensors from so called "unit objects" [6]. Similarly, we define the rules for operations between tensors, such as products, in terms of the operations between these unit objects. The basic operations between tensors are very similar to the usual ones between vectors and matrices, but extended to more than two subindices.

3.2.1 Unit Objects

A tensor of rank n is constructed as a linear combination of unit objects of that rank, e.g. a second rank tensor is built as a linear combination of the second rank unit objects or "dyads", and similarly for all ranks. Table 3.1 shows the unit objects up to rank four. The extension to higher rank unit objects is obvious.

Even though they look like, and can be interpreted as, some kind of product, the names of a unit object such as the dyad $\underline{\delta}_i\underline{\delta}_j$ is *not* "delta i times delta j", but "delta i delta j". Analogously, the name of a tetrad is "delta i delta j delta k delta l" and *not* "delta i times delta j times delta k times delta l".

The particular letters chosen as subindices in the sums in Table 3.1 are not important, but their position is essential. We can always change the name of a subindex at will, even in the middle of a computation, so long as we change it systematically, i.e. everywhere it appears and so long as we respect its position. Such subindices are often called *dummy* indices.

The first two of the following expressions have the same meaning, but they are different from the third one:

$$d_{ijk}\underline{\delta}_i\underline{\delta}_j\underline{\delta}_k = d_{twb}\underline{\delta}_t\underline{\delta}_w\underline{\delta}_b \neq d_{twb}\underline{\delta}_w\underline{\delta}_t\underline{\delta}_b$$

Table 3.1 Tensors as linear combination of unit objects

Rank	Unit object	Example
0	1 (the *scalar* unit, the number 1)	$n = n1$
1	$\underline{\delta}_i$ $(\underline{\delta}_1, \underline{\delta}_2, \ldots)$ (*monads*, there are $d^1 = 3^1 = 3$ in $d = 3$ dimensions)	$\underline{u} = u_1\underline{\delta}_1 + u_2\underline{\delta}_2 + u_3\underline{\delta}_3 = \sum_{i=1}^{3} u_i\underline{\delta}_i$
2	$\underline{\delta}_i\underline{\delta}_j$ $(\underline{\delta}_1\underline{\delta}_1, \underline{\delta}_1\underline{\delta}_2, \ldots \underline{\delta}_3\underline{\delta}_3)$ (*unit dyads*, there are $d^2 = 3^2 = 9$ in $d = 3$ dimensions)	$\underline{\underline{\epsilon}} = \sum_{i=1}^{3}\sum_{j=1}^{3} \epsilon_{ij}\underline{\delta}_i\underline{\delta}_j$
3	$\underline{\delta}_i\underline{\delta}_j\underline{\delta}_k$ $(\underline{\delta}_1\underline{\delta}_1\underline{\delta}_1, \underline{\delta}_1\underline{\delta}_1\underline{\delta}_2, \ldots \underline{\delta}_3\underline{\delta}_3\underline{\delta}_3)$ (*unit triads*, there are $d^3 = 3^3 = 27$ in $d = 3$ dimensions)	$\underline{\underline{d}} = \sum_{i=1}^{3}\sum_{j=1}^{3}\sum_{k=1}^{3} d_{ijk}\underline{\delta}_i\underline{\delta}_j\underline{\delta}_k$
4	$\underline{\delta}_i\underline{\delta}_j\underline{\delta}_k\underline{\delta}_l$ $(\underline{\delta}_1\underline{\delta}_1\underline{\delta}_1\underline{\delta}_1, \underline{\delta}_1\underline{\delta}_1\underline{\delta}_1\underline{\delta}_2, \ldots \underline{\delta}_3\underline{\delta}_3\underline{\delta}_3\underline{\delta}_3)$ (*unit tetrads*, there are $d^3 = 3^4 = 81$ in $d = 3$ dimensions)	$\underline{\underline{c}} = \sum_{i=1}^{3}\sum_{j=1}^{3}\sum_{k=1}^{3}\sum_{l=1}^{3} c_{ijkl}\underline{\delta}_i\underline{\delta}_j\underline{\delta}_k\underline{\delta}_l$

You will have noticed how bulky the multiple sums in Table 3.1 can be, the more so the higher the rank. An elegant and very practical shortcut due to A. Einstein gets rid of all superfluous matter in these expressions. This shortcut is called Einstein's *sum convention* over repeated indices: in sums such as $\sum_{i=1}^{3} u_i\underline{\delta}_i$ in which an index is repeated (i in this example appears twice: once in u_i and again in $\underline{\delta}_i$) the summation symbol will be omitted, but it will be understood that the sum exists and is to be carried out. Or viceversa: when an index is repeated in an expression like $u_i v_i$, a summation over that dummy index is always assumed to exist and to be carried out:

$$u_i v_i = \sum_{i=1}^{3} u_i v_i.$$

The convention is applied not just to one repeated index, but consistently to *all* repeated indices appearing in an expression. Keep in mind that the repeated index may appear as subindex of the component (u_i) or as subindex of the unit object ($\underline{\delta}_i$). Not only does the Einstein convention considerably reduce clutter, it also focuses our attention on the important part of an expression.

The expressions of Table 3.1 can be written more compactly using the repeated index convention:

$$\underline{u} = \sum_{i=1}^{3} u_i \underline{\delta}_i = u_i \underline{\delta}_i \qquad \underline{\underline{\epsilon}} = \sum_{i=1}^{3} \sum_{j=1}^{3} \epsilon_{ij} \underline{\delta}_i \underline{\delta}_j = \epsilon_{ij} \underline{\delta}_i \underline{\delta}_j$$

$$\underline{\underline{d}} = \sum_{i=1}^{3} \sum_{j=1}^{3} \sum_{k=1}^{3} d_{ijk} \underline{\delta}_i \underline{\delta}_j \underline{\delta}_k = d_{ijk} \underline{\delta}_i \underline{\delta}_j \underline{\delta}_k$$

$$\underline{\underline{c}} = \sum_{i=1}^{3} \sum_{j=1}^{3} \sum_{k=1}^{3} \sum_{l=1}^{3} c_{ijkl} \underline{\delta}_i \underline{\delta}_j \underline{\delta}_k \underline{\delta}_l = c_{ijkl} \underline{\delta}_i \underline{\delta}_j \underline{\delta}_k \underline{\delta}_l$$

Since each repeated index implies a sum over that index, a compact expression with several repeated indices may translate into a sum with a considerable number of terms. In $d = 3$, $c_{ijkl} \underline{\delta}_i \underline{\delta}_j \underline{\delta}_k \underline{\delta}_l$ is a quadruple sum containing a total of $3^4 = 81$ terms.

There are instances in which a subindex, e.g. k, is repeated in an expression without a sum over it being implied. In such relatively rare cases "no sum over k" will be explicitly stated.

It is also useful to interpret the unit objects as *pointers*:

- for a rank one tensor, $\underline{v} = \underline{\delta}_i v_i$ can also be read as "the tensor $\underline{v}$, viewed as a list of numbers, contains the value v_i in its ith position",
- similarly, $\underline{\underline{\epsilon}} = \underline{\delta}_i \underline{\delta}_j \epsilon_{ij}$ can be interpreted as "the tensor $\underline{\underline{\epsilon}}$, viewed as a table, contains the value ϵ_{ij} in the place pointed at by the dyad $\underline{\delta}_i \underline{\delta}_j$, or by the indices ij (ith row, jth column)", etc (see Fig. 3.2), and so on for higher ranks.

In general, the subindices $ijkl \ldots$ of the unit object $\underline{\delta}_i \underline{\delta}_j \underline{\delta}_k \underline{\delta}_l \ldots$ point to the location where the corresponding $T_{ijkl\ldots}$ component of a tensor is located.

3.2.2 Products

We will consider three types of products between tensors, each with its own product symbol:

- *dyadic product*: the factors are juxtaposed. No product symbol is required. The rank of the result is the sum of the ranks of the factors.
- *vector product*: this product is denoted by the usual "×" symbol. The rank of the result is the sum of the ranks of the factors minus 1.
- *contraction*: the symbol for this product is "·". There can be more than one contraction: thus, ":" represents two contractions, "⫶" three, and so forth. The rank of the result is the sum of the ranks of the factors minus 2 for each contraction, i.e. for each "·" appearing between the factors (see Table 3.2).

Table 3.2 Tensor multiplication

Symbol	Rank of the result	Example	Rank
None	Sum of ranks	$\underline{u}\,\underline{v}$	$1 + 1 = 2$
$\times$	Sum of ranks–1	$\underline{u} \times \underline{v}$	$1 + 1 - 1 = 1$
$\cdot$	Sum of ranks–2	$\underline{\underline{\sigma}} \cdot \underline{v}$	$2 + 1 - 2 = 1$
		$\underline{\underline{\sigma}} \cdot \underline{\underline{\tau}}$	$2 + 2 - 2 = 2$
$:$	Sum of ranks–4	$\underline{\underline{\sigma}} : \underline{\underline{\tau}}$	$2 + 2 - 2 \cdot 2 = 0$
		$\underline{\underline{c}} : \underline{\underline{\epsilon}}$	$4 + 2 - 2 \cdot 2 = 2$
$\vdots$	Sum of ranks–6	$\underline{d} \vdots \underline{\underline{d}}$	$3 + 3 - 3 \cdot 2 = 0$

In general, none of these products is commutative[1]: $\underline{a} \times \underline{b} \neq \underline{b} \times \underline{a}$, and $\underline{\underline{a}} \cdot \underline{\underline{b}} \neq \underline{\underline{b}} \cdot \underline{\underline{a}}$. The vector product is not associative and all three products are distributive with respect to the sum:

$$\underline{u}\,(\underline{v} + \underline{w}) = \underline{u}\,\underline{v} + \underline{u}\,\underline{w}$$
$$\underline{u} \times (\underline{v} + \underline{w}) = \underline{u} \times \underline{v} + \underline{u} \times \underline{w}$$
$$\underline{u} \cdot (\underline{v} + \underline{w}) = \underline{u} \cdot \underline{v} + \underline{u} \cdot \underline{w}$$

The product can be broken up in steps:

1. the factors are written as "component times unit objects". The dummy indices used in each of the factors have to be different; the scalar parts (i.e. the components) of the factors are multiplied as numbers. The unit objects of the factors (dyads, etc) are multiplied among them preserving the order. Examples:

$$\underline{u}\,\underline{v} = \left(u_i \underline{\delta}_i\right)\left(v_j \underline{\delta}_j\right) = \underline{\delta}_i \underline{\delta}_j u_i v_j$$
$$\underline{u} \times \underline{v} = \left(u_i \underline{\delta}_i\right) \times \left(v_j \underline{\delta}_j\right) = \underline{\delta}_i \times \underline{\delta}_j u_i v_j$$
$$\underline{\underline{\sigma}} \cdot \underline{\underline{\tau}} = \left(\sigma_{nm} \underline{\delta}_n \underline{\delta}_m\right) \cdot \left(\tau_{pq} \underline{\delta}_p \underline{\delta}_q\right) = \underline{\delta}_n \underline{\delta}_m \cdot \underline{\delta}_p \underline{\delta}_q \sigma_{nm} \tau_{pq}$$
$$\underline{v} \cdot \underline{u} = \left(v_i \underline{\delta}_i\right) \cdot \left(u_j \underline{\delta}_j\right) = \underline{\delta}_i \cdot \underline{\delta}_j v_i u_j$$

2. the products of unit objects are defined by:

$$\underline{\delta}_i \underline{\delta}_j \equiv \underline{\delta}_i \underline{\delta}_j \tag{3.1}$$
$$\underline{\delta}_i \times \underline{\delta}_j \equiv \epsilon_{ijk} \underline{\delta}_k \tag{3.2}$$
$$\underline{\delta}_i \cdot \underline{\delta}_j \equiv \delta_{ij} \tag{3.3}$$

---

[1] Although in specific cases such as $\underline{a} \cdot \underline{b} = \underline{b} \cdot \underline{a}$ the order of the factors can be irrelevant.

The first definition (3.1) may look tautological. Its meaning is "the dyadic product of $\underline{\delta}_i$ y $\underline{\delta}_j$ is the dyad $\underline{\delta}_i\underline{\delta}_j$". Even though both sides look identical, the left hand side is a product, the right hand side is a dyad. The fact that both sides look identical is deliberate.

The *permutation symbol* ϵ_{ijk} appears in the rule (3.2). It is defined by[2]:

$$\epsilon_{ijk} \equiv \begin{cases} +1 & \text{if } ijk = 123, 231 \text{ o } 312 \\ -1 & \text{if } ijk = 132, 321 \text{ o } 213 \\ 0 & \text{if two indices are identical} \end{cases} \tag{3.4}$$

Notice that the three sequences $123, 231, 312$ and $132, 321, 213$ are even and odd circular permutations respectively. In the right hand side of (3.2) the index k appears twice so that a sum is implied:

$$\underline{\delta}_i \times \underline{\delta}_j = \epsilon_{ijk}\underline{\delta}_k = \sum_{k=1}^{3} \epsilon_{ijk}\underline{\delta}_k = \epsilon_{ij1}\underline{\delta}_1 + \epsilon_{ij2}\underline{\delta}_2 + \epsilon_{ij3}\underline{\delta}_3$$

with ϵ_{ijk} given by (3.4). In this expression i y j are not repeated and may both *independently* take the values 1, 2, 3, e.g. for $i = 1$, $j = 2$,

$$\underline{\delta}_1 \times \underline{\delta}_2 = \epsilon_{12k}\underline{\delta}_k = \epsilon_{121}\underline{\delta}_1 + \epsilon_{122}\underline{\delta}_2 + \epsilon_{123}\underline{\delta}_3 = 0\underline{\delta}_1 + 0\underline{\delta}_2 + 1\underline{\delta}_3 = \underline{\delta}_3$$

It follows from its definition that out of the 27 components of ϵ_{ijk}, 21 are 0. Of the remaining six, three are $+1$ and the other three -1. The signs of the non-zero components correspond to those needed in the calculation of a determinant, so it is possible to write:

$$\begin{vmatrix} a_{11} & a_{12} & a_{13} \\ a_{21} & a_{22} & a_{23} \\ a_{31} & a_{32} & a_{33} \end{vmatrix} = \sum_{i=1}^{3}\sum_{j=1}^{3}\sum_{k=1}^{3} \epsilon_{ijk}a_{1i}a_{2j}a_{3k} = \epsilon_{ijk}a_{1i}a_{2j}a_{3k}$$

which is the reason ϵ_{ijk} can be used to write the vector product as a determinant:

$$\underline{u} \times \underline{v} = \begin{vmatrix} \underline{\delta}_1 & \underline{\delta}_2 & \underline{\delta}_3 \\ u_1 & u_2 & u_3 \\ v_1 & v_2 & v_3 \end{vmatrix} = \epsilon_{ijk}\underline{\delta}_i u_j v_k$$

---

[2]The permutation or Levi-Civita symbol provides the componentes of the alternating or totally antisymmetric tensor $\underset{\equiv}{\epsilon}$, which is a third rank, axial, antisymmetric tensor. The values of ϵ_{ijk} are the components of $\underset{\equiv}{\epsilon}$, just as the values of Kronecker's δ_{ij} are the components of the second rank unit or isotropic tensor $\underset{\equiv}{\delta}$.

The vector product of $\underline{u}$ and $\underline{v}$ can also be (loosely) written as the double contraction:

$$\underline{u} \times \underline{v} = \underline{\underline{\epsilon}} : \underline{v}\,\underline{u} \quad \text{or in component form} \quad (\underline{u} \times \underline{v})_i = \epsilon_{ijk} v_k u_j = \epsilon_{ijk} u_j v_k$$

The following holds for ϵ_{ijk}:

$$\epsilon_{ijk} = \frac{1}{2}(i-j)(j-k)(k-i)$$

$$\epsilon_{ijk}\epsilon_{mnk} = \delta_{im}\delta_{jn} - \delta_{in}\delta_{jm} \qquad \text{(sum over the index } k)$$

$$\epsilon_{ijk}\epsilon_{mjk} = 2\delta_{im} \qquad\qquad\quad \text{(sum over the indices } j, k)$$

$$\epsilon_{ijk}\epsilon_{ijk} = 6 \qquad\qquad\qquad\quad \text{(sum over all three indices)}$$

The third definition (3.3) makes use of *Kronecker's delta* δ_{ij}, whose meaning is:

$$\delta_{ij} \equiv \begin{cases} 1 & \text{if } i = j \\ 0 & \text{if } i \neq j \end{cases} \tag{3.5}$$

i.e. its value is 1 if both subindices are identical, 0 otherwise.

3. the product rules (3.1), (3.2), (3.3) are applied to the $\underline{\delta}$s immediately adjacent (from the left and from the right) to the product. The unit objects not immediately adjacent to the product are left untouched. If more than one contraction is present, they are carried out one after the other, starting by the $\underline{\delta}$'s closest to the contraction symbol(s) as shown in these examples:

$$\underline{\delta}_i\underline{\delta}_j \cdot \underline{\delta}_k = \underline{\delta}_i\left(\underline{\delta}_j \cdot \underline{\delta}_k\right) = \underline{\delta}_i\delta_{jk}$$

$$\underline{\delta}_i \cdot \underline{\delta}_j\underline{\delta}_k = \left(\underline{\delta}_i \cdot \underline{\delta}_j\right)\underline{\delta}_k = \underline{\delta}_k\delta_{ij}$$

$$\underline{\delta}_i\underline{\delta}_j \cdot \underline{\delta}_k\underline{\delta}_l = \underline{\delta}_i\left(\underline{\delta}_j \cdot \underline{\delta}_k\right)\underline{\delta}_l = \underline{\delta}_i\underline{\delta}_l\delta_{jk}$$

$$\left(\underline{\delta}_i\underline{\delta}_j : \underline{\delta}_k\underline{\delta}_l\right) = \left(\underline{\delta}_i \cdot \underline{\delta}_l\right)\left(\underline{\delta}_j \cdot \underline{\delta}_k\right) = \delta_{il}\delta_{jk}$$

4. in products in which a Kronecker's δ_{ij} appears, the terms having different values of i and j are cancelled by the rule $\delta_{ij} = 0$ if $i \neq j$. Only those with $i = j$ remain. The effect of a δ_{ij} in an expression is therefore to make the indices i and j equal in the product. Either *all* the i are changed to j's in the product or viceversa. Examples:

• contraction (or scalar product) of two rank one tensors:

$$\underline{u} \cdot \underline{v} = (\underline{\delta}_i u_i) \cdot (\underline{\delta}_j v_j) = (\underline{\delta}_i \cdot \underline{\delta}_j)u_i v_j = \delta_{ij}u_i v_j = u_i v_i$$

which matches the usual definition of the scalar product of two vectors.

- vector product of two rank one tensors:

$$\underline{u} \times \underline{v} = (\underline{\delta}_i u_i) \times (\underline{\delta}_j v_j) = (\underline{\delta}_i \times \underline{\delta}_j) u_i v_j = \epsilon_{ijk} \underline{\delta}_k u_i v_j = \begin{vmatrix} \underline{\delta}_1 & \underline{\delta}_2 & \underline{\delta}_3 \\ u_1 & u_2 & u_3 \\ v_1 & v_2 & v_3 \end{vmatrix}$$

$$= \underline{\delta}_1 (u_2 v_3 - u_3 v_2) + \underline{\delta}_2 (u_3 v_1 - u_1 v_3) + \underline{\delta}_3 (u_1 v_1 - u_1 v_3)$$

The same comment as for the scalar product applies.
- contraction of a second and rank one tensors:

$$\underline{\underline{\sigma}} \cdot \underline{v} = (\underline{\delta}_i \underline{\delta}_j \sigma_{ij}) \cdot (\underline{\delta}_k v_k) = \underline{\delta}_i (\underline{\delta}_j \cdot \underline{\delta}_k) \sigma_{ij} v_k = \underline{\delta}_i \delta_{jk} \sigma_{ij} v_k = \underline{\delta}_i \sigma_{ij} v_j$$

The result is a rank one tensor, the ith component of which (we know it is the ith component because of the subindex of the $\underline{\delta}_i$) is $\sigma_{ij} v_j = \sum_{j=1}^{3} \sigma_{ij} v_j$. You can check that this expression is the usual product rule of matrix times column vector.
- contraction of a first and a second rank tensors:

$$\underline{v} \cdot \underline{\underline{\sigma}} = (\underline{\delta}_k v_k) \cdot (\underline{\delta}_i \underline{\delta}_j \sigma_{ij}) = \underline{\delta}_j (\underline{\delta}_k \cdot \underline{\delta}_i) v_k \sigma_{ij} = \underline{\delta}_j \delta_{ki} v_k \sigma_{ij} = \underline{\delta}_j v_i \sigma_{ij}$$

The result is a rank one tensor, the jth component of which is

$$v_i \sigma_{ij} = \sum_{i=1}^{3} v_i \sigma_{ij}$$

This expression corresponds to the usual product rule of row vector times matrix.
- contraction of two second rank tensors:

$$\underline{\underline{\sigma}} \cdot \underline{\underline{\tau}} = (\underline{\delta}_i \underline{\delta}_j \sigma_{ij}) \cdot (\underline{\delta}_k \underline{\delta}_l \tau_{kl}) = \underline{\delta}_i \underline{\delta}_l (\underline{\delta}_j \cdot \underline{\delta}_k) \sigma_{ij} \tau_{kl} = \underline{\delta}_i \underline{\delta}_l \delta_{jk} \sigma_{ij} \tau_{kl} = \underline{\delta}_i \underline{\delta}_l \sigma_{ij} \tau_{jl}$$

The result is a second rank tensor, the i, lth component of which is

$$\sigma_{ij} \tau_{jl} = \sum_{j=1}^{3} \sigma_{ij} \tau_{jl}$$

We know this is the i, lth component because of the subindices in the surviving dyad $(\underline{\delta}_i \underline{\delta}_l)$. This expression matches the usual product rule of two square matrices.
- scalar times second rank tensor:

$$s\underline{\underline{\sigma}} = s(\underline{\delta}_i \underline{\delta}_j \sigma_{ij}) = \underline{\delta}_i \underline{\delta}_j s \sigma_{ij}$$

All components of the second rank tensor are multiplied by the scalar. The same applies to the product of a scalar by a tensor of any rank.

- double contraction of a third rank and a second rank tensor:

$$\underline{\underline{d}} : \underline{\underline{\tau}} = (\underline{\delta}_i \underline{\delta}_j \underline{\delta}_k d_{ijk}) : (\underline{\delta}_m \underline{\delta}_n \tau_{mn}) = \underline{\delta}_i \delta_{km} \delta_{jn} d_{ijk} \tau_{mn} = \underline{\delta}_i d_{ijk} \tau_{kj}$$

The result is a rank one tensor, the ith component of which is:

$$d_{ijk}\tau_{kj} = \sum_{j=1}^{3}\sum_{k=1}^{3} d_{ijk}\tau_{kj}$$

- contraction of a rank one and a third rank tensor:

$$\underline{E} \cdot \underline{\underline{d}} = (\underline{\delta}_m E_m) \cdot (\underline{\delta}_i \underline{\delta}_j \underline{\delta}_k d_{ijk}) = \underline{\delta}_j \underline{\delta}_k (\underline{\delta}_m \cdot \underline{\delta}_i) E_m d_{ijk} = \underline{\delta}_j \underline{\delta}_k \delta_{mi} E_m d_{ijk} =$$

$$\underline{\delta}_j \underline{\delta}_k E_m d_{mjk}$$

The result is a second rank tensor, the j, kth component of which is:

$$E_m d_{mjk} = \sum_{m=1}^{3} E_m d_{mjk}$$

- fourth rank tensor contracted twice with second rank tensor:

$$\underline{\underline{c}} : \underline{\underline{\epsilon}} = (\underline{\delta}_i \underline{\delta}_j \underline{\delta}_k \underline{\delta}_l c_{ijkl}) : (\underline{\delta}_m \underline{\delta}_n \epsilon_{mn}) = \underline{\delta}_i \underline{\delta}_j \delta_{lm} \delta_{kn} c_{ijkl} \epsilon_{mn} = \underline{\delta}_i \underline{\delta}_j c_{ijkl} \epsilon_{lk}$$

The result is a second rank tensor, the i, jth component of which is:

$$c_{ijkl}\epsilon_{lk} = \sum_{k=1}^{3}\sum_{l=1}^{3} c_{ijkl}\epsilon_{lk}$$

Notice how this double sum implies that for each of the nine components of the result, nine multiplications have to be performed and the results added.

3.2.3 Other Definitions

- *modulus* of a rank one tensor:

$$|\underline{v}| \equiv \sqrt{\underline{v} \cdot \underline{v}} = \sqrt{v_i v_i}$$

This definition coincides again with the usual one for vectors. $|\underline{v}|$ is a measure of the "size" of a first rank tensor. Notice that $\sqrt{v_i^2} \neq \sqrt{v_i v_i}$. The reason is that there is no repeated index in v_i^2 and therefore no sum. The meaning of v_i^2 is "the square of the ith component of $\underline{v}$", the value of i being unspecified. On the other hand,

$$\sqrt{v_i v_i} = \sqrt{\sum_{i=1}^{3} v_i^2}, \text{ which is the correct } |\underline{v}|.$$

- the *transpose* of a second rank tensor $\underline{\underline{T}} = \underline{\delta}_i \underline{\delta}_j T_{ij}$ is defined as:

$$\underline{\underline{T}}^T \equiv \underline{\delta}_i \underline{\delta}_j T_{ji}$$

i.e., the ijth component of $\underline{\underline{T}}$ occupies the ji position of $\underline{\underline{T}}^T$.

For higher rank tensors there are several possibilities to transpose indices. A simple notation that is typographically practical up to rank four consists in placing the transpose symbol "^{T}" in the following way: for any fourth rank tensor $\underline{\underline{c}} = \underline{\delta}_i \underline{\delta}_j \underline{\delta}_k \underline{\delta}_l c_{ijkl}$,

$\underline{\underline{c}}^T = \underline{\delta}_i \underline{\delta}_j \underline{\delta}_k \underline{\delta}_l c_{ijlk}$ is the transpose in the last two indices,

$\underline{\underline{c}}^T = \underline{\delta}_i \underline{\delta}_j \underline{\delta}_k \underline{\delta}_l c_{ikjl}$ is the transpose in two middle indices,

$^T\underline{\underline{c}} = \underline{\delta}_i \underline{\delta}_j \underline{\delta}_k \underline{\delta}_l c_{jikl}$ is the transpose in the first two indices.

- *modulus* of a second rank tensor:

$$|\underline{\underline{\tau}}| \equiv \sqrt{\frac{1}{2}\underline{\underline{\tau}} : \underline{\underline{\tau}}^T} = \sqrt{\frac{1}{2}\tau_{ij}\tau_{ij}}$$

The two repeated indices in $\tau_{ij}\tau_{ij}$ imply a double sum, i.e. nine terms in 3D, each of which is the square of the component τ_{ij}. Just as above, $\sqrt{\frac{1}{2}\tau_{ij}\tau_{ij}} \neq \sqrt{\frac{1}{2}\tau_{ij}^2}$ because in the second root there are no repeated indices. Also in analogy to the modulus of a rank one tensor $|\underline{\underline{\tau}}|$ is a measure of the "size" of a second rank tensor.

- a tensor is *symmetric* in two indices kl if the following holds:

$$\tau_{...jklm...} = \tau_{...jlkm...} \qquad \text{symmetric in } kl$$

For a second rank tensor, $\underline{\underline{\tau}}^T = \underline{\underline{\tau}} \Leftrightarrow \underline{\underline{\tau}}$ is symmetric.

- a tensor is *antisymmetric* or *skew symmetric* in two indices kl if the following holds:

$$\tau_{...jklm...} = -\tau_{...jlkm...} \qquad \text{antisymmetric in } kl$$

A tensor is *totally* antisymmetric if it is antisymmetric in all its pairs of indices.

- the *trace* of a second rank tensor:

$$\text{tr}(\underline{\underline{\tau}}) \equiv \tau_{ii} \tag{3.6}$$

is the sum of the "diagonal" elements of the tensor, i.e. of those elements that would appear on the diagonal if the tensor were represented as a matrix. You can easily check that $\text{tr}(\underline{\underline{\tau}}) = \underline{\underline{\delta}} : \underline{\underline{\tau}} = \underline{\underline{\tau}} : \underline{\underline{\delta}}$, i.e. double contraction with $\underline{\underline{\delta}}$ works as a tool for extracting the trace of a second rank tensor. The trace satisfies:

$$\text{tr}(\underline{\underline{\tau}} + \underline{\underline{\sigma}}) = \text{tr}(\underline{\underline{\tau}}) + \text{tr}(\underline{\underline{\sigma}})$$

- *double bracket* notation: it is often necessary to explicitly write the components of a tensor. This makes practical sense when the tensor is of rank one (it will be represented by a vector) or two (by a square matrix). When doing so, the name of the tensor will be enclosed in double brackets. For general first and second rank tensors, their vector or matrix forms are written:

$$[\![\underline{v}]\!] = \begin{bmatrix} v_1 \\ v_2 \\ v_3 \end{bmatrix} \qquad [\![\underline{\underline{T}}]\!] = \begin{bmatrix} T_{11} & T_{12} & T_{13} \\ T_{21} & T_{22} & T_{23} \\ T_{31} & T_{32} & T_{33} \end{bmatrix}$$

The reason for the distinction between first/second rank tensors and vector/matrices was already hinted at on Sect. 3.1. Depending on the application, some authors strictly enforce the distinction, while others consider it less of a necessity. In this book Voigt notation (Sect. 3.5) and conventional reference frames play a major role, so clear separation between a tensor and its matrix representation is a must. A useful habit to acquire is to ask about the reference frame as soon as specific components of a tensor are used, or as soon as a tensor is presented as a vector or a matrix.[3]

- the second rank *unit* tensor or *isotropic* tensor is:

$$\underline{\underline{\delta}} \equiv \underline{\delta}_i \underline{\delta}_j \delta_{ij} \tag{3.7}$$

Its components take on the values 0 or 1 according to the value of δ_{ij}. Notice the appearance of the letter "delta" δ with three different meanings in the previous equation: "$\underline{\underline{\delta}}$" is the unit second rank tensor; δ_{ij} are its components (whose name coincides, not by chance, with Kronecker's δ_{ij}); finally $\underline{\delta}_i\underline{\delta}_j$ is a unit dyad. Consistent usage of the underline/index notation will not lead to confusion. In d dimensions: $\text{tr}(\underline{\underline{\delta}}) = \delta_{ii} = d$. The reason for calling $\underline{\underline{\delta}}$ isotropic will be explained in Sect. 3.4.

---

[3]It is not unusual to find notations like $\mathbf{T}_{\hat{e}}$, where the subscript $\underline{\hat{e}}$ explicitly indicates the frame to which the tensor $\mathbf{T}$ is referred.

- we define the *inverse* of a second rank tensor as the tensor $\underline{\underline{T}}^{-1}$ that satisfies:

$$[\![\,\underline{\underline{T}}\cdot\underline{\underline{T}}^{-1}\,]\!] = [\![\,\underline{\underline{\delta}}\,]\!] = \underline{1} = \begin{bmatrix} 1\,0\,0 \\ 0\,1\,0 \\ 0\,0\,1 \end{bmatrix}$$

where $\underline{1}$ is the unit matrix ($\underline{1} \equiv [\![\,\underline{\underline{\delta}}\,]\!]$).

- for the purposes of this book we define the *inverse* of a fourth rank tensor $\underline{\underline{A}}$ only for tensors that have the following symmetries, written in component form:

$$A_{ijkl} = A_{ijlk} = A_{jikl} = A_{klij} \quad \forall i, j, k, l = 1, 2, \ldots d$$

The first two equalities are sometimes called the *minor* symmetries of $\underline{\underline{A}}$, and can be also be written in underline notation as $\underline{\underline{A}} = \underline{\underline{A}}^T = {}^T\!\underline{\underline{A}}$. The last one is the *major* symmetry (try to write it in underline notation using the transpose symbol only), and in addition have $\det(A) \neq 0$ when written as matrix in Voigt notation (Sect. 3.5). If these two conditions are satisfied, the inverse $\underline{\underline{A}}^{-1}$ is defined as the tensor that, when also written in matrix form using Voigt notation, satisfies:

$$A A^{-1} = \underline{1}$$

This roundabout definition obscures how simple it is to obtain $\underline{\underline{A}}^{-1}$ for a tensor $\underline{\underline{A}}$ that fulfills the two conditions above:

1. write it in Voigt notation, A,
2. invert the matrix A to obtain A^{-1},
3. undo Voigt notation for A^{-1} to obtain $\underline{\underline{A}}^{-1}$.

In all applications discussed in this book, the symmetry and $\det(A) \neq 0$ conditions for $\underline{\underline{A}}$ will be fulfilled.

- the third rank *isotropic* tensor is:

$$\underline{\underline{\epsilon}} \equiv \underline{\delta}_i \underline{\delta}_j \underline{\delta}_k \epsilon_{ijk}$$

Its components take on the values 0, -1 o $+1$ according to the value of the Levi-Civita symbol ϵ_{ijk}. The tensor $\underline{\underline{\epsilon}}$ is also known as the alternating or totally antisymmetric third rank tensor.

- there are three *unit* or *isotropic* fourth rank tensors[4]:

$$\underset{\equiv}{\delta}\,\underset{\equiv}{\delta} \equiv \underline{\delta}_i\underline{\delta}_j\underline{\delta}_k\underline{\delta}_l\delta_{ij}\delta_{kl}$$

$$\underset{\equiv}{I} \equiv \underline{\delta}_i\underline{\delta}_j\underline{\delta}_k\underline{\delta}_l\delta_{il}\delta_{jk}$$

$$\underset{\equiv}{I}^T \equiv \underline{\delta}_i\underline{\delta}_j\underline{\delta}_k\underline{\delta}_l\delta_{ik}\delta_{jl}$$

Notice how they differ in the index sequence of the components. Again not by chance, the name of the first one looks like a dyadic product, just as a dyad looks like the dyadic product of two monads. The effect of multiplying a second rank tensor by $\underset{\equiv}{\delta}\,\underset{\equiv}{\delta}$ is:

$$\underline{A} : \underset{\equiv}{\delta}\,\underset{\equiv}{\delta} = \mathrm{tr}(\underline{A})\underline{\delta} \qquad [\![\underline{A} : \underset{\equiv}{\delta}\,\underset{\equiv}{\delta}]\!] = \begin{bmatrix} \mathrm{tr}(\underline{A}) & 0 & 0 \\ 0 & \mathrm{tr}(\underline{A}) & 0 \\ 0 & 0 & \mathrm{tr}(\underline{A}) \end{bmatrix}$$

to produce an isotropic second rank tensor, i.e. a multiple of the unit $\underline{\delta}$. Multiplying a second rank tensor by $\underset{\equiv}{I}$ or $\underset{\equiv}{I}^T$ produces:

$$\underline{A} : \underset{\equiv}{I} = \underset{\equiv}{I} : \underline{A} = \underline{A} \qquad \underline{A} : \underset{\equiv}{I}^T = \underline{A}^T \qquad \underset{\equiv}{I}^T : \underline{A} = \underline{A}^T$$

For any $\underline{A}$:

- double contraction with $\underset{\equiv}{I}$ leaves a second rank tensor unaltered,
- double contraction with $\underset{\equiv}{I}^T$ transposes its indices,

If a higher rank tensor is multiplied by $\underset{\equiv}{I}$ or $\underset{\equiv}{I}^T$:

- left or right double contraction with $\underset{\equiv}{I}$ leaves the tensor unaltered,
- left double contraction with $\underset{\equiv}{I}^T$ transposes the first two indices of the tensor and leaves the other indices unaltered,
- right double contraction with $\underset{\equiv}{I}^T$ transposes the last two indices of the tensor and leaves the other indices unaltered.

Try to verify the equalities above as a good exercise in index gymnastics:

---

[4]All rank zero tensors (scalars) are isotropic; there is no rank one isotropic tensor. The number of isotropic tensors for ranks $n = 0, 1, 2, 3, 4, 5, 6, \ldots$ is $1, 0, 1, 1, 3, 6, 15 \ldots$ but we will not need them for $n > 4$.

$$\underline{\underline{\delta}}\,\underline{\underline{\delta}} : \underline{\underline{A}} = \underline{\delta}_i\underline{\delta}_j\underline{\delta}_k\underline{\delta}_l\delta_{ij}\delta_{kl} : \underline{\delta}_m\underline{\delta}_n A_{mn}$$

$$= \underline{\delta}_i\underline{\delta}_j\delta_{ij}\underbrace{\delta_{kl}\delta_{lm}\delta_{kn}}_{\Rightarrow k=l=m=n} A_{mn} = \underline{\delta}_i\underline{\delta}_j\delta_{ij}\underbrace{A_{kk}}_{\text{tr}(\underline{\underline{A}})} = \text{tr}(\underline{\underline{A}})\underbrace{\underline{\underline{\delta}}}_{\underline{\underline{\delta}}}$$

$$\underline{\underline{I}} : \underline{\underline{A}} = \underline{\delta}_i\underline{\delta}_j\underline{\delta}_k\underline{\delta}_l\delta_{il}\delta_{jk} : \underline{\delta}_m\underline{\delta}_n A_{mn} = \underline{\delta}_i\underline{\delta}_j\underbrace{\delta_{il}\delta_{jk}\delta_{lm}\delta_{kn}}_{\Rightarrow i=l=m,\ j=k=n} A_{mn}$$

$$= \underline{\delta}_i\underline{\delta}_j A_{ij} = \underline{\underline{A}}$$

$$\underline{\underline{A}} : \underline{\underline{I}} = \underline{\delta}_m\underline{\delta}_n A_{mn} : \underline{\delta}_i\underline{\delta}_j\underline{\delta}_k\underline{\delta}_l\delta_{il}\delta_{jk} = \underline{\delta}_k\underline{\delta}_l\delta_{il}\delta_{jk}\delta_{jm}\delta_{in} A_{mn} = \underline{\delta}_k\underline{\delta}_l A_{kl} = \underline{\underline{A}}$$

$$\underline{\underline{I}}^T : \underline{\underline{A}} = \underline{\delta}_i\underline{\delta}_j\underline{\delta}_k\underline{\delta}_l\delta_{ik}\delta_{jl} : \underline{\delta}_m\underline{\delta}_n A_{mn} = \underline{\delta}_i\underline{\delta}_j\delta_{ik}\delta_{jl}\delta_{lm}\delta_{kn} A_{mn} = \underline{\delta}_i\underline{\delta}_j A_{ji} = \underline{\underline{A}}^T$$

$$\underline{\underline{A}} : \underline{\underline{I}}^T = \underline{\delta}_m\underline{\delta}_n A_{mn} : \underline{\delta}_i\underline{\delta}_j\underline{\delta}_k\underline{\delta}_l\delta_{ik}\delta_{jl} = \underline{\delta}_k\underline{\delta}_l\delta_{ik}\delta_{jl}\delta_{ni}\delta_{mj} A_{mn} = \underline{\delta}_k\underline{\delta}_l A_{lk} = \underline{\underline{A}}^T$$

The unit tensor $\underline{\underline{\delta}}\,\underline{\underline{\delta}}$ is symmetric in the first and in the second pair of subindices:
$\delta_{ij}\delta_{kl} = \delta_{ji}\delta_{kl} = \bar{\delta}_{ij}\delta_{lk} = \delta_{kl}\delta_{ij}$, or in underline notation:

$$\underline{\underline{\delta}}\,\underline{\underline{\delta}} = (\underline{\underline{\delta}}\,\underline{\underline{\delta}})^T = {}^T(\underline{\underline{\delta}}\,\underline{\underline{\delta}}) = {}^T(\underline{\underline{\delta}}\,\underline{\underline{\delta}})^T$$

It is also symmetric under exchange of the first pair of indices by the second pair of indices $\delta_{ij}\delta_{kl} = \delta_{kl}\delta_{ij}$. It has both minor and major symmetries.
Neither $\underline{\underline{I}}$ nor $\underline{\underline{I}}^T$ alone have these symmetries, but the linear combination:

$$\underline{\underline{I}} + \underline{\underline{I}}^T = \underline{\delta}_i\underline{\delta}_j\underline{\delta}_k\underline{\delta}_l(\delta_{il}\delta_{jk} + \delta_{ik}\delta_{jl})$$

does.

- the *invariants* of a tensor are scalars that are obtained from the tensor and which, being scalars, have the same value in all reference frames. Their great usefulness stems from their frame independence: any expression that is written in terms of invariants will automatically be valid in all reference frames.
Depending on its rank, one or more independent[5] invariants can be obtained from a tensor:

  - zero rank: a scalar is its own invariant.
  - rank one tensors have only one invariant, typically its modulus $|\underline{u}|$. If the tensor $\underline{u}$ is expressed in two different reference frames, the values of the components will be different, say u_i and u'_i, but $|\underline{u}| = \sqrt{u_i u_i} = \sqrt{u'_i u'_i} = |\underline{u}'|$ will be fulfilled. This idea is very natural for a vector because its modulus can be thought of as its length, which is "obviously" independent of the frame of reference. i.e. invariant.

---

[5]An infinite number of invariants can be defined from a given one: if I is an invariant, any function containing only I and constants will be invariant as well, although only one of them will be independent.

– second rank tensors have three invariants, e.g.:

$$I = \text{tr}(\underline{\underline{T}}) = T_{ii} \tag{3.8}$$

$$II = \text{tr}(\underline{\underline{T}}^2) = \text{tr}(\underline{\underline{T}} \cdot \underline{\underline{T}}) = T_{ij} T_{ji} \tag{3.9}$$

$$III = \text{tr}(\underline{\underline{T}}^3) = \text{tr}\left((\underline{\underline{T}} \cdot \underline{\underline{T}}) \cdot \underline{\underline{T}}\right) = T_{ij} T_{jk} T_{ki} \tag{3.10}$$

Another popular choice is:

$$I_1 = I$$

$$I_2 = \frac{1}{2}(I^2 - II)$$

$$I_3 = \frac{1}{6}(I^3 - 3I \cdot II + 2III) = \det(\llbracket \underline{\underline{T}} \rrbracket)$$

where det( ) is the determinant of a matrix.
– it is also possible to define *joint invariants* of two (or more) tensors. For rank one tensors, $\underline{u}$, $\underline{v}$ the contraction:

$$\underline{u} \cdot \underline{v} = u_i v_i$$

is invariant. For second rank tensors, these expressions are also invariant:

$$\underline{\underline{\sigma}} : \underline{\underline{T}} = \text{tr}(\underline{\underline{\sigma}} \cdot \underline{\underline{T}}) \quad \underline{\underline{\sigma}}^2 : \underline{\underline{T}} = \text{tr}(\underline{\underline{\sigma}}^2 \cdot \underline{\underline{T}}) \quad \underline{\underline{\sigma}} : \underline{\underline{T}}^2 = \text{tr}(\underline{\underline{\sigma}} \cdot \underline{\underline{T}}^2) \quad \underline{\underline{\sigma}}^2 : \underline{\underline{T}}^2 = \text{tr}(\underline{\underline{\sigma}}^2 \cdot \underline{\underline{T}}^2)$$

– similarly, mixed invariants of first and second rank invariants can be obtained:

$$\underline{u} \cdot \underline{\underline{\sigma}} \cdot \underline{v} = \underline{\underline{\sigma}} : \underline{v}\,\underline{u} = \sigma_{ij} u_i v_j$$

$$\underline{v} \cdot \underline{\underline{\sigma}} \cdot \underline{u} = \underline{\underline{\sigma}} : \underline{u}\,\underline{v} = \sigma_{ij} v_i u_j$$

If $\underline{\underline{\sigma}}$ is symmetric, both forms are identical. For the particular case $\underline{u} = \underline{v}$ and independently of the symmetry of $\underline{\underline{\sigma}}$:

$$\underline{u} \cdot \underline{\underline{\sigma}} \cdot \underline{u} = \underline{\underline{\sigma}} : \underline{u}\,\underline{u} = \underline{u}\,\underline{u} : \underline{\underline{\sigma}} = \sigma_{ij} u_i u_j$$

which will appear often in later chapters.

• a second rank tensor can be decomposed in a *symmetric* and a *skew symmetric* part. Adding and subtracting $\frac{1}{2}\underline{\underline{T}}^T$ to $\underline{\underline{T}}$:

$$\underline{\underline{T}} = \underbrace{\frac{1}{2}\left(\underline{\underline{T}} + \underline{\underline{T}}^T\right)}_{\text{symmetric}} + \underbrace{\frac{1}{2}\left(\underline{\underline{T}} - \underline{\underline{T}}^T\right)}_{\text{antisymmetric}} \tag{3.11}$$

- a second rank tensor can be decomposed in a *deviatoric* and an *isotropic* part. In d dimensions:

$$\underline{\underline{\tau}} = \underbrace{\underline{\underline{\tau}} - \frac{1}{d} \operatorname{tr}(\underline{\underline{\tau}})\underline{\underline{\delta}}}_{\text{deviatoric}} + \underbrace{\frac{1}{d} \operatorname{tr}(\underline{\underline{\tau}})\underline{\underline{\delta}}}_{\text{isotropic}}$$

The isotropic part is a scalar multiple of $\underline{\underline{\delta}}$, while the deviatoric part is traceless:

$$\operatorname{tr}\left(\underline{\underline{\tau}} - \frac{1}{d} \operatorname{tr}(\underline{\underline{\tau}})\underline{\underline{\delta}} \right) = \operatorname{tr}(\underline{\underline{\tau}}) - \frac{1}{d} \operatorname{tr}(\underline{\underline{\tau}}) \operatorname{tr}(\underline{\underline{\delta}}) = \operatorname{tr}(\underline{\underline{\tau}}) - \frac{1}{d} \operatorname{tr}(\underline{\underline{\tau}})d = 0$$

- the "nabla" or "del" operator in Cartesian coordinates is:

$$\underline{\nabla} \equiv \underline{\delta}_i \frac{\partial}{\partial x_i} \tag{3.12}$$

It is a rank one tensor operator, meaning that it will always appear prefixed to an expression on which it operates by taking partial derivatives with respect to the d spatial coordinates x_i. As far as products involving $\underline{\nabla}$ are concerned, the same rules apply to it as to any other rank one tensor. In addition, the usual rules for derivatives also apply.[6]

Notice that up to now the components of a tensor of any rank $\underline{\underline{T}}$ were named by the same letter: $T_{ij\ldots}$. One would then expect the components of $\underline{\nabla}$ to be named ∇_i:

$$\underline{\nabla} \equiv \underline{\delta}_i \nabla_i$$

which is perfectly fine, except that ∇_i remains undefined. Thus, (3.12) can also be understood as $\underline{\nabla} = \underline{\delta}_i \nabla_i$ with $\nabla_i \equiv \frac{\partial}{\partial x_i}$.[7]

The operator $\underline{\nabla}$ allows us to evaluate:

- the gradient of a scalar field, which is a rank one tensor field:

$$\underline{\nabla}s = \underline{\delta}_i \frac{\partial s}{\partial x_i}$$

where $s(x_1, x_2, x_3)$ is a scalar field (temperature inside a material, concentration of a chemical species, ...).
- the gradient of a vector field (of a rank one tensorial field) which is a second rank tensor field:

---

[6]We strictly follow [6] left-to-right rule: indices are assigned to the terms appearing in an expression in the same order in which they appear when the expression is read from left to right; see (3.13) and (3.14). If $\underline{v}$ in (3.13) is a velocity field, the left-to-right rule leads to a velocity gradient which is the transpose of that defined by other authors: $\underline{\delta}_i \underline{\delta}_j \frac{\partial v_i}{\partial x_j}$.

[7]It is also common practice to call x_i the components of $\underline{r}$, so that $\underline{r} = \underline{\delta}_i x_i$.

$$\underline{\nabla}\,\underline{v} = \left(\underline{\delta}_i \frac{\partial}{\partial x_i}\right)(\underline{\delta}_j v_j) = \underline{\delta}_i \underline{\delta}_j \frac{\partial v_j}{\partial x_i} \tag{3.13}$$

– the divergence of a rank one tensorial field, which is a scalar field:

$$\underline{\nabla} \cdot \underline{v} = \left(\underline{\delta}_i \frac{\partial}{\partial x_i}\right) \cdot (\underline{\delta}_j v_j) = \delta_{ij} \frac{\partial v_j}{\partial x_i} = \frac{\partial v_i}{\partial x_i}$$

– the divergence of a second rank tensorial field, which is a rank one tensor field:

$$\underline{\nabla} \cdot \underline{\underline{\tau}} = \left(\underline{\delta}_i \frac{\partial}{\partial x_i}\right) \cdot (\underline{\delta}_j \underline{\delta}_k \tau_{jk}) = \underline{\delta}_k \delta_{ij} \frac{\partial \tau_{jk}}{\partial x_i} = \underline{\delta}_k \frac{\partial \tau_{ik}}{\partial x_i} \tag{3.14}$$

3.3 Transformation of Tensors

As already anticipated, a tensorial magnitude has an existence independent of any frame of reference. But depending on the choice of reference frame (the "unit base vectors" of our Cartesian space), its components will have different numerical values. In this section we will see how the components of a tensor in *two* different frames are related to each other.

The goal of this section is to answer the question:

if we know the components $T_{ij...}$ of a tensor $\underline{T}$ in a given frame ①②③, and the position/orientation of a second frame ①'②'③' with respect to ①②③, which are its components $T'_{ij...}$ in this second frame?

In the following, different frames will be distinguished by primes or asterisks. A second frame can differ from the original one by a translation without rotation with respect to the original one. In this case, the transformation rule cannot be simpler: the components of a tensor in both frames are the same.

If the second frame is related to the original one by a rotation, the situation is a bit more involved (and you probably already know the answer for rank one tensors). The reasoning to obtain the transformation rule can be split in simple steps:

1. tensors will not appear isolated, but in equations relating several of them. Take as example the (linear, microscopic) inverse Ohm's relation: $\underline{J} = \underline{\underline{\sigma}} \cdot \underline{E}$ where $\underline{J}$ is the electric current density, $\underline{\underline{\sigma}}$ is the symmetric second rank tensor of electric conductivity, and $\underline{E}$ the electric field.
2. for this equation to have physical meaning, it must hold in all reference frames. If Ohm's law looks like $\underline{J} = \underline{\underline{\sigma}} \cdot \underline{E}$ in the frame of reference ①②③, it must be of the form $\underline{J}' = \underline{\underline{\sigma}}' \cdot \underline{E}'$ in a different frame defined by axes ①'②'③'.
3. if we accept that all components of $\underline{J}$, $\underline{\underline{\sigma}}$ and $\underline{E}$ change when these tensors are referred to the new frame, the 12 numbers (three J_i, six σ_{ij}, three E_i) cannot change in an arbitrary way, but must obviously follow some quite specific rule if

Fig. 3.3 Components of a
rank one tensor (e.g. electric
field $\underline{E}$) in two different
reference frames

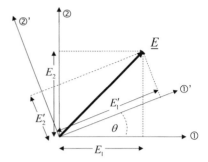

$\underline{J}' = \underline{\underline{\sigma}}' \cdot \underline{E}'$ is to hold. This requirement univocally determines the transformation
rule we are seeking.

We will now make the previous reasoning concrete for tensors of increasingly
higher rank. It is possible to obtain the general transformation rule of Sect. 3.3 directly,
but it is a good practical exercise to work through the ranks.

- transformation for 0th rank tensors: the first case is trivially simple: a scalar like the
  mass density ρ has the same value in all frames of reference. The transformation
  rule is as simple as:

$$\rho' = \rho \tag{3.15}$$

- transformation for rank one tensors: this probably is a familiar case. If we accept
  that first order tensors can be represented by arrows, a geometrical reasoning,
  especially in two-dimensional Cartesian coordinates, is clear enough.
  In Figure 3.3 we have represented the electric field and two reference frames. The
  original or "old" frame is defined by ①②, the destination or "new" one by ①'②'.
  This new systems is rotated a known angle θ with respect to the old one. The
  components E_i of the field in the old frame are known and we wish to calculate
  the values of the components E_i' in the new reference.[8]
  The new components can be obtained by adding up the projections of the two old
  components E_1, E_2 onto the new axes (Fig. 3.4):

$$
\begin{aligned}
E_1' &= \cos\theta E_1 + \sin\theta E_2 \\
E_2' &= -\sin\theta E_1 + \cos\theta E_2
\end{aligned}
\tag{3.16}
$$

which is a very familar expression. It is less probable that you have seen a similar
geometric construction in three dimensions, as it becomes rather unsightly. A
clearer and more efficient notation is based on the angles between the old and the
new axes. Let's define:

---

[8]The use of a "prime" to tag the new frame in this particular case does *not* imply that the old frame
will always be unprimed and the new frame always be primed.

Fig. 3.4 The new components E_1', E_2' are the sum of the projections of the old components E_1, E_2 onto the new axes ①'②'

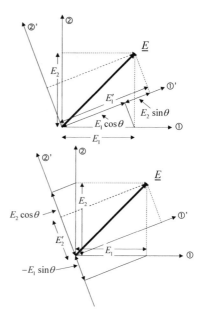

$$\theta_{ij} \equiv \text{angle between the new } i \text{ axis and the old } j \text{ axis}$$
$$l_{ij} \equiv \cos\theta_{ij}$$

In the example of Figs.3.3 and 3.4 these angles are θ_{11}, θ_{12}, θ_{21} and θ_{22}. They are not all independent, in fact: $\theta_{11} = \theta$, $\theta_{12} = \frac{\pi}{2} - \theta$, $\theta_{21} = \frac{\pi}{2} + \theta$, $\theta_{22} = \theta$. The advantage of using four angles when apparently one suffices is a clearer notation that can be easily extended. Equation (3.16) can now be written as:

$$E_1' = l_{11}E_1 + l_{12}E_2$$
$$E_2' = l_{21}E_1 + l_{22}E_2$$

which easily lends itself to generalization to three dimensions:

$$E_1' = l_{11}E_1 + l_{12}E_2 + l_{13}E_3$$
$$E_2' = l_{21}E_1 + l_{22}E_2 + l_{23}E_3 \qquad (3.17)$$
$$E_3' = l_{31}E_1 + l_{32}E_2 + l_{33}E_3$$

This can be written more compactly using the repeated index convention:

$$E_i' = l_{ij}E_j \qquad (3.18)$$

Equation (3.18) is the transformation law for rank one tensors we have sought.[9] The expression $E'_i = l_{ij} E_j$ guarantees that $\underset{\sim}{E}$ (e.g. the electric field or any other rank one tensor) keeps its physical significance when we refer it to different frames of reference.

The cosines appearing in (3.17) and (3.18) can be arranged in a matrix:

$$\underset{\sim}{L} \equiv \begin{pmatrix} l_{11} & l_{12} & l_{13} \\ l_{21} & l_{22} & l_{23} \\ l_{31} & l_{32} & l_{33} \end{pmatrix} \tag{3.19}$$

which has some special properties: it is an *orthogonal* matrix, i.e. if we interpret each of its three rows and three columns as vectors, these six vectors have unit modulus.

Furthermore, the three columns are orthogonal to each other; the same is true for the three rows. Its determinant $\det(\underset{\sim}{L})$ is equal to either 1 or -1, depending on whether $\underset{\sim}{L}$ represents a proper or an improper rotation. In addition, inverting an orthogonal matrix is very simple: $\underset{\sim}{L}^{-1} = \underset{\sim}{L}^T$, i.e. the inverse of an orthogonal matrix is its transpose.

In addition, the rows and columns of the transformation matrix $\underset{\sim}{L}$ are not just *any* unit vectors, but:

- its rows contain the components of the unit vectors of the *new* frame of reference, expressed in the old frame,
- its columns contain the components of the unit vectors of the *old* frame of reference, expressed in the new frame.

The previous two statements will come in handy when writing or interpreting specific instances of $\underset{\sim}{L}$.

For the purposes of this book, $\underset{\sim}{L}$ is a matrix, not a tensor (hence the "tilde" underline), that contains information about both the old and the new frames of reference. As an additional but not very surprising bonus, the matrix that transforms back from the new to the old frame of reference is the inverse of $\underset{\sim}{L}$, i.e. its transpose $\underset{\sim}{L}^{-1} = \underset{\sim}{L}^T$. The transformation rule that gives us the old components in terms of the new ones is:

$$E_i = l_{ji} E'_j.$$

---

[9] Again, the nine angles θ_{ij} (or their nine cosines l_{ij}) are not independent. Only three are. This is a consequence of the fact that in order to orient a frame of reference with respect to another in two dimensions, a single angle suffices, whereas three are required in three dimensions (for example three Euler angles). However, to orient a *single* axis or vector with respect to a frame of reference, one angle is enough in two dimensions, and two in three dimensions. We will come back to this point in later chapters.

Since $\underset{\sim}{L}$ is a matrix, we can use the double bracket notation to write the transformation rule (3.18) as:

$$[\![\underline{E}']\!] = \underset{\sim}{L} [\![\underline{E}]\!] \tag{3.20}$$

and the inverse transformation:

$$[\![\underline{E}]\!] = \underset{\sim}{L}^T [\![\underline{E}']\!] \tag{3.21}$$

The meaning of (3.20) is clear: if we place the components of a rank one tensor in the slots of a vector $[\![\underline{E}']\!]$ and multiply the transformation matrix by this vector (3.20), we obtain another vector $[\![\underline{E}']\!]$, the components of which are the components of the transformed rank one tensor $\underline{E}'$.

Since clearly there was nothing specific to the electric field $\underline{E}$ in the previous argument, the transformation law (3.18) is valid for all rank one tensors.

From the point of view of the operations (multiplications and additions) to be performed, (3.18) and (3.20) are identical. The shortest explanation of why $[\![\underline{E}']\!] = \underset{\sim}{L} [\![\underline{E}]\!]$ is correct, while $\underline{E}' = \underset{\sim}{L}\underline{E}$ is incorrect, is that the product of matrices and tensors is not even defined.

- transformation of second rank tensors: we will now follow a different route to obtain the transformation law for second rank tensors: we will demand that requirement 3 on Sect. 3.3 be fulfilled. We take a constitutive law, e.g. the inverse Ohm's law $(J_i = \sigma_{ij} E_j)$, in which tensors of both rank one, the electric field $\underline{E}$ and the electric current density $\underline{J}$, and second rank, the electric conductivity $\underset{=}{\sigma}$ (symmetric), appear.

Let's assume two different observers A and B whose frames of reference, unprimed for A and primed for B, are related by a rotation given by a known $\underset{\sim}{L}$.

Both observe one and the same experiment in which the electric field, the electric current density and the conductivity are measured. Since the components of rank one tensors like $\underline{E}$ and $\underline{J}$ depend on the reference, A and B will measure different components for them: J_i, E_j for A, J'_m, E'_l for B. They will also obtain different values for the six independent components of the conductivity: A will measure σ_{ij}, B will measure σ'_{ml}.

Imposing the condition that Ohm's inverse law be valid for both A and B implies:

Ohm's inverse law for A: $\quad \underline{E} = \underset{=}{\sigma} \cdot \underline{J} \quad (E_i = \sigma_{ij} J_j)$

Ohm's inverse law for B: $\quad \underline{E}' = \underset{=}{\sigma}' \cdot \underline{J}' \quad (E'_m = \sigma'_{ml} J'_l)$

Since we already know that the components of rank one tensors for A and B are related by (3.18), we can do the following:

- write Ohm's inverse law in B's frame: $J'_m = \sigma'_{ml} E'_l$,
- use the transformation law for $\underline{J}'$: $J'_m = l_{mn} J_n$,
- use Ohm's inverse law in A's frame $\underline{J}$: $J_n = \sigma_{nj} E_j$,
- transform $\underline{E}$ to B's frame: $E_j = l_{lj} E'_l$.

Taken together, the three last steps yield:

$$J'_m = l_{mn} J_n = l_{mn} \sigma_{nj} E_j = l_{mn} l_{lj} \sigma_{nj} E'_l \tag{3.22}$$

if we now compare (3.22) with the first expression in the list above:

$$J'_m = \sigma'_{ml} E'_l \tag{3.23}$$

we see that both the LHS and E'_l on the RHS of (3.22) and (3.23) are identical, so we have to conclude that the rest of (3.22) and (3.23) must also be equal:

$$\sigma'_{ml} = l_{mn} l_{lj} \sigma_{nj} \tag{3.24}$$

Equation 3.24 is the transformation rule for second rank tensors. Again, there is nothing special about $\underline{\underline{\sigma}}$ in the previous argument, so Eq. 3.24 must be valid for any second rank tensor.[10]

Notice the two repeated indices n and j on the RHS. They imply a double sum with nine terms (in 3D): in order to obtain *one* of the transformed components (like σ'_{ml}), we have to carry out nine multiplications and add up the results.

The inverse transformation is given by:

$$\sigma_{ml} = l_{nm} l_{jl} \sigma'_{nj} \tag{3.25}$$

The key difference between (3.24) and (3.25) is that in one case the subindices of $\underline{\underline{\sigma}}'$ on the LHS appear in the first position in the two "l"s, whereas in the inverse transformation the subindices of $\underline{\underline{\sigma}}$ appear in the second position of the "l"s.

The transformation law can (3.24) also be written in terms of vectors and matrices:

$$[\![\underline{\underline{\sigma}}']\!] = \underset{\sim}{L} [\![\underline{\underline{\sigma}}]\!] \underset{\sim}{L}^T \tag{3.26}$$

and the inverse transformation (3.25):

$$[\![\underline{\underline{\sigma}}]\!] = \underset{\sim}{L}^T [\![\underline{\underline{\sigma}}']\!] \underset{\sim}{L} \tag{3.27}$$

Expressions (3.24) and (3.26), and expressions (3.25) and (3.27) are equivalent from the point of view of the operations to be carried out.

- transformation of third and fourth rank tensors: an identical argument can be applied to constitutive laws in which third and fourth rank tensors appear in combination with lower rank tensors, the transformation rules of which are already known.

---

[10]The fact that $\underline{\underline{\sigma}}$ is a symmetrical tensor did not play any role in this derivation. Equation 3.24 is valid for asymmetric tensors as well.

Taking the piezoelectric moduli $\underset{=}{d}$ as a typical third rank tensor and the constitutive law of direct piezoelectricity $\underline{P} = \underset{=}{d} : \underline{\underline{\tau}}$:

$$P'_i = l_{ij}P_j = l_{ij}d_{jkm}\tau_{mk} = l_{ij}d_{jkm}l_{pm}l_{qk}\tau'_{pq} \tag{3.28}$$

$$= \underbrace{l_{ij}l_{qk}l_{pm}d_{jkm}}_{d'_{iqp}}\tau'_{pq} \Rightarrow d'_{iqp} = l_{ij}l_{qk}l_{pm}d_{jkm} \tag{3.29}$$

For fourth rank we can take the elastic compliance and Hooke's law $\underset{=}{\epsilon} = \underset{=}{s} : \underline{\underline{\tau}}$ and use the same argument:

$$\epsilon'_{ij} = l_{im}l_{jn}\epsilon_{mn} = l_{im}l_{jn}s_{mnpq}\tau_{qp} = l_{im}l_{jn}l_{sq}l_{tp}s_{mnpq}\tau'_{st}$$

$$= \underbrace{l_{im}l_{jn}l_{tp}l_{sq}s_{mnpq}}_{s'_{ijts}}\tau'_{st} \Rightarrow s'_{ijts} = l_{im}l_{jn}l_{tp}l_{sq}s_{mnpq} \tag{3.30}$$

which are the transformation laws for third and fourth rank tensors. There exist matrix-vector expressions analogous to (3.20) or (3.26) for tensors of rank higher than two [11, 12] but they are not essential for our purposes. When necessary, we will transform the components of third and fourth rank tensors by means of (3.28) or (3.30).

It should by now be quite obvious that the transformation rule for a Cartesian tensor $\underset{\vdots}{\underline{T}}$ of any rank is:

$$T'_{ijkmn}\ldots = l_{ip}l_{jq}l_{kr}l_{ms}l_{nt}\ldots T_{pqrst}\ldots \tag{3.31}$$

where:

- the T_{pqrst} are the (generally) known components to be transformed,
- the T'_{ijkmn} are the (generally) unknown transformed components,
- the factors $l_{ip}, l_{jq}, \ldots$ are the elements of the orthogonal transformation matrix $\underset{\smile}{L}$ (3.19),
- there are as many l's as the rank of the tensor,
- the subindices of T' on the LHS are the first subindices of the l's,
- the subindices of T on the RHS are the second subindices of the l's,

The transformation rules (3.15), (3.18), (3.24), (3.28), (3.30) are but particular cases of (3.31).

The rule for the inverse transformation is as expected:

$$T_{ijkmn}\ldots = l_{pi}l_{qj}l_{rk}l_{sm}l_{tn}\ldots T'_{pqrst}\ldots \tag{3.32}$$

At this point we can already understand why $\underset{\smile}{\delta}$ is called isotropic. Let's transform it by an arbitrary $\underset{\smile}{L}$. Being a second rank tensor, its new components are given by:

$$\delta'_{ij} = l_{ip}l_{jq}\delta_{pq} = l_{ip}l_{jp} \tag{3.33}$$

Written in full, the product $l_{ip}l_{jp}$ (sum over p) is $l_{i1}l_{j1} + l_{i2}l_{j2} + l_{i3}l_{j3}$, which is the scalar product of the ith and jth rows of $\underset{\smile}{L}$, which we know to be unit and orthogonal, hence the product is either 1 if the rows are the same ($i = j$), or 0 if the rows are different ($i \neq j$). This result can be expressed as $l_{ip}l_{jp} = \delta_{ij}$, so that (3.33) is:

$$\delta'_{ij} = l_{ip}l_{jp} = \delta_{ij}$$

that is, the new components of $\underline{\delta}$ are the same as the old ones. The tensor $\underline{\delta}$ has the same components in all frames of reference:

$$[\![\underline{\delta}]\!] = \begin{bmatrix} 1 & 0 & 0 \\ 0 & 1 & 0 \\ 0 & 0 & 1 \end{bmatrix}$$

and is therefore invariant under any rotation, proper or improper. This invariance is the hallmark of isotropy (Sect. 2.1), hence the name "isotropic" for $\underline{\delta}$.

It is important to realize that transforming a tensor becomes increasingly cumbersome as its rank increases[11]:

- transforming each of the three components of a rank one tensor implies carrying out three products and adding the results. Since there are three components, a total of 3×3 products are required.
- for second rank, two repeated indices appear in Eq. 3.24 and nine products are necessary to transform each one of the nine components of the tensor, yielding a total of $9 \times 9 = 81$ products.
- for third and fourth rank tensors, with three (3.28) and four (3.30) repeated indices respectively, a total of $27 \times 27 = 729$ and $81 \times 81 = 6561$ products have to be carried out.
- in general, transforming all components of an n rank tensor in d dimensions implies d^{2n} products.

The following table sums up the transformation laws for ranks one through four:

rank	transformation law	equivalent matrix expression
0	$T' = T$	$- \; - \; -$
1	$T'_i = l_{ij}T_j$	$[\![\underline{T'}]\!] = \underset{\smile}{L}\,[\![\underline{T}]\!]$
2	$T'_{ij} = l_{im}l_{jn}T_{mn}$	$[\![\underline{T'}]\!] = \underset{\smile}{L}\,[\![\underline{T}]\!]\underset{\smile}{L}^T$
3	$T'_{iqp} = l_{ij}l_{qk}l_{pm}T_{jkm}$	$- \; - \; -$
4	$T'_{ijts} = l_{im}l_{jn}l_{tp}l_{sq}T_{mnpq}$	$- \; - \; -$

---

[11] The following applies to tensors without any special symmetry. Symmetry reduces the number of independent components.

3.4 Diagonalization of a Second Rank Tensor; Principal Directions

Many important constitutive equations involve a linear relation between two field tensors, the proportionality constant being a second rank matter tensor. Ohm's law $\underline{E} = \underline{\underline{\rho}} \cdot \underline{J}$ is a typical example in which the material property is the second rank, symmetric resistivity tensor $\underline{\underline{\rho}}$. From Ohm's law, written either in tensor or in double bracket notation:

$$\underline{E} = \underline{\underline{\rho}} \cdot \underline{J} \quad [\![\underline{E}]\!] = [\![\underline{\underline{\rho}}]\!][\![\underline{J}]\!] \tag{3.34}$$

it is clear that, unless $\underline{\underline{\rho}}$ has a very special form, the tensors $\underline{E}$ and $\underline{J}$, interpreted as "arrows", will point in different directions: cause and effect will not be colinear (Fig. 3.5). The only way for $\underline{E}$ and $\underline{J}$ to be colinear in *all* cases is for $\underline{\underline{\rho}}$ to be a) diagonal and b) have the same value in all diagonal components, i.e. for $\underline{\underline{\rho}}$ being a scalar multiple of the second rank isotropic tensor $\underline{\underline{\rho}} = \rho \underline{\underline{\delta}}$ so that:

$$\underline{E} = \rho \underline{\underline{\delta}} \cdot \underline{J} = \rho \underline{J}$$

so that $\underline{J}$ is a scalar multiple of $\underline{E}$. This result is not specific for the electric resistivity; any second rank property $\underline{\underline{T}}$ for an isotropic material will be of the form:

$$\underline{\underline{T}} = T \underline{\underline{\delta}} \quad \text{or} \quad T_{ij} = T \delta_{ij} \quad \text{or} \quad [\![\underline{\underline{T}}]\!] = T [\![\underline{\underline{\delta}}]\!] = \begin{bmatrix} T & 0 & 0 \\ 0 & T & 0 \\ 0 & 0 & T \end{bmatrix}$$

There are however some directions along which cause and effect are colinear, even for an anisotropic material. These directions are called *principal directions* and can be found by demanding colinearity of cause and effect. We want the product

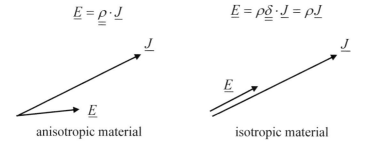

$$\underline{E} = \underline{\underline{\rho}} \cdot \underline{J} \qquad\qquad \underline{E} = \rho \underline{\underline{\delta}} \cdot \underline{J} = \rho \underline{J}$$

anisotropic material   isotropic material

Fig. 3.5 An anisotropic material (left) deviates effect from cause, (electric field $\underline{E}$ and electric current density $\underline{J}$). For an isotropic material (right), cause and effect are colinear

$\underline{\rho} \cdot \underline{J}$ to be a scalar multiple of $\underline{J}$ with an a priori unknown proportionality constant $\overline{\overline{\lambda}}$:

$$\underline{E} = \underline{\underline{\rho}} \cdot \underline{J} = \lambda \underline{J} = \lambda \underline{\underline{\delta}} \cdot \underline{J} \quad \Rightarrow \quad (\underline{\underline{\rho}} - \lambda \underline{\underline{\delta}}) \cdot \underline{J} = \underline{0}$$

where the value of λ is to be determined together with the direction of $\underline{J}$. In matrix-vector notation:

$$\left([\![\underline{\underline{\rho}}]\!] - \lambda \mathbb{1} \right) \cdot [\![\underline{J}]\!] = [\![\underline{0}]\!] \qquad \begin{bmatrix} \rho_{11} - \lambda & \rho_{12} & \rho_{13} \\ \rho_{12} & \rho_{22} - \lambda & \rho_{23} \\ \rho_{13} & \rho_{23} & \rho_{33} - \lambda \end{bmatrix} \begin{bmatrix} J_1 \\ J_2 \\ J_3 \end{bmatrix} = \begin{bmatrix} 0 \\ 0 \\ 0 \end{bmatrix}$$

This is a homogeneous system of linear equations, the solutions of which are obtained from the condition:

$$\begin{vmatrix} \rho_{11} - \lambda & \rho_{12} & \rho_{13} \\ \rho_{12} & \rho_{22} - \lambda & \rho_{23} \\ \rho_{13} & \rho_{23} & \rho_{33} - \lambda \end{vmatrix} = 0$$

which is a cubic polynomial in λ. Since the matrix $[\![\underline{\underline{\rho}}]\!]$ is symmetric and its components are real numbers, the polynomial is guaranteed to have three real roots, called the *eigenvalues* or *principal* values of $[\![\underline{\underline{\rho}}]\!]$, and at least three principal directions. Depending on how many of the principal values are distinct, a minimum of three and at most a double infinity[12] of principal directions exist.

Principal directions have specific physical meaning:

- along the principal directions, cause and effect are colinear for all materials,
- if expressed in a reference frame formed by three unit vectors $\underline{n}^{①}$, $\underline{n}^{②}$, $\underline{n}^{③}$ along the principal directions (having associated eigenvalues λ_1, λ_2, λ_3),

$$\underbrace{L}_{\underline{n}^{①}\underline{n}^{②}\underline{n}^{③}} = \begin{bmatrix} n_1^{①} & n_2^{①} & n_3^{①} \\ n_1^{②} & n_2^{②} & n_3^{②} \\ n_1^{③} & n_2^{③} & n_3^{③} \end{bmatrix}$$

a second rank property is diagonal,
- the eigenvalues appear along the diagonal in the same sequence as the order in which the eigenvectors are assigned as unit vectors:

$$[\![\underline{\underline{\rho}}]\!]_{\underline{n}^{①}\underline{n}^{②}\underline{n}^{③}} = \begin{bmatrix} \lambda_1 & 0 & 0 \\ 0 & \lambda_2 & 0 \\ 0 & 0 & \lambda_3 \end{bmatrix} \qquad \begin{matrix} \lambda_1 \Leftrightarrow \underline{n}^{①} \\ \lambda_2 \Leftrightarrow \underline{n}^{②} \\ \lambda_3 \Leftrightarrow \underline{n}^{③} \end{matrix}$$

---

[12]Why double?

- for all classes in which the crystallographic axes are orthogonal (i.e. orthorhombic, tetragonal and cubic systems), both conventional and crystallographic axes are defined so as to coincide with the principal directions,
- for classes in the tetragonal, trigonal and hexagonal system, plus the conical and cylindrical limit classes, the crystallographic axis z and the conventional axis ③ coincide with one principal direction. The eigenvalues associated with the other two directions are equal (double eigenvalue),
- for classes in the monoclinic system, the crystallographic axis y and the conventional axis ② coincide with a principal direction, the conventional axes ① and ③ can be chosen at will,
- in classes in the triclinic system (in which all three conventional axes can be chosen at will), neither crystallographic nor conventional axes coincide with any principal directions,
- the largest eigenvalue (say λ_1 if we take $\lambda_1 > \lambda_2 > \lambda_3$) is the maximum value the projection of the second rank property can have. Similarly λ_3 is the minimum value of the projection.

3.5   Voigt Notation

That tensors of rank one and two can be represented by means of vectors and matrices is at least intuitive, since their components can be arranged in one- and two-dimensionaless arrays in a natural way. Tensors of third and fourth rank would correspond to three- and four-dimensional arrays (Fig. 3.2) which cannot easily be displayed let alone manipulated on two-dimensional paper.

There is however a neat trick invented by Voigt [13] that condenses two indices into one and thus "flattens" third and fourth rank tensors down to matrices, provided some of their indices have certain symmetries.

Voigt condensation can be applied to a pair of subindices of any tensor *if and only if* the tensor is symmetric in these two indices, a case that comes up in practice very often. Table 3.3 shows how two indices are condensed into one.

There is also a very simple mnemotechnic rule. Take a walk along the dotted line as indicated by the arrow and renumber the pair of indices as you come across them in the following scheme:

Table 3.3 Voigt notation

indices of the tensor →	11	22	33	23 and 32	13 and 31	12 and 21
	↓	↓	↓	↓	↓	↓
single, condensed index →	1	2	3	4	5	6

$$\begin{pmatrix} 11 & 12 & 13 \\ & 22 & 23 \\ & & 33 \end{pmatrix}$$

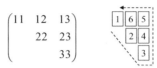

Since the tensor must be symmetric for the condensation to be applicable, pairs of subindices like 21 and 12 condense into the same index. Notice that while each of the original indices ranges from 1 to 3, the condensed index runs from 1 to 6.

Indices in which the tensor is not symmetric are not affected by Voigt condensation. A tensor like the piezoelectric moduli $\underset{=}{d}$, which is symmetric in the last two indices, can be represented in Voigt notation by a 3×6 matrix: its first index runs from 1 to 3, the remaining two are condensed into one which runs from 1 to 6.

The resulting 3×6 array of components is a matrix and not a tensor. It does not transform according to any of the transformation rules for tensors. Remember we use the "under-tilde" symbol to denote matrices (square or not, like $\underset{\sim}{d}$) and distinguish them from tensors $\underset{=}{d}$

The question may arise as to how to tell apart a component of a tensor from a component of a matrix: is ρ_i the ith component of a rank one tensor ρ, or is it the ith *condensed* component of a symmetric second rank tensor written in Voigt notation? The answer is that, in general, more information is needed to answer the question. If a sufficiently large number of properties (symbols) are used, as it is bound to happen in a book about material properties, a clash of symbols will likely happen.[13]

The property names of Table 1.1 have been chosen so that no confusion of this kind can arise. Taking the symbol ρ as an example:

- ρ (no subindices) is the mass density, a zero rank tensor,
- ρ_{ij} is the i, jth component of the electric resistivity $\underset{=}{\rho}$, in tensor subindex notation,
- ρ_i is the ith component of the same electric resistivity written in Voigt notation $\underset{\sim}{\rho}$,
- ρ_{ijkl}^{mag} is the i, j, k, l component of the magnetoresistivity $\underset{\equiv}{\rho}^{mag}$,
- ρ_{ij}^{mag} is the i, jth component of the same magnetoresistivity written in Voigt notation $\underset{\sim}{\rho}^{mag}$.

Table 3.5 shows some selected properties together with the symmetry of their indices, in both tensor and Voigt notation.

The symmetries listed in the third column of Table 3.4 can also be expressed using the transposition symbol "T". For example, the stress tensor $\underset{=}{\tau}$ is symmetric in its only pair of subindices:

$$\underset{=}{\tau} = \underset{=}{\tau}^T \qquad \tau_{ij} = \tau_{ji}$$

The Kerr coefficient is symmetric in its first two indices and, separately, in its last two indices, but not by interchange of the first and last pair of indices:

---

[13]This is the reason for naming the stress tensor $\underset{=}{\tau}$ instead of $\underset{=}{\sigma}$, (that is reserved for the electric conductivity) and the dielectric permittivity $\underset{=}{\kappa}$ instead of $\underset{=}{\epsilon}$ (which is reserved for the deformation or displacement gradient tensor).

Table 3.4 Examples of factors of 2 and 4 to be used in index condensation. Subindices m y n are condensed indices obtained from $i, j, \ldots$, according to Table 3.3

Tensor		
Stress (no * in Table 3.5)	$\tau_m = \tau_{ij}$	In all cases
Deformation (one *)	$\epsilon_m = \epsilon_{ij}$	If m is 1, 2 o 3
	$\epsilon_m = 2\epsilon_{ij}$	If m is 4, 5 o 6
Pöckels coeff. (no *)	$r_{mk} = r_{ijk}$	In all cases
Piezoelectric moduli (one *)	$d_{im} = d_{ijk}$	If m is 1, 2 o 3
	$d_{im} = 2d_{ijk}$	If m is 4, 5 o 6
Elastic stiffness (no *)	$c_{mn} = c_{ijkl}$	In all cases
Kerr coeff. (one *)	$K_{mn} = K_{ijkl}$	If n is 1, 2 o 3 (for any m)
	$K_{mn} = 2K_{ijkl}$	If n is 4, 5 o 6 (for any m)
Elastic compliance (two *)	$s_{mn} = s_{ijkl}$	If m is 1, 2 o 3 and if n is 1, 2 o 3
	$s_{mn} = 2s_{ijkl}$	If m or n but not both are 4, 5 o 6
	$s_{mn} = 4s_{ijkl}$	If both m y n are 4, 5 o 6

$$\underline{\underline{K}} = {}^T\underline{\underline{K}} = \underline{\underline{K}}^T \quad (\Rightarrow = {}^T\underline{\underline{K}}^T) \qquad K_{ijmn} = K_{jimn} = K_{jimn} = K_{jinm} \neq K_{mnij}$$

In these two examples, both the transpose symbol and subindices are notationally roughly equally efficient. The elastic compliance $\underline{\underline{s}}$ and stiffness $\underline{\underline{c}}$ have the same symmetries as the Kerr coefficient, but, in addition, are symmetric under interchange of the first and second pairs of subindices. For these very symmetric tensors, the use of the transposition symbol borders on the ridiculous. Compare the use of subindices:

$$s_{ijmn} = s_{jimn} = s_{ijnm} \quad (\Rightarrow = s_{jinm}) \qquad \text{(minor symmetries)}$$
$$= s_{mnij} \qquad \text{(major symmetry)}$$

with the use of "T":

$$\underline{\underline{s}} = {}^T\underline{\underline{s}} = \underline{\underline{s}}^T \quad (\Rightarrow = {}^T\underline{\underline{s}}^T) \qquad \text{(minor symmetries)}$$

$$= {}^T\left[\left\{ {}^T\left[\left({}^T\underline{\underline{s}}^T \right) \right]^T \right\}^T \right]^T \qquad \text{(major symmetry)}$$

Voigt's notation is a practical method to display the components of tensors of third and fourth rank, but its interest would be rather limited if this were its only function. Index condensation is valuable because tensorial constitutive equations can be re-

Table 3.5 Index symmetries, tensor and Voigt notation for selected properties

Tensor	Rank	Symmetry	Tensorial	Voigt		Size
Stress	2	$1 \leftrightarrow 2$	$\underline{\underline{T}}\, \tau_{ij}$	$\vec{\tau}$	τ_m	6×1
Strain*	2	$1 \leftrightarrow 2$	$\underline{\underline{\epsilon}}\, \epsilon_{ij}$	$\vec{\epsilon}$	ϵ_m	6×1
Thermal expansion*	2	$1 \leftrightarrow 2$	$\underline{\underline{\alpha}}\, \alpha_{ij}$	$\vec{\alpha}$	α_m	6×1
Rel. diel. imperm.	2	$1 \leftrightarrow 2$	$\underline{\underline{B}}\, B_{ij}$	$\vec{B}$	B_m	6×1
Piezoelectric mod.*	3	$2 \leftrightarrow 3$	$\underline{\underline{\underline{d}}}\, d_{ijk}$	$\underset{\smile}{d}$	d_{im}	3×6
Pöckels coeff.	3	$1 \leftrightarrow 2$	$\underline{\underline{r}}\, r_{ijk}$	$\underset{\smile}{r}$	r_{mk}	6×3
Elasto-optic coeff.	4	$1 \leftrightarrow 2,\, 3 \leftrightarrow 4$	$\underline{\underline{\underline{p^{opt}}}}\, p^{opt}_{ijkl}$	$\underset{\smile}{p^{opt}}$	p^{opt}_{mn}	6×6
Piezo-optic coeff.*	4	$1 \leftrightarrow 2,\, 3 \leftrightarrow 4$	$\underline{\underline{\underline{\pi^{opt}}}}\, \pi^{opt}_{ijkl}$	$\underset{\smile}{\pi^{opt}}$	π^{opt}_{mn}	6×6
Piezoresistive coeff.*	4	$1 \leftrightarrow 2,\, 3 \leftrightarrow 4$	$\underline{\underline{\underline{\pi}}}\, \pi_{ijkl}$	$\underset{\smile}{\pi}$	π_{mn}	6×6
Electrostric. coeff**	4	$1 \leftrightarrow 2,\, 3 \leftrightarrow 4$	$\underline{\underline{\underline{M}}}\, M_{ijkl}$	$\underset{\smile}{M}$	M_{mn}	6×6
Magnetores. coef.*	4	$1 \leftrightarrow 2,\, 3 \leftrightarrow 4$	$\underline{\underline{\underline{\rho^{mag}}}}\, \rho^{mag}_{ijkl}$	$\underset{\smile}{\rho}$	ρ_{mn}	6×6
Coef. Kerr*	4	$1 \leftrightarrow 2,\, 3 \leftrightarrow 4$	$\underline{\underline{\underline{K}}}\, K_{ijkl}$	$\underset{\smile}{K}$	K_{mn}	6×6
Elastic stiffness	4	$1 \leftrightarrow 2,\, 3 \leftrightarrow 4,$ $(12) \leftrightarrow (34)$	$\underline{\underline{\underline{c}}}\, c_{ijkl}$	$\underset{\smile}{c}$	c_{mn}	6×6
Elastic compliance**	4	$1 \leftrightarrow 2,\, 3 \leftrightarrow 4,$ $(12) \leftrightarrow (34)$	$\underline{\underline{\underline{s}}}\, s_{ijkl}$	$\underset{\smile}{s}$	s_{mn}	6×6

written in Voigt notation, i.e. using matrices and vectors, and still maintain the same formal aspect. If we take the elastic compliance $\underline{\underline{\underline{s}}}$ and Hooke's law as examples:

- the tensorial form of Hooke's law states that deformation equals the product (double contraction) of the elastic compliance and the stress tensor: $\underline{\underline{\epsilon}} = \underline{\underline{\underline{s}}} : \underline{\underline{\tau}}$,

- written in component form, $\epsilon_{ij} = s_{ijkl}\tau_{lk}$, the two repeated indices k, l imply that in order to calculate one of the nine components of $\underline{\underline{\epsilon}}$, it is necessary to perform nine products and add them together,

- the nifty thing about index condensation is that if the three tensors $\underline{\underline{\epsilon}}, \underline{\underline{\underline{s}}}, \underline{\underline{\tau}}$ are written in Voigt notation, the equality $\vec{\epsilon} = \underset{\smile}{s}\,\vec{\tau}$ still holds. In this case, Voigt notation simplifies a double contraction to a simple matrix times vector product in which a component of the result is computed by performing only three products and adding them.

In short, constitutive equations like Hooke's can be used both in tensor notation and in Voigt's condensed notation. This is by no means obvious. If we consider the major shuffling of components caused by condensation, it seems almost miraculous that the condensed form of Hooke's law $\vec{\epsilon} = \underset{\sim}{s}\,\vec{\tau}$ ($\epsilon_i = s_{ij}\tau_j$) still works.

As a matter of fact, *it does not*, unless a final correction is performed. For Voigt notation to work it is necessary to include a factor of 2 or 4 when condensing the indices of some tensors. Table 3.4 shows a few selected examples in more detail. Table 3.5 lists which tensors require no factor, which require a factor of 2 and which a factor of 4.

In Table 3.5:

- components of properties without asterisk need no correction factor,
- components of properties of second and third rank with an asterisk (*) take a factor of 2 when the (only) condensed index of the component is ≥ 4,
- components of properties of fourth rank with an asterisk (*) take a factor of 2 when the *second* (of the two) condensed indices of the component is ≥ 4,
- components of properties of fourth rank with two asterisks (**) take a factor of 2 when any one of the two condensed indices of the component is ≥ 4, and take a factor of 4 when both condensed indices are ≥ 4.

It is straightforward to see why these factors must arise. Let us take, as an example, the piezoelectric moduli $\underset{\sim}{d}$ that appear in the constitutive equation for direct piezolectricity $\underline{P} = \underset{\equiv}{d} : \underset{\equiv}{\tau}$. The goal is for the equation written in condensed notation $\vec{P} = \underset{\sim}{d}\,\vec{\tau}$ to also be satisfied.

For Voigt notation to work, the components of the electric polarization $\underline{P}$ must have the same values when calculated via $\underline{P} = \underset{\equiv}{d} : \underset{\equiv}{\tau}$ and when calculated via $\vec{P} = \underset{\sim}{d}\,\vec{\tau}$.

- in tensor notation, the component P_1 is the sum of the nine terms $P_1 = d_{1jk}\tau_{kj}$ for $j = 1, 2, 3$ and $k = 1, 2, 3$:

$$P_1 = d_{111}\tau_{11} + d_{112}\tau_{21} + d_{113}\tau_{31} + d_{121}\tau_{12} + d_{122}\tau_{22} \tag{3.35}$$

$$+ d_{123}\tau_{32} + d_{131}\tau_{13} + d_{132}\tau_{23} + d_{133}\tau_{33} = \tag{3.36}$$

$$= d_{111}\tau_{11} + d_{122}\tau_{22} + d_{133}\tau_{33} + 2d_{123}\tau_{23} + 2d_{113}\tau_{13} + 2d_{112}\tau_{12} \tag{3.37}$$

where the symmetries of $\underset{\equiv}{d}$ ($\underset{\equiv}{d}^T = \underset{\equiv}{d}$) and of $\underset{\equiv}{\tau}$ ($\underset{\equiv}{\tau}^T = \underset{\equiv}{\tau}$) have been used.

- in Voigt notation, the component P_1 is the sum of only six terms obtained from $P_1 = d_{1j}\tau_j$ for $j = 1, \ldots, 6$:

$$P_1 = d_{11}\tau_1 + d_{12}\tau_2 + d_{13}\tau_3 + d_{14}\tau_4 + d_{15}\tau_5 + d_{16}\tau_6$$

If we undo index condensation in this expression without including any factor of 2, we would obtain:

$$P_1 = d_{111}\tau_{11} + d_{122}\tau_{22} + d_{133}\tau_{33} + d_{123}\tau_{23} + d_{113}\tau_{13} + d_{112}\tau_{12} \quad \text{(wrong!)} \tag{3.38}$$

We see that Eqs. (3.37) and y (3.38) are not identical unless a factor of 2 is introduced in the terms: $d_{14} = 2d_{123}$, $d_{15} = 2d_{113}$, $d_{16} = 2d_{112}$.

Thanks to Voigt notation we will be able to use:

$$\vec{P} = \underset{\sim}{d}\,\vec{\tau} \qquad \vec{\epsilon}^T = \vec{E}^T \underset{\sim}{d} \qquad \vec{\tau} = \underset{\sim}{c}\,\vec{\epsilon} \qquad \vec{\epsilon} = \underset{\sim}{s}\,\vec{\tau}q \qquad \text{Voigt notation}$$

$$P_i = d_{ij}\tau_j \qquad \epsilon_i = E_j d_{ji} \qquad \tau_i = c_{ij}\epsilon_j \qquad \epsilon_i = s_{ij}\tau_j$$

instead of:

$$\underline{P} = \underset{=}{d} : \underline{\tau} \qquad \underset{=}{\epsilon} = \underset{=}{E} \cdot \underset{=}{d} \qquad \underset{=}{\tau} = \underset{=}{c} : \underset{=}{\epsilon} \qquad \underset{=}{\epsilon} = \underset{\equiv}{s} : \underset{=}{\tau} \qquad \text{tensor notation}$$

$$P_i = d_{ijk}\tau_{kj} \qquad \epsilon_{ij} = E_k d_{kij} \qquad \tau_{ij} = c_{ijkl}\epsilon_{lk} \qquad \epsilon_{ij} = s_{ijkl}\tau_{lk} \quad \text{index notation}$$

whenever required.

In some of the above expressions we have written $\vec{P}$ instead of $\underline{P}$, despite the fact the $\underline{P}$ only has one subindex and Voigt notation does not even apply to it. The reason is that it is not allowed to mix tensors and matrices in the same expression. Thus $\underline{P} = \underset{\sim}{d}\,\vec{\tau}$ is not correct. The meaning of $\vec{P}$ is just $[\![\,\underline{P}\,]\!]$, so that $\vec{P} = [\![\,\underline{P}\,]\!] = \underset{\sim}{d}\vec{\tau}$, and $\vec{P}$ is used instead of $[\![\,\underline{P}\,]\!]$ for conciseness.

Voigt notation is practical and it is extensively used in practice. There is however a key issue to bear in mind about index condensation: once a tensor is written in Voigt notation, it is converted into a vector or a matrix and does not obey the transformation rules (3.31) for tensors any more. Thus, it cannot be directly transformed to another reference frame.

On the other hand, the numerical values of the components of tensorial properties found in the literature are invariably expressed in Voigt notation. The obvious question is then: to which frame are these values referred to? The choice is in principle arbitrary, but a very general consensus exists: properties in Voigt notation are referred to the conventional axes explained in Sect. 2.4.1.

Therefore, when using numerical values for the components of a tensorial property like $\underset{\sim}{s}$ in Voigt notation we are tied to the conventional reference frame. If it is necessary to use Voigt notation in a different frame, extra work is needed:

- given the values of the components of the property, say $\underset{\sim}{s}$, first undo index condensation: $s_{ij} \to s_{mnpq}$ ($\underset{\sim}{s} \to \underset{=}{s}$),
- since $\underset{=}{s}$ is a bona fide tensor, use (3.31) to transform each of its components to the new reference frame: $s_{mnpq} \to s'_{mnpq}$ ($\underset{=}{s} \to \underset{=}{s'}$)
- apply index condensation again $s'_{mnpq} \to s'_{ij}$ ($\underset{=}{s'} \to \underset{\sim}{s'}$) to get the values of the components in Voigt notation in the new frame.

The computational effort to perform the change of reference frame can be considerable, the more so the higher the tensorial rank (Sect. 3.3).

3.6 Geometrical Symmetry and Independence of Components

Chapter 2 dealt with the subject of geometrical symmetry. It primarily relied on visual/spatial intuition and was largely equation-free. The present chapter has focused up to now on symbol manipulation for Cartesian tensors and is largely image-free. It is not easy to see how both chapters could be related, and yet there is very direct link between both that we will now explore.

The key idea is simple: geometric symmetry in the material may enforce relations between the components of its properties. Two or more components may be equal, whereas others or even all components of a property may cancel, etc.

The most intuitive way to see how geometric symmetry establishes relationships between the components of a tensorial property is to carry out the calculation for a specific example.

The starting point is the knowledge of the point group G to which the material belongs: for each of the elements $g \in G$ of the point group of the material the representation matrix $\underset{\sim}{L_g}$ is known. The reasoning can be split in steps:

1. the geometry of the material is invariant under g: transforming it by means of $\underset{\sim}{L_g}$ leaves it invariant. As g produces no geometrical change in the material, all properties (which of course depend on the geometry) must also remain invariant under $\underset{\sim}{L_g}$.
2. on the other hand, all tensor properties of the material must transform according to (3.31).
3. from the previous two statements we conclude that the properties transformed by g and the properties of the original material before the transformation, must be equal. When this condition is enforced, numerical relationships among components will appear.

The previous steps are then repeated for all $\underset{\sim}{L_g}$ representing the elements $g \in G$. In practice, it is seldom necessary to repeat the procedure for all $g \in G$.

One important and simple example is the effect of an inversion center on the piezoelectric moduli $\underset{\equiv}{d}$. Let's assume a material belongs to point group $\bar{1}$, i.e. it is invariant under the unit E ("do nothing"), and under an inversion. The unit E leads to the trivial conclusion that every component of $\underset{\equiv}{d}$ must be equal to itself.

For the inversion center $\bar{1}$:

1. the representation matrix $\underset{\sim}{L_{\bar{1}}}$ is obtained immediately from its action on the coordinate axes as shown in Fig. 3.6.
   The matrix $\underset{\sim}{L_{\bar{1}}}$ is then (its rows are the new base vectors expressed in the old reference frame):

$$\underset{\sim}{L_{\bar{1}}} = \begin{pmatrix} -1 & 0 & 0 \\ 0 & -1 & 0 \\ 0 & 0 & -1 \end{pmatrix} \tag{3.39}$$

Fig. 3.6 An inversion center
transforms axes ①②③ into
①'②'③'

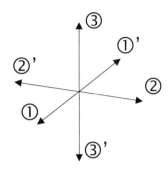

2. since the material is invariant under $\bar{1}$, nothing must change when the geometry
   of the material is subjected to inversion. Therefore:

$$d'_{ijk} = d_{ijk} \tag{3.40}$$

   must hold.
3. independently of the previous step, each of the components of $\underset{=}{d}$ is transformed
   by the $\bar{1}$ according to $d'_{ijk} = l_{im}l_{jn}l_{kp}d_{mnp}$, where the l_{ij} are the components of
   $\underset{\smile}{L_{\bar{1}}}$ (3.39). The component ①①① must therefore transform as:

$$d'_{111} = l_{1m}l_{1n}l_{1p}d_{mnp}$$

   There are three repeated indices on the RHS, hence three sums and 27 terms in
   all. Many of them however are zero. As a matter of fact only one is non-zero,
   because from the three l_{1i} two are zero and only $l_{11} = -1$:

$$d'_{111} = l_{11}^3 d_{111} = (-1)^3 d_{111} = -d_{111} \tag{3.41}$$

4. the only possibility for (3.40) and (3.41) to be simultaneously fulfilled is $d_{111} =$
   $d'_{111} = 0$. The component ①①① must hence be zero.

   It turns out that if the same procedure is now applied to the ①①② component, the
result is again the same. In fact, it is easy to verify that the cancellation takes place
for *all* components of $\underset{=}{d}$, that is: $d_{ijk} = d'_{ijk} = 0$, $\forall i, j, k$.

   It is possible to save time by noticing that the elements of $\underset{\smile}{L_{\bar{1}}}$ are $l_{ij} = -\delta_{ij}$.
Applying the transformation rule to a generic component d_{mnp}:

$$d'_{ijk} = (-\delta_{im})(-\delta_{jn})(-\delta_{kp})d_{mnp} = (-1)^3 d_{ijk} = -d_{ijk}$$
$$\Rightarrow \quad d_{ijk} = 0 \ \forall i, j, k \quad \Rightarrow \quad \underset{=}{d} = \underset{=}{0}$$

   We thus conclude that any material whose structure possesses an inversion center
(a centrosymmetric material) cannot be used as piezoelectric. It is not just that the

numerical values of its $\underline{\underline{d}}$ are very small, they are exactly zero due to the presence of an inversion center.

It is also clear that the previous argument is independent of the name of the property, the only important aspect is that the property be third rank.

The result is even more general: if we consider a generic fifth rank property, say $\underline{\underline{T}}$, and the same procedure is applied we obtain:

$$T'_{ijkmn} = (-\delta_{ip})(-\delta_{jq})(-\delta_{kr})(-\delta_{ms})(-\delta_{nt})T_{pqrst} = (-1)^5 T_{ijkmn} = -T_{ijkmn}$$
$$\Rightarrow \quad T_{ijkmn} = 0 \; \forall i, j, k, m, n \quad \Rightarrow \quad \underline{\underline{T}} = \underline{\underline{0}}$$

The essential point is that the appearance of an odd number of -1 factors, as in (3.41), sets all components of *any* odd rank property to zero. Hence:

All odd rank tensor properties for centrosymmetric materials vanish.

Let's apply the previous argument to *even* rank tensor properties for centrosymmetric materials. In this case, there will be an even number of -1 factors. For fourth rank we have:

$$s'_{ijkl} = \delta_{im}\delta_{jn}\delta_{kp}\delta_{lq} s_{mnpq} = (-1)^4 s_{ijkl} = s_{ijkl}$$

which is always fulfilled. Thus, a center of symmetry imposes no restrictions whatsoever on the components of even rank tensor properties.

The point group of the material may contain other symmetry elements apart from the inversion center. These other elements may give rise to further relations among components of both odd and even rank properties, as we will now see.

Let us consider a generic second rank property $\underline{T}$, not necessarily symmetric, for a monoclinic material of class m and see the influence of the point group symmetry on the values of the components of $\underline{T}$.

The geometric symmetry elements belonging to the monoclinic point group m are the unit E and a plane m_y perpendicular to the conventional ② axis. The first action m'_y of the plane m_y[14] on the conventional axes is represented in Fig. 3.7. Hence, the point group contains two elements, whose representation matrices are:

$$\underset{\sim}{L}_E = \begin{bmatrix} 1 & 0 & 0 \\ 0 & 1 & 0 \\ 0 & 0 & 1 \end{bmatrix} \qquad \underset{\sim}{L}_{m'_y} = \begin{bmatrix} 1 & 0 & 0 \\ 0 & -1 & 0 \\ 0 & 0 & 1 \end{bmatrix}$$

While the unit E imposes no relationships between the components of $\underline{T}$, the results of carrying out steps 1 to 4 above for $\underset{\sim}{L}_{m'_y}$ and each of the components of $\underline{\underline{T}}$ are shown in Table 3.6.

---

[14] What is the second action of the plane?

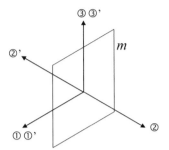

Fig. 3.7 First action m_y^1 of the symmetry plane m_y on the conventional axes

Table 3.6 Structures for rank one properties (such as the pyroelectric modulus p). Only materials belonging to one of these 12 polar classes can have first rank properties. The integer to the right of str( ) is the number of independent components of the property

·	zero component
●	non-zero component

system(s)	class(es)	str ()	
triclinic	1	$\begin{bmatrix} • \\ • \\ • \end{bmatrix}$	3
monoclinic	2	$\begin{bmatrix} · \\ • \\ · \end{bmatrix}$	1
	m	$\begin{bmatrix} • \\ · \\ • \end{bmatrix}$	2
orthorhombic	$mm2$	$\begin{bmatrix} · \\ · \\ • \end{bmatrix}$	1
tetragonal	$4, 4mm$	$\begin{bmatrix} · \\ · \\ • \end{bmatrix}$	1
trigonal	$3, 3m$		
hexagonal	$6, 6mm$		
limit classes	$\infty, \infty m$		

In this table, the first column shows the result of applying the transformation law for rank one tensors (step 3), the second column states that nothing should change when the material is subjected to $\underset{\sim}{L}_{m_y^1}$ because the material is invariant under this

Table 3.7 Structures for second rank symmetric properties. The integer to the right of str( ) is the number of independent components of the property

.	zero component
•	non-zero component
•——•	components equal

system(s)	class(es)	str ()
triclinic	all	$\begin{bmatrix} • & • & • \\ & • & • \\ & & • \end{bmatrix}$ 6
monoclinic	all	$\begin{bmatrix} • & . & • \\ & • & . \\ & & • \end{bmatrix}$ 4
orthorhombic	all	$\begin{bmatrix} • & . & . \\ & • & . \\ & & • \end{bmatrix}$ 3
tetragonal	all	
trigonal	all	
hexagonal	all	$\begin{bmatrix} • \diagdown & . & . \\ & • & . \\ & & • \end{bmatrix}$ 2
limit classes	$\infty, \infty m, \infty/m$	
	$\infty 2, \infty/mm$	
cubic	all	$\begin{bmatrix} • \diagdown & . & . \\ & • \diagdown & . \\ & & • \end{bmatrix}$ 1
limit classes	$\infty\infty m, \infty\infty\infty$	

operation; making these first two columns compatible leads to the results of the third column.

by transformation law	by symmetry	result
$\rho'_{11} = l_{1i}l_{1j}\rho_{ij} = 1 \times 1 \times \rho_{11}$	$\rho'_{11} = \rho_{11}$	$\rho_{11} \neq 0$ (free)
$\rho'_{22} = l_{2i}l_{2j}\rho_{ij} = 1 \times 1 \times \rho_{22}$	$\rho'_{22} = \rho_{22}$	$\rho_{22} \neq 0$ (free)
$\rho'_{33} = l_{3i}l_{3j}\rho_{ij} = 1 \times 1 \times \rho_{33}$	$\rho'_{33} = \rho_{33}$	$\rho_{33} \neq 0$ (free)
$\rho'_{12} = l_{1i}l_{2j}\rho_{ij} = -1 \times 1 \times \rho_{12}$	$\rho'_{12} = \rho_{12}$	$\rho_{12} = 0$
$\rho'_{21} = l_{2i}l_{1j}\rho_{ij} = -1 \times 1 \times \rho_{21}$	$\rho'_{21} = \rho_{21}$	$\rho_{21} = 0$
$\rho'_{13} = l_{1i}l_{3j}\rho_{ij} = 1 \times 1 \times \rho_{13}$	$\rho'_{13} = \rho_{13}$	$\rho_{13} \neq 0$ (free)
$\rho'_{31} = l_{3i}l_{1j}\rho_{ij} = 1 \times 1 \times \rho_{31}$	$\rho'_{31} = \rho_{31}$	$\rho_{31} \neq 0$ (free)
$\rho'_{23} = l_{2i}l_{3j}\rho_{ij} = -1 \times 1 \times \rho_{23}$	$\rho'_{23} = \rho_{23}$	$\rho_{23} = 0$
$\rho'_{32} = l_{3i}l_{2j}\rho_{ij} = -1 \times 1 \times \rho_{32}$	$\rho'_{32} = \rho_{32}$	$\rho_{32} = 0$

Compare these results with the entry for monoclinic materials in Table 9.1. Tables 3.6, 3.7, 3.8, 3.9, 3.10, 3.11, 3.12, 3.13 and 3.14 have been obtained either by this kind of reasoning or by more sophisticated methods which are in essence similar to it. As a general rule, the higher the symmetry of the material (i.e. the larger

Table 3.8 Structures of the Pöckels coefficient r. The integer to the right of str( ) is the number of independent components of the property

·	zero component
•	non-zero component
•—•	components equal
•—○	same absolute value, opposite signs

system(s)	class(es)	str ()	system(s)	class(es)	str ()
triclinic	1	$\begin{bmatrix} \bullet & \bullet & \bullet \\ \bullet & \bullet & \bullet \\ \bullet & \bullet & \bullet \\ \bullet & \bullet & \bullet \\ \bullet & \bullet & \bullet \\ \bullet & \bullet & \bullet \end{bmatrix}$ 18	monoclinic	2	$\begin{bmatrix} \cdot & \bullet & \cdot \\ \cdot & \bullet & \cdot \\ \cdot & \bullet & \cdot \\ \bullet & \cdot & \bullet \\ \bullet & \cdot & \bullet \\ \bullet & \cdot & \cdot \end{bmatrix}$ 8
monoclinic	m	$\begin{bmatrix} \bullet & \cdot & \bullet \\ \bullet & \cdot & \bullet \\ \bullet & \cdot & \bullet \\ \cdot & \bullet & \cdot \\ \cdot & \bullet & \cdot \\ \cdot & \bullet & \cdot \end{bmatrix}$ 10	orthorhombic	222	$\begin{bmatrix} \cdot & \cdot & \cdot \\ \cdot & \cdot & \cdot \\ \cdot & \cdot & \cdot \\ \bullet & \cdot & \cdot \\ \cdot & \bullet & \cdot \\ \cdot & \cdot & \bullet \end{bmatrix}$ 3
orthorhombic	$mm2$	$\begin{bmatrix} \cdot & \cdot & \bullet \\ \cdot & \cdot & \bullet \\ \cdot & \cdot & \bullet \\ \cdot & \bullet & \cdot \\ \bullet & \cdot & \cdot \\ \cdot & \cdot & \cdot \end{bmatrix}$ 5	trigonal	3	6
trigonal	32	2	trigonal	$3m$	4
tetragonal hexagonal limit classes	4 6 ∞	4	tetragonal	$\bar{4}$	4
tetragonal hexagonal limit classes	422 622 $\infty2$	1	tetragonal hexagonal limit classes	$4mm$ $6mm$ ∞m	3
tetragonal	$\bar{4}2m$	2	hexagonal	$\bar{6}$	2
hexagonal	$\bar{6}m2$	1	cubic	$\bar{4}3m, 23$	1
cubic limit class	432 $\infty\infty$	0			

Table 3.9 Structures of the piezoelectric moduli $\underset{\sim}{d}$. The integer to the right of str( ) is the number of independent components of the property

·	zero component
●	non-zero component
●——●	components equal
●——○	same absolute value, opposite signs
⊙	-2 the ● to which it is joined by a line

system(s)	class(es)	str ( )	system(s)	class(es)	str ( )
triclinic	1	18	monoclinic	2	8
monoclinic	m	10	orthorhombic	222	3
orthorhombic	mm2	5	tetragonal	4	4
tetragonal	$\bar{4}$	4	tetragonal	422	1
tetragonal	4mm	3	tetragonal	$\bar{4}2m$	2
trigonal	3	6	trigonal	32	2
trigonal	3m	4	hexagonal limit classes	6 ∞	4
hexagonal limit classes	6mm ∞m	3	hexagonal limit classes	622 ∞2	1
hexagonal	$\bar{6}$	2	hexagonal	$\bar{6}m2$	1
cubic limit classes	432 ∞∞	0	cubic	$\bar{4}3m$ 23	1

Table 3.10 Structures of the piezo-optic π^{opt} and elasto-optic p^{opt} coefficients, piezoresistivity π, magnetoresistivity ρ^{mag} and Kerr effect K. The integer to the right of str( ) is the number of independent components of the property

$\cdot$	zero component
$\bullet$	non-zero component
$\bullet\!\!-\!\!\bullet$	components equal
$\bullet\!\!-\!\!\circ$	same absolute value, opposite signs
en π^{opt} (y similar en π, ρ, K): $\quad\times$	$(\pi^{opt}_{11} - \pi^{opt}_{12})$
$\odot$	2 the $\bullet$ to which it is joined by a line
$\otimes$	-2 the $\bullet$ to which it is joined by a line
en p^{opt}: $\quad\times$	$\frac{1}{2}(p^{opt}_{11} - p^{opt}_{12})$
$\odot$	same as the $\bullet$ to which it is joined by a line
$\otimes$	sign opposite to $\bullet$ to which it is joined by a line

system(s)	class(es)	str ()
triclinic	all	$\begin{bmatrix} \bullet & \bullet & \bullet & \bullet & \bullet & \bullet \\ \bullet & \bullet & \bullet & \bullet & \bullet & \bullet \\ \bullet & \bullet & \bullet & \bullet & \bullet & \bullet \\ \bullet & \bullet & \bullet & \bullet & \bullet & \bullet \\ \bullet & \bullet & \bullet & \bullet & \bullet & \bullet \\ \bullet & \bullet & \bullet & \bullet & \bullet & \bullet \end{bmatrix}$ 36
monoclinic	all	$\begin{bmatrix} \bullet & \bullet & \bullet & \cdot & \bullet & \cdot \\ \bullet & \bullet & \bullet & \cdot & \bullet & \cdot \\ \bullet & \bullet & \bullet & \cdot & \bullet & \cdot \\ \cdot & \cdot & \cdot & \bullet & \cdot & \bullet \\ \bullet & \bullet & \bullet & \cdot & \bullet & \cdot \\ \cdot & \cdot & \cdot & \bullet & \cdot & \bullet \end{bmatrix}$ 20
orthorhombic	all	$\begin{bmatrix} \bullet & \bullet & \bullet & \cdot & \cdot & \cdot \\ \bullet & \bullet & \bullet & \cdot & \cdot & \cdot \\ \bullet & \bullet & \bullet & \cdot & \cdot & \cdot \\ \cdot & \cdot & \cdot & \bullet & \cdot & \cdot \\ \cdot & \cdot & \cdot & \cdot & \bullet & \cdot \\ \cdot & \cdot & \cdot & \cdot & \cdot & \bullet \end{bmatrix}$ 12
trigonal	$3, \bar{3}$	12
	$32, 3m, \bar{3}m$	8

Table 3.11 Structures of the piezo-optic π^{opt} and elasto-optic p^{opt} coefficients, piezoresistivity $\underset{\sim}{\pi}$, magnetoresistivity ρ^{mag} and Kerr effect $\underset{\sim}{K}$ (continued). The integer to the right of str( ) is the number of independent components of the property

system(s)	class(es)	str ()
tetragonal	$4, \bar{4}, 4/m$	10
	$422, 4mm,$ $\bar{4}2m, 4/mmm$	7
hexagonal limit classes	$6, \bar{6}, 6/m,$ $\infty, \infty/m$	8
hexagonal limit classes	$622, 6mm,$ $\bar{6}m2, 6/mmm,$ $\infty 2, \infty m,$ ∞/mm	6
cubic	$23, m3$	4
cubic	$432, \bar{4}3m,$ $m3m$	3
limit classes	$\infty\infty, \infty\infty m$	2

Table 3.12 Structures of the electrostrictive coefficients M. The integer to the right of str( ) is the number of independent components of the property

·	zero component
•	non-zero component
•——•	components equal
•——○	same absolute value, opposite signs
×	$2(M_{11} - M_{12})$
⊙	2 the • to which it is joined by a line

system(s)	class(es)	str ( )
triclinic	all	$\begin{bmatrix} \bullet & \bullet & \bullet & \bullet & \bullet & \bullet \\ \bullet & \bullet & \bullet & \bullet & \bullet & \bullet \\ \bullet & \bullet & \bullet & \bullet & \bullet & \bullet \\ \bullet & \bullet & \bullet & \bullet & \bullet & \bullet \\ \bullet & \bullet & \bullet & \bullet & \bullet & \bullet \\ \bullet & \bullet & \bullet & \bullet & \bullet & \bullet \end{bmatrix}$ 36
monoclinic	all	$\begin{bmatrix} \bullet & \bullet & \bullet & \cdot & \bullet & \cdot \\ \bullet & \bullet & \bullet & \cdot & \bullet & \cdot \\ \bullet & \bullet & \bullet & \cdot & \bullet & \cdot \\ \cdot & \cdot & \cdot & \bullet & \cdot & \bullet \\ \bullet & \bullet & \bullet & \cdot & \bullet & \cdot \\ \cdot & \cdot & \cdot & \bullet & \cdot & \bullet \end{bmatrix}$ 20
orthorhombic	all	$\begin{bmatrix} \bullet & \bullet & \bullet & \cdot & \cdot & \cdot \\ \bullet & \bullet & \bullet & \cdot & \cdot & \cdot \\ \bullet & \bullet & \bullet & \cdot & \cdot & \cdot \\ \cdot & \cdot & \cdot & \bullet & \cdot & \cdot \\ \cdot & \cdot & \cdot & \cdot & \bullet & \cdot \\ \cdot & \cdot & \cdot & \cdot & \cdot & \bullet \end{bmatrix}$ 12
trigonal	$3, \bar{3}$	(matrix with cross-linked components) 12
	$32, 3m, \bar{3}m$	(matrix with cross-linked components) 8

the set of symmetry operation under which it is invariant), the higher the number of relationships that appear among its components.

The same cancelling and/or equality of components applies to field tensors as well, even if they are not properties of materials. In this case, the point group is not that of the material, but of the physical magnitude under consideration.

Table 3.13 Structures of the electrostrictive coefficients $\underset{\sim}{M}$ (continued). The integer to the right of str( ) is the number of independent components of the property

system(s)	class(es)	str ()
tetragonal	$4, \bar{4}, 4/m$	10
	$422, 4mm,$ $\bar{4}2m, 4/mmm$	7
hexagonal limit classes	$6, \bar{6}, 6/m,$ $\infty, \infty/m$	8
hexagonal limit classes	$622, 6mm,$ $\bar{6}m2, 6/mmm,$ $\infty 2, \infty m,$ ∞/mm	6
cubic	$23, m3$	4
cubic	$432, \bar{4}3m,$ $m3m$	3
limit classes	$\infty\infty, \infty\infty m$	2

Table 3.14 Structures of the elastic compliance s and stiffness c. The integer to the right of str( )
is the number of independent components of the property

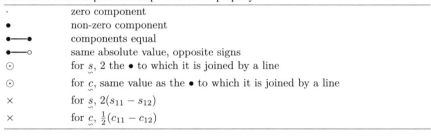

·	zero component
•	non-zero component
•——•	components equal
•——○	same absolute value, opposite signs
⊙	for s, 2 the • to which it is joined by a line
⊙	for c, same value as the • to which it is joined by a line
×	for s, $2(s_{11} - s_{12})$
×	for c, $\frac{1}{2}(c_{11} - c_{12})$

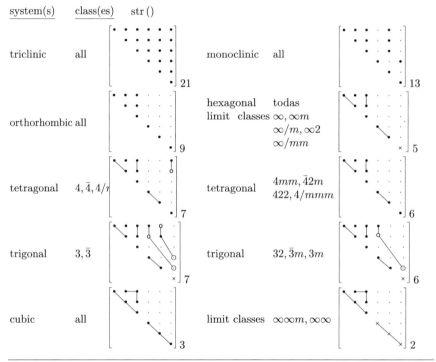

system(s)   class(es)   str ()

triclinic   all   21

monoclinic   all   13

orthorhombic all   9

hexagonal   todas
limit classes $\infty, \infty m$
              $\infty/m, \infty 2$
              ∞/mm 5

tetragonal $4, \bar{4}, 4/m$ 7

tetragonal $4mm, \bar{4}2m$
              $422, 4/mmm$ 6

trigonal $3, \bar{3}$ 7

trigonal $32, \bar{3}m, 3m$ 6

cubic   all   3

limit classes $\infty\infty m, \infty\infty$ 2

3.7   Structure of a Property

The purpose of the two examples of the previous section is to illustrate how geometric symmetry imposes relationships on the components of tensorial properties. The complete task of determining these relationships between components, for all point

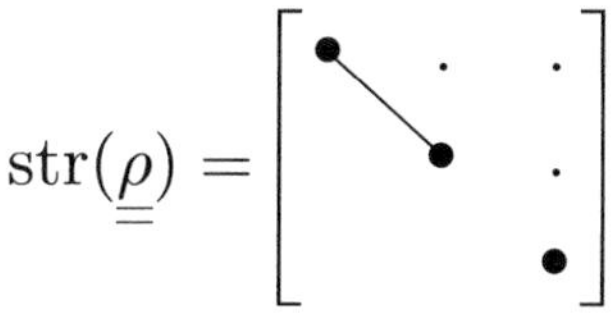

$$\text{str}(\underline{\underline{\rho}}) = \begin{bmatrix} \bullet & \diagdown & \cdot & \cdot \\ & \bullet & \cdot \\ & & \bullet \end{bmatrix}$$

Fig. 3.8 An example of the structure of a second rank symmetric property, like the electric resistivity $\underline{\underline{\rho}}$. The black circles correspond to non-zero values, dots to zero values. In this example, two circles connected by a line are numerically equal. Since the property is symmetric, only the diagonal and the upper triangle are represented

groups and for properties of varying rank[15] can be carried out by several methods [11, 12, 15] the knowledge of which is not essential for the purposes of this book.

But is is important to get acquainted with the way such relationships between components are presented in the literature. Although a table like Table 3.6 would do in principle, there is a more efficient, graphical way, which we will call the *structure* of a property and denote by str( ). This is a one- or two-dimensional graphical representation: str( ) looks like a vector or a matrix whose components are graphical symbols, possibly connected by lines. It is thus immediately suited to represent relationships between components of tensors of first and second ranks.

Figure 3.8 is a typical example: for materials belonging to any of the classes of the tetragonal system, all symmetric, second rank tensorial properties have the same structure (Fig. 3.8).

According to this str($\underline{\underline{\rho}}$) two components of $\underline{\underline{\rho}}$ are identical ($\rho_{11} = \rho_{22}$) while the third is different. All other components are zero. It is important to keep in mind that the structures of the properties are always referred to conventional axes. In a different set of axes the str( ) will almost always look different.

For tensors of third and fourth rank with minor or major symmetries under index permutation, Voigt notation is used to flatten them. For tensors of higher ranks, str( ) cannot be used.

We will take as an example the fourth rank coefficient of piezoresistivity $\underline{\underline{\pi}}$, which has minor, but not major index symmetry:

$$\underline{\underline{\pi}}^T = {}^T\underline{\underline{\pi}} = \underline{\underline{\pi}} \qquad (\pi_{ijkl} = \pi_{jikl} = \pi_{ijlk} = \pi_{jilk})$$

Voigt condensation can be applied to the first and to the second pair of indices, so that the resulting matrix $\underline{\underline{\pi}} \to \underset{\smile}{\pi}$ is 6×6, asymmetric and has a maximum of 36 independent coefficients.

For a fourth rank property with higher symmetry (minor and major), such as the elastic stiffness $\underline{\underline{c}}$, which fulfills:

---

[15]The extensive compilation [14] goes up to rank seven.

$$c_{ijkl} = c_{jikl} = c_{ijlk} = c_{jilk} = c_{klij} = c_{lkij} = c_{klji} = c_{lkji}$$

Voigt condensation $\underline{\underline{c}} \to \underline{c}$ produces a symmetric 6×6 matrix with at most 21 independent coefficients.

Tables 3.6 through 3.14 present all structures for properties up to fourth rank, for all 32 crystallographic and 7 limit classes. The str( ) of classes not appearing in the tables contains only zeros, i.e. the material does not have the property.

References

1. Jeffreys, H.: Cartesian Tensors. Cambridge University Press, Cambridge (1969)
2. Halmos, P.R.: Finite Dimensional Vector Spaces. Van Nostrand-Reinhold, New York (1958)
3. Gel'fand, I.M.: Lectures on Linear Algebra. Dover Publications Inc., Mineola (1989)
4. Knowles, J.K.: Linear Vector Spaces and Cartesian Tensors. Oxford University Press, Oxford (1998)
5. Arfken, G.B., Weber, H.J.: Mathematical Methods for Physicists. Harcourt Academic Press, San Diego (2001)
6. Bird, B., Stewart, W.E., Lightfoot, E.N.: Transport Phenomena, 2nd edn. Wiley, Hoboken (2002)
7. Dodson, C.T.J., Poston, T.: Tensor Geometry: The Geometric Viewpoint and Its Uses. Springer, Berlin (2013)
8. Das, A.J.: Tensors: The Mathematics of Relativity Theory and Continuum Mechanics. Springer, Berlin (2007)
9. Jeevanjee, N.: An Introduction to Tensors and Group Theory for Physicists. Springer, Berlin (2011)
10. Abraham, R., Marsden, J.E., Ratiu, T.: Manifolds, Tensor Analysis, and Applications. Springer, Berlin (2012)
11. Newnham, R.E.: Properties of Materials: Anisotropy, Symmetry. Structure. Oxford University Press, Oxford (2005)
12. Tinder, R.F.: Tensor Properties of Solids: Phenomenological Development of the Tensor Properties of Crystals. Morgan & Claypool Publishers, San Rafael (2008)
13. Voigt, W.: Lehrbuch der Kristallphysik. Teubner (1928)
14. Popov, S.V., Svirko, Y.P., Zheludev, N.I.: CD Encyclopedia of Material Tensors. Wiley, Hoboken (1999)
15. Nye, J.F.: Physical Properties of Crystals. Oxford University Press, Oxford (1985)

Chapter 4
Representation Surfaces

A representation surface is a graphical way of visualizing the dependence of a given tensorial magnitude (both material and field tensors) on the direction in which it is measured. In loose terms, the RS is the polar spherical representation of the projection of a tensor in one (or more) directions.

The defining property of the RS is that the modulus of the vector radius from the origin to the RS in a given direction is equal to the value of the property projected in that direction (Fig. 4.1). We will indicate the direction of the vector radius $\underline{r}$ by the unit vector $\underline{n}$:

$$\underline{n} = \frac{\underline{r}}{|\underline{r}|}$$

As the vector $\underline{r}$ scans all possible orientations in 3D, its modulus shortens and lengthens according to the value of the projection of the property in the direction of $\underline{n}$. The RS is the set of points the tip of the vector visits.

The equation of the RS can be written in a polar (scalar) way as:

$$r(\underline{n}) = r(n_1, n_2, n_3) = r(\varphi, \theta) \tag{4.1}$$

where n_i is the ith director cosine, or component, of $\underline{n}$ and φ, θ are two angular parameters that define the orientation of the radius vector $\underline{r}$. They may, but must not necessarily be the usual polar spherical angles.

Irrespective of its rank, the measurement of a tensorial property along a given direction(s) $\underline{n}$ yields a scalar; this scalar is the value of $r(\underline{n})$.

Depending on the situation it may be convenient to make explicit the dependence on the n_i's, or on angles such as φ, θ. The components n_i of the direction unit vector are functions of the angles:

$$n_i = n_i(\varphi, \theta), \quad i = 1, 2, 3 \tag{4.2}$$

© Springer Nature Switzerland AG 2020
M. Laso and N. Jimeno, *Representation Surfaces for Physical Properties
of Materials*, Engineering Materials, https://doi.org/10.1007/978-3-030-40870-1_4

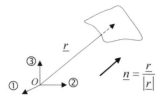

Fig. 4.1 The modulus of the vector $\underline{r}$ (the radius vector) from the origin to the representation surface of a tensorial property is equal to the scalar value of that property in the direction of the vector

and satisfy $\sqrt{n_1^2 + n_2^2 + n_3^2} = 1$. Only two of them (as many as angles) are independent.

If a vector version of Eq. (4.1) is required[1]:

$$\underline{r}(\underline{n}) = r(n_1, n_2, n_3)\underline{n} = r(\varphi, \theta)\underline{n}$$

or in components:

$$\underline{r}(\underline{n}) = \begin{cases} r_1 = r(n_1, n_2, n_3)n_1 \\ r_2 = r(n_1, n_2, n_3)n_2 \\ r_3 = r(n_1, n_2, n_3)n_3 \end{cases}$$

Whether the equation of the RS is written in terms of direction cosines n_1, n_2, n_3 or in terms of spherical polar angles φ, θ depends on the situation. Some symmetries of the RS are easiest to detect when the RS is written in terms of direction cosines, in other occasions it is more convenient to write the RS in terms of angles.

As we will make plausible below, the operation that corresponds to the measurement is a projection. Unit vectors are basic ingredients in a projection, for this reason we will quickly review the most common ways of defining a unit vector.

4.1   The Unit Vector

Depending on the application, it is usual to find that the unit direction vector $\underline{n}$ is defined in different ways:

- if $\underline{n}$ is required to point in the same direction as another, known, non-unit vector, e.g. a given electric field $\underline{E}$, then:

$$\underline{n} = \frac{\underline{E}}{|\underline{E}|}$$

---

[1]Using the same symbol for a vector $\underline{r}$ and for its modulus r; for a function of the components $r(n_1, n_2, n_3)$ and for a function of the angles $r(\varphi, \theta)$ is an abuse of notation.

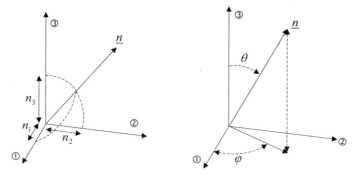

Fig. 4.2 Unit vector $\underline{n}$ defined by its projections on the conventional axes (left) or by its spherical angles azimuth and colatitude

- if the projections n_i of $\underline{n}$ on three or two (say on ① and ②) of the conventional axes ①②③ are known (Fig. 4.2, left):

$$[\![\underline{n}]\!] = \begin{bmatrix} n_1 \\ n_2 \\ n_3 = \sqrt{1 - n_1^2 - n_2^2} \end{bmatrix}$$

- if the spherical angles azimuth $\varphi \in [0, 2\pi)$ and co-latitude $\theta \in [0, \pi]$ are given (Fig. 4.2 right):

$$[\![\underline{n}]\!] = \begin{bmatrix} \sin\theta \cos\varphi \\ \sin\theta \sin\varphi \\ \cos\theta \end{bmatrix} \tag{4.3}$$

- on occasion, it is necessary to define a unit vector $\underline{n}$ along a crystallographic direction, given by its usual direction indices $[ijk]$. Taking an orthorhombic structure as an example, the unit vector that points in the cyrstallographic direction $[ijk]$ is:

$$[\![\underline{n}]\!] = \begin{bmatrix} \dfrac{i}{\sqrt{i^2 + j^2 + k^2}} \\ \dfrac{j}{\sqrt{i^2 + j^2 + k^2}} \\ \dfrac{k}{\sqrt{i^2 + j^2 + k^2}} \end{bmatrix}$$

In all previous cases, only two components of $\underline{n}$ are independent, since $|\underline{n}| = \sqrt{n_1^2 + n_2^2 + n_3^2} = 1$ and only two independent numbers are necessary to orient a single direction or a single axis.

Fig. 4.3 Orienting a slab.
The three unit vectors $\underline{n}^A$,
$\underline{n}^B$ and $\underline{n}^C$ are perpendicular
to the faces of the slab

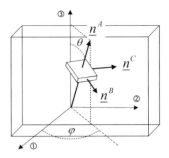

If it is required to orient not just a unit vector but an orthonormal reference frame, Fig. 4.3 shows a possible choice that is sometimes found in the literature. The three orthonormal vectors $\underline{n}^A$, $\underline{n}^B$ and $\underline{n}^C$ are given in this case by:

$$[\![\underline{n}^A]\!] = \begin{bmatrix} \sin\theta\cos\varphi \\ \sin\theta\sin\varphi \\ \cos\theta \end{bmatrix} \quad [\![\underline{n}^B]\!] = \begin{bmatrix} \cos\theta\cos\varphi \\ \cos\theta\sin\varphi \\ -\sin\theta \end{bmatrix} \quad [\![\underline{n}^C]\!] = \begin{bmatrix} -\sin\varphi \\ \cos\varphi \\ 0 \end{bmatrix} \quad (4.4)$$

This is however a choice of inherent limitation: although $\underline{n}^A$ still points in every possible direction when the polar angles are varied through their ranges, $\underline{n}^B$ will only sweep the lower hemisphere, including the equator. The situation for $\underline{n}^C$ is even more precarious.[2]

Two polar angles are insufficient to orient a reference frame in all possible directions, a third degree of freedom (angle) is required. These three numbers can be the usual spherical angles plus an extra rotation around the axis defined by $\underline{n}^A$ in Fig. 4.3, or the usual Euler angles φ, θ, ψ.[3]

The overall rotation caused by φ, θ, ψ is obtained by three successive rotations in a specific sequence, each about a different axis, as explained in Figs. 4.4, 4.5 and 4.6. The orthogonal matrix $\underset{\smile}{L}$ for this standard choice of Euler angles is obtained as the product of the three individual rotation matrices, one for each of the rotations displayed in these figures.

The transformation from the reference system ①②③ to the reference system formed by ①***②***③*** is given by:

$$\underset{\smile}{L} = \begin{bmatrix} \cos\psi\cos\varphi - \cos\theta\sin\varphi\sin\psi & \cos\psi\sin\varphi + \cos\theta\cos\varphi\sin\psi & \sin\psi\sin\theta \\ -\sin\psi\cos\varphi - \cos\theta\sin\varphi\cos\psi & -\sin\psi\sin\varphi + \cos\theta\cos\varphi\cos\psi & \cos\psi\sin\theta \\ \sin\theta\sin\varphi & -\sin\theta\cos\varphi & \cos\theta \end{bmatrix}$$

---

[2] Along which directions does $\underline{n}^C$ of Fig. 4.3 point when the spherical angles are varied in their ranges $\varphi \in [0, 2\pi)$ and $\theta \in [0, \pi]$?

[3] For a discussion of other possibilities see [1].

Fig. 4.4 Rotation by Euler angle φ about the ③ axis yields intermediate axes ①*②*③*. For this rotation, axes ③ and ③* coincide

Fig. 4.5 Rotation by Euler angle θ about the ①* axis yields intermediate axes ①**②**③**. For this rotation, axes ①* and ①** coincide

Fig. 4.6 Rotation by Euler angle ψ about the ③** axis yields the final axes ①***②***③***. For this rotation, axes ③** y ③*** coincide

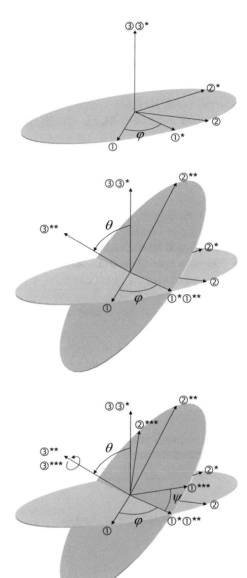

As already discussed on Sect. 3.3, the rows of L are the unit vectors ①***, ②*** and ③***, expressed in the reference system ①②③. Conversely, the columns of $\overset{\smile}{L}$ are the unit vectors ①②③, expressed in the reference system ①***②***③***.

All unit vectors formed by the rows and the columns of $\overset{\smile}{L}$ sample all possible orientations in 3D space with neither gaps nor overlaps when all three Euler angles sweep their ranges $\theta \in [0, \pi]$, $\varphi \in [0, 2\pi)$, $\psi \in [0, 2\pi)$.

4.2 Projection

The general idea of projection is a minor generalization of the familiar decomposition of a vector in its components, as so often encountered in elementary physics, and which we will briefly review now.

Taking a force $\underline{F}$ as a prototypical rank one field tensor it can be decomposed in a component along a given direction, defined by a unit vector $\underline{n}$, and a component perpendicular to it (Fig. 4.7).

The modulus of the component of $\underline{F}$ parallel to $\underline{n}$ is given by the projection of $\underline{F}$ along $\underline{n}$:

$$|\underline{F}^{\|}| = \underline{F} \cdot \underline{n} = F_i n_i \tag{4.5}$$

while the vector component of $\underline{F}$ parallel to $\underline{n}$ is:

$$\underline{F}^{\|} = (\underline{F} \cdot \underline{n})\underline{n} = \delta_j (F_i n_i) n_j$$

and the perpendicular vector component is the difference between the original vector and its parallel component:

$$\underline{F}^{\perp} = \underline{F} - \underline{F}^{\|} = \underline{F} - (\underline{F} \cdot \underline{n})\underline{n} = \delta_j \left[F_j - (F_i n_i) n_j \right] \tag{4.6}$$

It is also possible to write (4.6) as:

$$\underline{F}^{\perp} = (\underline{\underline{\delta}} - \underline{n}\,\underline{n}) \cdot \underline{F}$$

Fig. 4.7 Decomposition of a force in two components, parallel and perpendicular, to a direction defined by the unit vector $\underline{n}$

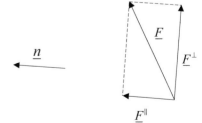

As $\underline{n}\,\underline{n}$ projects along $\underline{n}$, it is usual to call it a *longitudinal* projector, while $(\underline{\underline{\delta}} - \underline{n}\,\underline{n})$ projects perpendicularly to $\underline{n}$. The perpendicular component to $\underline{F}$ is obtained as the difference (4.6), which depends on the particular $\underline{F}$.

Which does *not* depend on $\underline{F}$ but only on $\underline{n}$ are the terms $\underline{n}\,\underline{n}$ and $(\underline{\underline{\delta}} - \underline{n}\,\underline{n})$, which yield the projections parallel and perpendicular to any vector we apply them to. The projector $(\underline{\underline{\delta}} - \underline{n}\,\underline{n})$ is thus not just a projector "perpendicular" to a given rank one tensor, but a *transversal* projector, emphasizing that it projects onto the plane formed by the infinite directions perpendicular to $\underline{n}$. These directions lie on the plane transversal to $\underline{n}$.

Notice that $\underline{n}\,\underline{n}\cdot$ and $(\underline{\underline{\delta}} - \underline{n}\,\underline{n})\cdot$ yield *vector* components when applied to a rank one tensor. Scalar projections, i.e. the moduli of $\underline{F}^{\parallel}$ and $\underline{F}^{\perp}$ are obtained through (4.5) and as the modulus of (4.6).

It is also possible to decompose the force $\underline{F}$ in its three components F^u, F^v and F^w along any three non-coplanar, not necessarily orthogonal vectors $\underline{u}$, $\underline{v}$ and $\underline{w}$ in a unique way, by computing its scalar products with these unit vectors:

$$F^u = \underline{F} \cdot \underline{u} \qquad F^v = \underline{F} \cdot \underline{v} \qquad F^w = \underline{F} \cdot \underline{w}$$

so that

$$\underline{F} = F^u \underline{u} + F^v \underline{v} + F^w \underline{w}$$

If the vectors $\underline{u}$, $\underline{v}$ and $\underline{w}$ are of unit modulus and orthogonal to each other, they can be used as the basis of a Cartesian reference frame. They may, in particular, coincide with the conventional axes ①②③.

4.2.1 Tensors as Multilinear Maps

Expressions such as:

$$\underline{R} \cdot \underline{u} = R_i u_i \qquad \underline{\underline{S}} \cdot \underline{u}\,\underline{v} = S_{ij} u_j v_i \qquad \underline{\underline{\underline{T}}} : \underline{u}\,\underline{v}\,\underline{w} = T_{ijk} u_k v_j w_i$$

where $\underline{R}$, $\underline{\underline{S}}$, $\underline{\underline{\underline{T}}}$ are generic tensors, can be interpreted as multilinear maps. Viewing a tensor such as $\underline{\underline{\underline{T}}}$, as a multilinear map $\underline{\underline{\underline{T}}}(\,\cdot\,,\,\cdot\,,\,\cdot\,)$, where the dots are placeholders for arguments (as many as the rank of the tensor), is never wrong and convenient in some situations. Index contraction is then equivalent to "using up" or "filling in" its argument slots with other tensors. Depending on how many of the slots are filled (i.e. how many of the indices are contracted), tensors of different ranks are obtained.

Thus, projecting a tensor along one or more directions is nothing but using up or filling in its argument slots with unit vectors along those directions.

We now apply these basic notions to representative tensors of increasing rank.

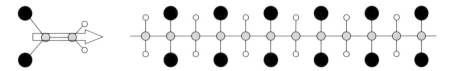

Fig. 4.8 Two views of the poly(vinylidene fluoride) molecule. The arrow indicates the electrical dipole

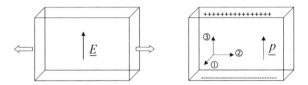

Fig. 4.9 Poling a PVDF sheet: en electric field is applied perpendicular to a longitudinal strain. Sometimes, the temperature is raised to enhance polymer chain mobility and facilitate orientation of the molecular dipole in the electric field. Poled PVDF belongs to class $mm2$

4.3 Pyroelectricity

Materials belonging to the polar classes (those appearing in Table 3.6) are pyroelectric, that is, they have a macroscopic polarization that changes when their temperature is changed. The pyroelectric coefficient appears in the linear constitutive law of pyroelectricity as the first rank tensor that couples the change in temperature and the polarization of the material:

$$\underline{P} = \underline{p}\,\Delta T \tag{4.7}$$

Poly(vinylidene fluoride) $\left[\begin{array}{c} F \ | \ -C- \ | \ F \end{array}\begin{array}{c} H \ | \ C- \ | \ H \end{array}\right]_n$ or PVDF is a prototypic pyroelectric and ferroelectric polymer [2]. Its molecular structure in the all-*trans* conformation is represented in Fig. 4.8. An individual PVDF molecule in the all-*trans* conformation belongs to the $mm2$ point group and due to the presence of the fluorine atoms, has a strong electric dipole. In its monocrystalline forms PVDF is strongly pyroelectric [3]. However, when freshly synthesized PVDF has a semicrystalline morphology: it consists of around 50–60% of crystalline domains distributed throughout an amorphous matrix at roughly random positions and with random orientations. As a consequence it is isotropic and non-pyroelectric [4].

For PVDF to be used as a pyroelectric material, it must undergo poling, i.e. an orientation or polarization processing step. A PVDF sheet or thin block is subjected to a strong static electric field ($\simeq$50 MV/m) $\underline{E}$ and simultaneously stretched transversely to $\underline{E}$; the process can also be carried out at increased temperature in order to enhance molecular mobility. The net effect is to produce a preferential alignment of the dipoles of the individual polymer chains along $\underline{E}$.

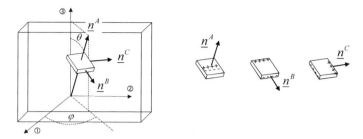

Fig. 4.10 For an arbitrary orientation of the slab, all three pairs of faces may develop surface charges that depend on their orientation (see Chap. 5)

The poling process renders PVDF pyroelectric, polar class $mm2$, the conventional axes of which are shown in Fig. 4.9. According to str(p) for class $mm2$, a temperature increase of the block will produce a macroscopic polarization given by (4.7) with $P_1 = P_2 = 0$, $P_3 \neq 0$. As a consequence the faces perpendicular to the axis ③ will polarize, while no surface charge will appear on the faces ① or ②.

If a small parallelepipedic sample is now cut out of the block with an arbitrary orientation (Fig. 4.10), all three pairs of faces will develop electric charge. The polarizations (surface charges) that appear on each pair of faces can, to first order, be obtained by projecting, $\underline{P}$ along the three orthogonal unit vectors, $\underline{n}^A$, $\underline{n}^B$ and $\underline{n}^C$ depicted in the figure:

$$P^A = \underline{P} \cdot \underline{n}^A \qquad P^B = \underline{P} \cdot \underline{n}^B \qquad P^C = \underline{P} \cdot \underline{n}^C$$

The polarization of each pair of faces will depend on the orientation of the slab.

It is important to keep in mind that varying the polar angles does not imply rotating a given sample. For every orientation to be sampled, a *new* slab with a different orientation must be cut out of the original block.

We will use the representation surface to visualize how the components of $\underline{P}$ change when samples with varying orientations are used, not when a sample is rotated. In the case of the pyroelectric effect, rotating a given sample in space has no effect whatsoever on the polarization of the sample.

4.4 Electrical Resistivity

Let's consider the measurement of the resistivity of a long, thin rod of an anisotropic material, like the layered material (limit class ∞/mm) of Fig. 4.11, where the conventional axes of the material ①②③ have also been drawn. The rod has length L, cross section A, and "long, thin" implies $L \gg \sqrt{A}$.

Although the resistivity $\underline{\underline{\rho}}$ is a second rank symmetric tensor, the measurement of the resistance and hence of the resistivity will yield a single number. A question that naturally arises is how this scalar resistivity is related to the six independent components of the (tensor) resistivity ρ.

The orientation of the rod is given by a unit vector $\underline{n}$. As in Sect. 4.3, different orientations $\underline{n}$ of the rod will be obtained cutting out different rods in different orientations from a given bulk sample of the material, *not* by rotating in space a single, given sample.

When an electric current density $\underline{J} = \dfrac{I}{A}\underline{n}$ flows along the anisotropic rod the electric field $\underline{E}$ is not colinear with $\underline{J}$. The scalar resistivity measured in such an experiment, which we will call $\rho_{\underline{n}}$, is the ratio of the modulus of the electric field along $\underline{n}$, $|\underline{E}^{\|}|$, and the modulus of the current density $|\underline{J}|$. The resistance of the rod will be measured as $R = \rho_{\underline{n}} \dfrac{L}{A}$:

- a current density $\underline{J}$ (A/m^2) of known modulus flows along the rod axis in the direction of the unit vector $\underline{n}$: $\underline{J} = J\underline{n}$ ($J_i = Jn_i$),
- the electric field in the rod is given by Ohm's law: $\underline{E} = \underline{\underline{\rho}} \cdot \underline{J}$ ($E_i = \rho_{ij}J_j$),
- the component $E^{\|}$ of the field $\underline{E}$ parallel to $\underline{J}$ is $E^{\|} = \underline{E} \cdot \underline{n}$ ($E^{\|} = E_i n_i$),
- the resistivity in the direction of the rod axis:

$$\rho_{\underline{n}} = \frac{E^{\|}}{|\underline{J}|} = \frac{E_i n_i}{J} = \frac{(\rho_{ij}J_j)n_i}{J} = \rho_{ij}n_i\frac{J_j}{J} = \rho_{ij}n_in_j = \rho_{\underline{n}} : \underline{n}\,\underline{n}$$

Unlike in an isotropic conductor, the electric field has a component transversal to the flow of electrons. The resistivity of the sample depends on the parallel component of $\underline{E}$ only. In the long, thin rod, the current density is aligned with the axis and hence with $\underline{n}$.

A similar experiment, which we will call the dual of the rod experiment, can be performed to measure the electric conductivity of a material. Instead of a long, thin rod, now a flat, thin sample (thickness δ is much smaller than the radius $\delta \ll R$.) of the material is placed between two conducting plates between which a potential difference exists. The equipotential lines are now perpendicular to $\underline{n}$, so that $\underline{E} = -\underline{\nabla}V$ is parallel to $\underline{n}$. According to Ohm's inverse law:

$$\underline{J} = -\underline{\underline{\sigma}} \cdot \underline{\nabla}V = \underline{\underline{\sigma}} \cdot \underline{E}$$

the current density will be deviated from the direction of $\underline{E}$. Just as in the previous experiment, the (scalar) value of the conductivity $\sigma_{\underline{n}}$ in the direction of $\underline{n}$ will be given by the ratio of the current density parallel to $\underline{E}$ divided by $|\underline{E}|$:

$$\sigma_{\underline{n}} = \frac{J^{\|}}{|\underline{E}|} = \frac{J_i n_i}{|\underline{E}|} = \frac{(\sigma_{ij}E_j)n_i}{|\underline{E}|} = \sigma_{ij}n_i\frac{E_j}{|\underline{E}|} = \sigma_{ij}n_in_j = \sigma_{\underline{n}} : \underline{n}\,\underline{n}$$

The previous results are general: the scalar value of a second rank tensor $\underline{\underline{T}}$ along a given direction $T_{\underline{n}}$ is given by its projection:

$$T_{\underline{n}} = T_{ij}n_j n_i \quad (= T_{ij}n_i n_j \text{ if } \underline{\underline{T}} \text{ symmetric}) \qquad T_{\underline{n}} = \underline{\underline{T}} : \underline{n}\,\underline{n} \qquad (4.8)$$

To make this more specific let's apply projection to an electrical resistance made of a laminar composite C (like that of Figs. 4.11 and 4.12) of two materials A and B of known resistivities ρ_A, ρ_B. The layers have known thicknesses δ_A, δ_B. The layers are inclined with respect to the rod axis, so that the normal to the layers forms an angle θ with the axis of the rod. From the point of view of the composite material and of its conventional axes, it is the rod that is inclined, as in Fig. 4.11.

The resistivity along the rod axis is obtained as the projection $\rho_{\underline{n}} = \underline{\underline{\rho}}_C : \underline{n}\,\underline{n}$ where $\underline{\underline{\rho}}_C$ is the resistivity of the composite. This resistivity has to be calculated first for the composite, independently of the shape of the sample, a rod in this example, and then projected along the direction of the rod axis (Fig. 4.13).

The calculation of $\underline{\underline{\rho}}_C$ for this $\infty\infty m$ composite is straightforward in the conventional axes of the composite (axes ①②③ in Figs. 4.11 and 4.12) because along the axes the layers are electrically either in series (along the longitudinal direction, ③) or in parallel (along the transversal directions, ① and ②).

As corresponds to a transversally isotropic material:

$$[\![\underline{\underline{\rho}}_C]\!] = \begin{bmatrix} (\rho_C)_{tt} & 0 & 0 \\ 0 & (\rho_C)_{tt} & 0 \\ 0 & 0 & (\rho_C)_{ll} \end{bmatrix} \text{ with } \begin{cases} (\rho_C)_{tt} = [V_A\rho_A^{-1} + V_B\rho_B^{-1}]^{-1} \\ (\rho_C)_{ll} = V_A\rho_A + V_B\rho_B \end{cases} \qquad (4.9)$$

with

$$V_A = \frac{\delta_A}{\delta_A + \delta_B} \qquad V_B = \frac{\delta_B}{\delta_A + \delta_B}$$

and where the resistivities along ③ have been added in series, those in the transversal ①② plane in parallel.

From Fig. 4.11 the simplest unit direction vector needed to project the resistivity $\underline{\underline{\rho}} : \underline{n}\,\underline{n}$ along the rod axis is[4]:

$$[\![\underline{n}]\!] = \begin{bmatrix} \sin\theta \\ 0 \\ \cos\theta \end{bmatrix}$$

so that:

$$\rho_{\underline{n}} = (\rho_C)_{tt} \sin^2\theta + (\rho_C)_{ll} \cos^2\theta$$

---

[4]In this choice $\underline{n}$ has been placed in the ①③ plane. Check that the same result is obtained if $\underline{n}$ is not in that plane, but still forms an angle θ with the ③ axis.

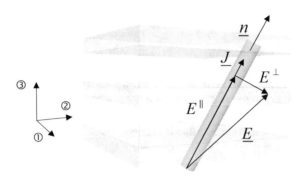

Fig. 4.11 The electric current density $\underline{J} = \dfrac{I}{A}\underline{n}$ and the electric field $\underline{E}$ are not colinear in a long, thin anisotropic conducting rod. The current density points in the direction of $\underline{n}$

Let's now compute the conductivity of the rod by the same method. An advantage of having used ①②③ is that $\underline{\underline{\rho}}_C$ is diagonal in this set of axes. Since the conductivity is $\underline{\underline{\sigma}}_C = \underline{\underline{\rho}}_C^{-1}$ we immediately obtain:

$$
[\![\underline{\underline{\sigma}}_C]\!] = \begin{bmatrix} (\sigma_C)_{tt} & 0 & 0 \\ 0 & (\sigma_C)_{tt} & 0 \\ 0 & 0 & (\sigma_C)_{ll} \end{bmatrix} \quad \text{with} \quad \begin{cases} (\sigma_C)_{tt} = [(\rho_C)_{tt}]^{-1} = [V_A\rho_A^{-1} + V_B\rho_B^{-1}] \\ (\sigma_C)_{ll} = [(\rho_C)_{ll}]^{-1} = [V_A\rho_A + V_B\rho_B]^{-1} \end{cases}
$$

The projection formula $\sigma_{\underline{n}} = \underline{\underline{\sigma}}_C : \underline{n}\,\underline{n}$ then leads to:

$$
\sigma_{\underline{n}} = (\sigma_C)_{tt}\sin^2\theta + (\sigma_C)_{ll}\cos^2\theta = [(\rho_C)_{tt}]^{-1}\sin^2\theta + [(\rho_C)_{ll}]^{-1}\cos^2\theta
$$

Since resistivity and conductivity are reciprocal to each other, let's check that the product $\rho_{\underline{n}}\sigma_{\underline{n}}$ is unity:

$$
\rho_{\underline{n}}\sigma_{\underline{n}} = [(\rho_C)_{tt}\sin^2\theta + (\rho_C)_{ll}\cos^2\theta]\left[\frac{1}{(\rho_C)_{tt}}\sin^2\theta + \frac{1}{(\rho_C)_{ll}}\cos^2\theta\right]
$$
$$
= \sin^4\theta + \cos^4\theta + \left[\frac{(\rho_C)_{tt}}{(\rho_C)_{ll}} + \frac{(\rho_C)_{ll}}{(\rho_C)_{t}}\right]\sin^2\theta\cos^2\theta
$$

The previous expression would be equal to one only if the value of the bracket were exactly 2. Substituting (4.9) in the previous expression yields the unexpected result that $\rho_{\underline{n}}\sigma_{\underline{n}} \neq 1$. This is somewhat counterintuitive and is a consequence of the resistivity $\rho_{\underline{n}}$ and conductivity $\sigma_{\underline{n}}$ being computed as:

$$
\sigma_{\underline{n}} = \underline{\underline{\sigma}}_n : \underline{n}\,\underline{n} \qquad \rho_{\underline{n}} = \underline{\underline{\rho}}_n : \underline{n}\,\underline{n}
$$

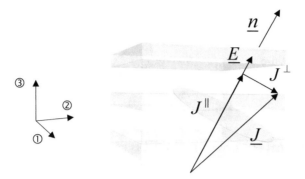

Fig. 4.12 The electric current density and the electric field $\underline{E}$ are not colinear in a flat, thin anisotropic conducting slab. The field points in the direction of $\underline{n}$

Fig. 4.13 Electrical resistance made of layered composite material with layers inclined an angle θ with respect to the axis of the rod

for one and the same sample, with one and the same $\underline{n}$. Whereas for an isotropic material, resistivity and conductivity are inverse, for an anisotropic material resistivity and conductivity *tensors* are inverse, but the projected values of these properties along *the same* direction are not inverse to each other:

$$\sigma_{\underline{n}} \neq \frac{1}{\rho_{\underline{n}}} \quad \text{with} \quad \sigma_{\underline{n}} = \underline{\sigma}_{\underline{n}} : \underline{n}\,\underline{n} \, , \, \rho_{\underline{n}} = \underline{\rho}_{\underline{n}} : \underline{n}\,\underline{n} \tag{4.10}$$

The reason for this apparent paradox is that in the measurement of $\rho_{\underline{n}}$ along a thin, long rod the flow of electrons (i.e. $\underline{J}$) is along the direction of $\underline{n}$, whereas in the expression $\sigma_{\underline{n}} = \underline{\sigma}_{\underline{n}} : \underline{n}\,\underline{n}$ it is the electric field $\underline{E}$ that points along $\underline{n}$, and the electron flow is at an angle to it. Since the flow of electrons in both cases is along different directions in the anisotropic material, and the properties are different along different directions, it is no wonder that $\sigma_{\underline{n}} \neq \frac{1}{\rho_{\underline{n}}}$.

As a matter of fact, in each of the previous cases one of the two situations described cannot be carried out in practice for an anisotropic material.[5] In the long thin rod

---

[5]"Long, thin" allows us to ignore the small regions of size $\sqrt{A}$ close to the ends of the rod. Similarly, "wide, flat", excludes a narrow zone of width δ at the periphery of the disc. In these transition regions, the fields are not homogeneous.

Fig. 4.14 Element of
surface. The exterior normal
points from the negative
material into the positive
material

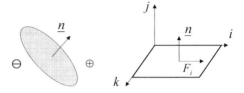

experiment it is not possible to align the electric field with the rod axis. Only an electron flow parallel to the axis can be imposed.

Conversely, in the flat, thin disk experiment it is not possible to align the electron flow perpendicular to the faces of the sample. Only the electric field parallel to $\underline{n}$ (perpendicular to the faces) can be imposed. Not only is (4.10) not surprising, it is a natural result:

Projections of reciprocal properties along the same direction are not reciprocal.

4.5 Stress

We will now present a minimal working definition of stress which is simple and useful enough for our purposes.[6] Consider a surface element of orientation given by the unit vector $\underline{n}$ (or *exterior normal*) within a material. If we define as positive side the side into which the exterior normal points, the component τ_{ij} of the stress is the force in the i-direction exerted by the positive material on the negative material on a unit area perpendicular to the j-direction (Fig. 4.14).

This definition uses a convention different from that used in many mechanical engineering treatises, which is based on the historical choice of sign, both in the order of the indices and in the sign. The reasons behind this choice are carefully discussed in [5], p. 7 and footnote 3.

According to this definition, the first index of the symmetric stress tensor refers to the direction in which the unit normal points, the second index to the direction of the force. The two indices in this definition correspond to two unit vectors in a given reference frame (Fig. 4.14, right):

- if the two indices are identical, i.e. the component is of the form τ_{ii} (no sum), it is called a *normal stress*. In a matrix representation, normal stresses appear on the diagonal of $[\![\tau]\!]$
- if the two indices are different, i.e. the component is of the form τ_{ij}, $i \neq j$, it is called a *shear stress*.

---

[6]This definition closely follows [5].

Fig. 4.15 Given the stress $\underline{\underline{\tau}}$ in a material, the force acting on a unit area whose exterior normal is $\underline{n}^{(A)}$ is
$$\underline{F} = \underline{\underline{\tau}}(\underline{n}^{(A)}, \cdot) = \underline{\underline{\tau}} \cdot \underline{n}^{(A)} = \underline{\underline{\tau}} \cdot \underline{n}^{(A)}(\cdot)$$

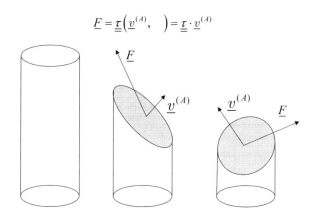

$$\underline{F} = \underline{\underline{\tau}}\left(\underline{v}^{(A)}, \quad\right) = \underline{\underline{\tau}} \cdot \underline{v}^{(A)}$$

Taking $\underline{\underline{\tau}}$ as a bilinear map, $\underline{\underline{\tau}}(\cdot, \cdot)$, in which the first argument (first index) corresponds to the area and the second one to the force, the expression:

$$\underline{F} = \underline{\underline{\tau}}(\underline{n}^{(A)}, \cdot) = \underline{\underline{\tau}} \cdot \underline{n}^A \tag{4.11}$$

where $\underline{n}^A$ is a unit vector, gives the *traction* vector, i.e. the force that acts within the material on a unit area whose normal is $\underline{n}^A$. In other words, if the material were to be actually cut along the surface, and the positive part of the material removed, a force $-\underline{\underline{\tau}} \cdot \underline{n}^A$ would have to be applied on the side of the removed material in order to maintain the same mechanical state in the negative material.[7]

Since $\underline{\underline{\tau}} \cdot \underline{n}^A$ is a force, if it is now desired to obtain the component $F_{\underline{n}}$ of this force in a direction defined by a second unit vector $\underline{n}$, all we have to do is project in the direction of the latter:

$$F_{\underline{n}} = \underline{\underline{\tau}}(\underline{n}^{(A)}, \underline{n}) = (\underline{\underline{\tau}} \cdot \underline{n}^{(A)}) \cdot \underline{n} = \underline{\underline{\tau}} : \underline{n}^{(A)}\underline{n} \tag{4.12}$$

As expected, if the map $\underline{\underline{\tau}}(\cdot, \cdot)$ is evaluated for two base vectors of a reference frame, $\underline{\delta}_i$ and $\underline{\delta}_j$, the result is just the ijth component of $\underline{\underline{\tau}}$ in that frame (Fig. 4.15):

$$\underline{\underline{\tau}}(\underline{\delta}_i, \underline{\delta}_j) = \underline{\underline{\tau}} : \underline{\delta}_i\underline{\delta}_j = \tau_{ij}$$

i.e. the force in direction j that acts on a unit area whose normal points in direction i.

As a consequence, the normal component (scalar!) of the force that acts on a unit area defined by its normal $\underline{n}^A$ is given by (Fig. 4.16):

$$F_{\underline{n}} = \underline{\underline{\tau}}(\underline{n}^{(A)}, \underline{n}^{(A)}) = \underline{\underline{\tau}} : \underline{n}^{(A)}\underline{n}^{(A)} \tag{4.13}$$

---

[7]By the linearity in all arguments, the force on an area A would be A times larger: $\underline{\underline{\tau}}(A\underline{v}^A,) = A\underline{\underline{\tau}} \cdot \underline{v}^A$.

Fig. 4.16 Using an arbitrary unit vector $\underline{n}$ as second argument $F_n = \underline{\tau}(\underline{n}^{(A)}, \underline{n}) = \underline{\underline{\tau}} : \underline{n}^{(A)}\underline{n}$ we obtain the (scalar) component of the force along $\underline{n}$

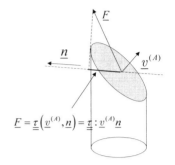

The tangential force is obtained as the difference between the total and the normal forces:

$$\underline{F} = \underline{\tau}(\underline{n}^{(A)}, \cdot) = \underline{\underline{\tau}} \cdot \underline{n}^{(A)} \qquad \text{(force on the surface)}$$

$$F_n = \underline{\tau}(\underline{n}^{(A)}, \underline{n}^{(A)}) = \underline{\underline{\tau}} : \underline{n}^{(A)}\underline{n}^{(A)} \qquad \text{(modulus of the normal force)}$$

$$\underline{F}_n = (\underline{\underline{\tau}} : \underline{n}^{(A)}\underline{n}^{(A)})\underline{n}^{(A)} \qquad \text{(normal force)}$$

$$\underline{F}^{\parallel} = \underline{F} - \underline{F}_n = \underline{\underline{\tau}} \cdot \underline{n}^{(A)} - (\underline{\underline{\tau}} : \underline{n}^{(A)}\underline{n}^{(A)})\underline{n}^{(A)} \qquad \text{(tangential force)}$$

The tangential force can also be written using the transversal projector as in (4.6):

$$\underline{F}^{\parallel} = (\underline{\underline{\delta}} - \underline{n}^A\,\underline{n}^A) \cdot \underline{F} = (\underline{\underline{\delta}} - \underline{n}^A\,\underline{n}^A) \cdot \underline{\underline{\tau}} \cdot \underline{n}^{(A)}$$

4.6   Strain

Only a basic discussion of the strain tensor $\underline{\underline{\epsilon}}$ is presented. Countless excellent references are available to the interested reader [6–8]. We consider a point P at position $\underline{r}$ within an undeformed solid S. The solid undergoes deformation due to some external cause, so that P moves to a new position P' $\underline{r}'$ (Fig. 4.17).

The *displacement* vector is defined as:

$$\underline{u} \equiv \underline{r}' - \underline{r}$$

and the infinitesimal strain or *displacement gradient* tensor is defined in terms of $\underline{u}$ as:

$$\underline{\underline{\epsilon}} \equiv \frac{1}{2}\left[\nabla\underline{u} + (\nabla\underline{u})^T\right] \qquad \epsilon_{ij} \equiv \frac{1}{2}\left(\frac{\partial u_j}{\partial x_i} + \frac{\partial u_i}{\partial x_j}\right) \qquad (4.14)$$

which is symmetric by construction.[8] Furthermore, infinitesimal strain implies $|\underline{\underline{\epsilon}}|\ll 1$, i.e. the components of $\underline{\underline{\epsilon}}$ are all $\ll 1$.

In general, the displacement $\underline{u}$ is a rank one tensor field that may depend on position and time $\underline{u}(x_1, x_2, x_3, t)$. But here we will deal with homogeneous, stationary situations only, in which $\underline{\underline{\epsilon}}$ is the same at every point and also time independent. Since $\underline{\underline{\epsilon}}$ is a constant, according to (4.14) the displacement $\underline{u}$ will be non-homogenous and a linear function of $\underline{r}$.

Geometrically, the definition (4.14) measures how different the displacements of two neighboring points P and Q ($\Delta\underline{u} = \underline{u}^P - \underline{u}^Q$) is, depending on the original difference of position ($\Delta\underline{x} = \underline{x}^P - \underline{x}^Q$) between P and Q.

Displacement alone does not necessarily produce deformation; it is the different displacements of neighboring points that cause strain (Fig. 4.18).

To better understand the meaning of $\underline{\underline{\epsilon}}$, let's see, step by step, its geometrical effect on a simple shape and in only 2D: a square of side $2L$ placed as in Fig. 4.19

- let's first see the effect of having one non-zero diagonal component in $\underline{\underline{\epsilon}}$:

$$[\![\underline{\underline{\epsilon}}]\!] = \begin{pmatrix} \epsilon_{11} & 0 \\ 0 & 0 \end{pmatrix} \qquad \epsilon_{11} = \text{const.}$$

Fig. 4.17 A deformable solid before (S) and after (S') a deformation

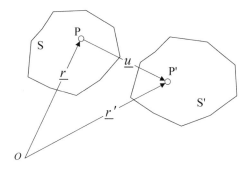

Fig. 4.18 Strain measures differences in the displacements of neighboring points

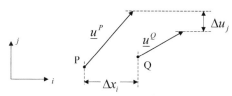

---

[8] As is obvious from its definition, strain is a purely geometric magnitude which may have different causes, stress in Hooke's law being one, but only one, of them. It is possible to have strain without stress and stress without strain.

Fig. 4.19 Undeformed
square of side $2L$

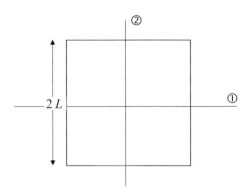

From the definition of $\underset{=}{\epsilon}$ (4.14) we can integrate:

$$\frac{\partial u_1}{\partial x_1} = \epsilon_{11} \Rightarrow u_1(x_1, x_2) = \epsilon_{11}x_1 + f_1(x_2) \qquad \frac{\partial u_2}{\partial x_2} = 0 \Rightarrow u_2(x_1, x_2) = f_2(x_1)$$

$$(4.15)$$

where $f_1(x_2)$, $f_2(x_1)$ are arbitrary functions, but only of the arguments x_2 y x_1 respectively. Substituting in the ①② component:

$$\frac{\partial u_2}{\partial x_1} + \frac{\partial u_1}{\partial x_2} = 0 \Rightarrow \frac{\partial f_2(x_1)}{\partial x_1} + \frac{\partial f_1(x_2)}{\partial x_2} = 0$$

Since $f_1(x_2)$, $f_2(x_1)$ are functions of only x_2 y x_1, so are their derivatives. The only way for the sum of two functions of two different variables to be zero is for both to be constant:

$$\frac{\partial f_2(x_1)}{\partial x_1} = K \quad \frac{\partial f_1(x_2)}{\partial x_2} = -K \Rightarrow f_2(x_1) = Kx_1 + C_2 \quad f_1(x_2) = -Kx_2 + C_1$$

and substituting in (4.15):

$$u_1(x_1, x_2) = \underbrace{\epsilon_{11}x_1}_{C} \underbrace{-Kx_2}_{B} + \underbrace{C_1}_{A} \qquad u_2(x_1, x_2) = \underbrace{Kx_1}_{B} + \underbrace{C_2}_{A} \qquad (4.16)$$

i.e. the displacement field is linear in $\underline{x}$. The two expressions in (4.16) give the two components of the u_1, u_2 displacement that a point of coordinates x_1, x_2 undergoes due to the component ϵ_{11}. Let's see the separate effects of the terms A, B and C in (4.16):

- evaluating (4.16) for the vertices of the square, we see it is just a rigid translation (Fig. 4.20).
- similarly, evaluating the B terms we get the effect shown in Fig. 4.21.
  Terms A y B in (4.16) produce rigid translation and rotation, but no displacement.

- term C (which contains the non-zero diagonal term ϵ_{11}) on the other hand, produces a longitudinal deformation in the direction of ①. The change in length of the $2L$ side is $2L\epsilon_{11}$, corner angles stay at $\frac{\pi}{2}$ (Fig. 4.22).

The same would be true for any term of the form ϵ_{ii} (no sum); it would produce the same type of deformation in direction i. For this reason the diagonal elements of $\underline{\underline{\epsilon}}$ are usually called *longitudinal* or elongational.

If there are more non-zero longitudinal components, the three contributions are superposed. Imagine (in 3D) that all diagonal elements are nonzero, and consider a cube whose edges L are parallel to the axes of the reference frame to which $\underline{\underline{\epsilon}}$ is referred.

The changes in the length of the sides of the cube would be: $L\epsilon_{11}, L\epsilon_{22}, L\epsilon_{33}$, so that the relative change in volume $\Delta V / V$ would be:

$$\frac{\Delta V}{V} = \frac{L^3(1 + \epsilon_{11})(1 + \epsilon_{22})(1 + \epsilon_{33}) - L^3}{L^3}$$
$$= 1 + \epsilon_{11} + \epsilon_{22} + \epsilon_{33} + \epsilon_{11}\epsilon_{22} + \cdots + \epsilon_{11}\epsilon_{22}\epsilon_{33} - 1 \qquad (4.17)$$
$$\simeq \epsilon_{11} + \epsilon_{22} + \epsilon_{33} = \mathrm{tr}(\underline{\underline{\epsilon}})$$

where we have neglected quadratic and cubic terms in ϵ_{ij}.[9]

- let's now see how the non-diagonal elements of $\underline{\underline{\epsilon}}$ act on the same square. Again in 2D let's assume $\underline{\underline{\epsilon}}$ has this form:

$$[\![\underline{\underline{\epsilon}}]\!] = \begin{pmatrix} 0 & \epsilon_{12} \\ \epsilon_{12} & 0 \end{pmatrix} \qquad \epsilon_{12} = \text{const.}$$

Integrating:

$$\frac{\partial u_1}{\partial x_1} = 0 \Rightarrow u_1(x_1, x_2) = f_1(x_2) \qquad \frac{\partial u_2}{\partial x_2} = 0 \Rightarrow u_2(x_1, x_2) = f_2(x_1) \quad (4.18)$$

where $f_1(x_2)$, $f_2(x_1)$ are again arbitrary function of x_2 y x_1 respectively. Substituting in the definition of component ϵ_{12}:

$$\frac{\partial u_2}{\partial x_1} + \frac{\partial u_1}{\partial x_2} = 2\epsilon_{12} \Rightarrow \frac{\partial f_2(x_1)}{\partial x_1} + \frac{\partial f_1(x_2)}{\partial x_2} = 2\epsilon_{12}$$

Similarly as for the longitudinal components:

$$\frac{\partial f_2(x_1)}{\partial x_1} = \epsilon_{12} + K \qquad \frac{\partial f_1(x_2)}{\partial x_2} = \epsilon_{12} - K \Rightarrow$$
$$f_2(x_1) = \epsilon_{12}x_1 + Kx_1 + C_2 \qquad f_1(x_2) = \epsilon_{12}x_2 - Kx_2 + C_1$$

---

[9]The result is the invariant $\mathrm{tr}(\underline{\underline{\epsilon}})$, which hints at the generality of the result. Equation (4.17) is indeed valid for any orientation of the cube with respect to the reference frame.

and substituting in (4.18):

$$u_1(x_1, x_2) = \underbrace{\epsilon_{12}x_2}_{C} - \underbrace{Kx_2}_{B} + \underbrace{C_1}_{A} \qquad u_2(x_1, x_2) = \underbrace{\epsilon_{12}x_1}_{C} + \underbrace{Kx_1}_{B} + \underbrace{C_2}_{A} \quad (4.19)$$

The terms B y C only produce rigid translation and rotation. However, the terms with ϵ_{12} do deform the square as shown in Fig. 4.23.

Therefore, the effect of an off-diagonal element (ϵ_{ij}, $i \neq j$) is to produce angular changes in the plane defined by the axes i, j (ϵ_{12} and ①② in this example). The kind of deformation of Fig. 4.23 is called *pure shear*. In addition:

$$\mathrm{tr}(\underline{\underline{\epsilon}}) = 0 \qquad (4.20)$$

which is another characteristic of (infinitesimal strain) angular deformations: they don't change areas (in 2D) nor volumes (in 3D).

The angular deformation for the corner angles $\frac{\pi}{2}$ is given to first order by:

$$2\theta \simeq 2\tan\theta = \frac{2\epsilon_{12}L}{L} = 2\epsilon_{12} \qquad (4.21)$$

i.e. opposite corners lose double the value of the ϵ_{12}, the other two grow by the same value (Fig. 4.24). Notice that the result Fig. 4.27 is not an invariant; ϵ_{12} has different values in different reference frames. This result is therefore not general. We will deduce the general result shortly.

Depending on the situation, there may exist geometrical restrictions that correspond to specific boundary conditions. A typical example is a fixed edge, as in Fig. 4.25. This situation is called *simple shear* and involves the same deformation, but different rigid translation and/or rotation.

Let's calculate the displacement field $\underline{u}$ for the 2D block which must remain in contact with a horizontal base and is subjected to $[\![\underline{\underline{\epsilon}}]\!] = \begin{bmatrix} 0 & \epsilon_{12} \\ \epsilon_{12} & 0 \end{bmatrix}$ (Fig. 4.25).

The displacement field is given by (4.19). The three integration constants that appear in (4.19) are determined to satisfy the geometrical restriction: the corners at $(0, 0)$ y $(L, 0)$ remain fixed; applying (4.19) to these points:

$$\left.\begin{array}{l} u_1(0, 0) = 0 \, , \, u_1(L, 0) = 0 \\ u_2(0, 0) = 0 \, , \, u_2(L, 0) = 0 \end{array}\right\} \quad \Rightarrow \quad C_1 = 0 \quad C_2 = 0 \quad K = -\epsilon_{12}$$

and the displacement field is:

$$\begin{cases} u_1(x_1, x_2) = 2\epsilon_{12}x_2 \\ u_2(x_1, x_2) = 0 \end{cases}$$

as shown in Fig. 4.26. Apart from this rigid rotation, the deformation is exactly the same as in Fig. 4.23.

Fig. 4.20 Terms A in (4.16) produce a rigid translation. The object translates, but its shape does not change. The original square is drawn in dashed line

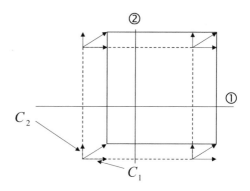

Fig. 4.21 Terms B in (4.16) produce a rigid rotation. The object rotates, but its shape does not change

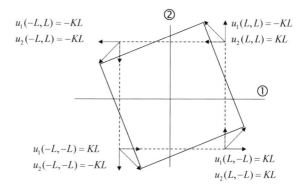

$u_1(-L,L) = -KL$
$u_2(-L,L) = -KL$

$u_1(L,L) = -KL$
$u_2(L,L) = KL$

$u_1(-L,-L) = KL$
$u_2(-L,-L) = -KL$

$u_1(L,-L) = KL$
$u_2(L,-L) = KL$

Fig. 4.22 Term C in (4.16) produces a longitudinal deformation. The original square is drawn in dashed line

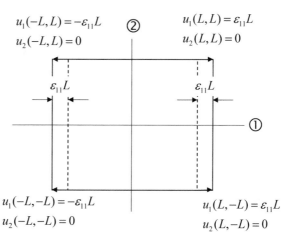

$u_1(-L,L) = -\varepsilon_{11}L$
$u_2(-L,L) = 0$

$u_1(L,L) = \varepsilon_{11}L$
$u_2(L,L) = 0$

$\varepsilon_{11}L$ $\varepsilon_{11}L$

$u_1(-L,-L) = -\varepsilon_{11}L$
$u_2(-L,-L) = 0$

$u_1(L,-L) = \varepsilon_{11}L$
$u_2(L,-L) = 0$

As off-diagonal components of $\underline{\underline{\varepsilon}}$ cause changes in some angles, they are often called the *angular* or *shear* components of the strain (Fig. 4.27).

Fig. 4.23 An off-diagonal element of $\underline{\underline{\epsilon}}$, like ϵ_{12}, produces angular deformation in the ①② plane. The original square is drawn in dashed line. Vertex displacements and angular changes are greatly exaggerated

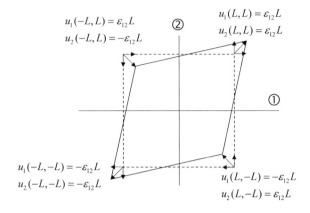

$$u_1(-L,L) = \varepsilon_{12}L$$
$$u_2(-L,L) = -\varepsilon_{12}L$$

$$u_1(L,L) = \varepsilon_{12}L$$
$$u_2(L,L) = \varepsilon_{12}L$$

$$u_1(-L,-L) = -\varepsilon_{12}L$$
$$u_2(-L,-L) = -\varepsilon_{12}L$$

$$u_1(L,-L) = -\varepsilon_{12}L$$
$$u_2(L,-L) = \varepsilon_{12}L$$

Fig. 4.24 The corners of the square of Fig. 4.19 change by $\pm 2\epsilon_{12}$

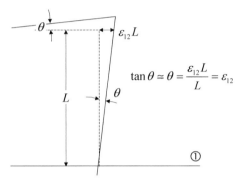

$$\tan\theta \simeq \theta = \frac{\varepsilon_{12}L}{L} = \varepsilon_{12}$$

Fig. 4.25 2D block with a fixed side

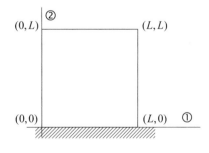

Fig. 4.26 Simple shear of the 2D block of Fig. 4.25. The displacements of the corners are shown. The corner angles increase or decrease by $2\epsilon_{12}$ rad, the same as in Figs. 4.23 and 4.24

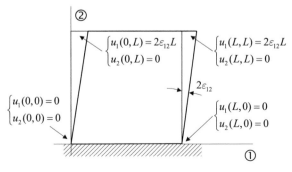

$$\begin{cases} u_1(0,L) = 2\varepsilon_{12}L \\ u_2(0,L) = 0 \end{cases}$$

$$\begin{cases} u_1(L,L) = 2\varepsilon_{12}L \\ u_2(L,L) = 0 \end{cases}$$

$$\begin{cases} u_1(0,0) = 0 \\ u_2(0,0) = 0 \end{cases}$$

$2\varepsilon_{12}$

$$\begin{cases} u_1(L,0) = 0 \\ u_2(L,0) = 0 \end{cases}$$

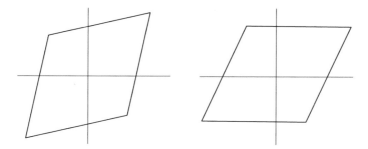

Fig. 4.27 The difference between pure (left) and simple shear is a rigid rotation, due to the K terms in (4.19). The deformation of the object is exactly the same. Only the integration constants are different in order to accommodate different boundary conditions: $K = 0$ for pure shear; $K = -\epsilon_{12}$ for simple shear

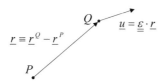

Fig. 4.28 The displacement of a point of coordinates $\underline{r}$ caused by a homogeneous strain $\underline{\underline{\epsilon}}$ is given by Eq. 4.22

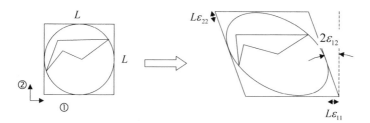

Fig. 4.29 The effect of a homogenous $\underline{\underline{\epsilon}}$ is to deform all space uniformly, including any objects embedded in it. The figure on the left is deformed as shown on the right by $[\![\underline{\underline{\epsilon}}]\!] = \begin{pmatrix} \epsilon_{11} & \epsilon_{12} \\ \epsilon_{12} & \epsilon_{22} \end{pmatrix}$. Deformations are very exaggerated in this figure

We can now put together all previous results in a general expression for the displacement field $\underline{u}$ caused by a homogeneous strain tensor $\underline{\underline{\epsilon}}$:

$$\underline{u} = \underline{\underline{\epsilon}} \cdot \underline{r} + \text{rigid rotation} + \text{rigid translation} \qquad (4.22)$$

Equation (4.22) (Fig. 4.28) is the general result that includes (4.16) y (4.19). Since $\underline{\underline{\epsilon}}$ is a constant and $\underline{u}$ is a first integral of $\underline{\underline{\epsilon}}$, Eq. (4.22) is the general linear expression for $\underline{u}$.

It is often useful to consider $\underline{\underline{\epsilon}}$ as a deformation of the space that also applies to all objects embedded in it (Fig. 4.29).

Fig. 4.30 Homogenous, isotropic deformation of a 2D ring

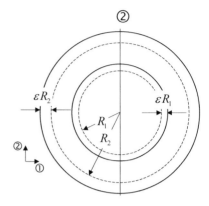

The following example should help understand the previous statement. Take a 2D ring of an isotropic material, of inner and outer radii R_1 and R_2 respectively. Imagine it really is a metal ring which fits a finger too tightly. As a way to remove it, someone suggests to heat it so that the hole gets larger. Someone else argues that the material of the ring itself will get wider and thus squeeze your finger even more tightly.

The starting point is the expansion law that gives the homogeneous strain in a material upon a change of temperature ΔT:

$$\underline{\underline{\epsilon}} = \underline{\underline{\alpha}} \Delta T$$

where $\underline{\underline{\alpha}} = \alpha \underline{\underline{\delta}}$ is the thermal expansion coefficient for an isotropic material and ΔT the change in temperature, both assumed known, $\alpha_{ii} > 0$ (no sum), $\Delta T > 0$, so that

$$\underline{\underline{\epsilon}} = \alpha \underline{\underline{\delta}} \Delta T \quad \Rightarrow \quad [\![\underline{\underline{\epsilon}}]\!] = \begin{bmatrix} \alpha \Delta T & 0 \\ 0 & \alpha \Delta T \end{bmatrix} \tag{4.23}$$

Before reading on, try to decide which argument is correct, using Fig. 4.29 without carrying out any calculation.

The answer is straightforward if we use $\underline{\underline{\epsilon}}$ to deform 2D space *and* the ring with it. The strain of Eq. 4.23 is an isotropic deformation, so that space and the ring grow identically in all directions. In particular, the inner radius also grows. The statement that heating the ring will increase its inner radius is correct.[10]

Figure 4.30 shows the increases in size calculated according to (4.22). The inner radius grows by $R_1 \epsilon$. The argument "the material of the the ring will expand and squeeze the finger even more tightly" has a correct first part (the material will indeed get wider, by $(R_2 - R_1)\epsilon$) and a wrong second part.

---

[10] A different matter is that in practice α is a very small number and ΔT cannot be very large if we are talking about a ring and a finger. The method is not practical; better use soapy water.

Fig. 4.31 A purely longitudinal

$$\llbracket \underline{\underline{\epsilon}} \rrbracket = \begin{pmatrix} \epsilon_{11} & 0 \\ 0 & 0 \end{pmatrix} \text{ does}$$

produce changes in the corner angles if the edges of the square are not parallel to the axes

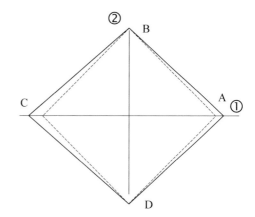

The $\underline{\underline{\epsilon}}$ of this example produces a homogenous, isotropic expansion of 2D space. If this is so, which is the center of this expansion?[11]

In the previous examples, longitudinal components of $\underline{\underline{\epsilon}}$ produce no changes in the corner angles, angular components of $\underline{\underline{\epsilon}}$ produce no changes in the lengths of the sides. This is so only because of the particular orientation of the square: its sides are parallel to the axes of the reference frame of $\underline{\underline{\epsilon}}$.

If we place the square as in Fig. 4.31 a purely longitudinal $\underline{\underline{\epsilon}}$ like $\llbracket \underline{\underline{\epsilon}} \rrbracket = \begin{pmatrix} \epsilon_{11} & 0 \\ 0 & 0 \end{pmatrix}$ does change the corner angles.

Similarly, an angular strain $\llbracket \underline{\underline{\epsilon}} \rrbracket = \begin{pmatrix} 0 & \epsilon_{12} \\ \epsilon_{12} & 0 \end{pmatrix}$ may modify lengths and leave the corner angles unchanged (Fig. 4.32).

There is no contradiction in this. If we transform $\llbracket \underline{\underline{\epsilon}} \rrbracket = \begin{pmatrix} 0 & \epsilon_{12} \\ \epsilon_{12} & 0 \end{pmatrix}$ to a new reference frame ①'②', rotated $\frac{\pi}{4}$ with respect to the original one, by means of $\underline{\underline{L}} = \begin{pmatrix} \sqrt{2}/2 & \sqrt{2}/2 \\ -\sqrt{2}/2 & \sqrt{2}/2 \end{pmatrix}$, we obtain:

$$\llbracket \underline{\underline{\epsilon}}' \rrbracket = \begin{pmatrix} \epsilon_{12} & 0 \\ 0 & -\epsilon_{12} \end{pmatrix} \tag{4.24}$$

In these new axes, the deformation "looks" purely longitudinal. Furthermore, it is a combination of two longitudinal deformations of the same modulus and opposite signs.

These changes of character (longitudinal vs. angular) in $\underline{\underline{\epsilon}}$ are just a consequence of the change of reference frame. But the geometrical effect of the strain tensor, i.e.

---

[11]The answer is not the apparently obvious.

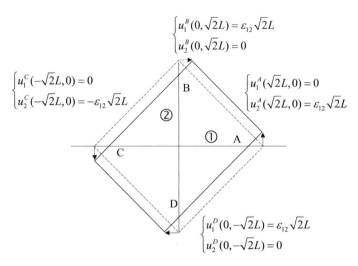

$$\begin{cases} u_1^B(0, \sqrt{2}L) = \varepsilon_{12}\sqrt{2}L \\ u_2^B(0, \sqrt{2}L) = 0 \end{cases}$$

$$\begin{cases} u_1^C(-\sqrt{2}L, 0) = 0 \\ u_2^C(-\sqrt{2}L, 0) = -\varepsilon_{12}\sqrt{2}L \end{cases}$$

$$\begin{cases} u_1^A(\sqrt{2}L, 0) = 0 \\ u_2^A(\sqrt{2}L, 0) = \varepsilon_{12}\sqrt{2}L \end{cases}$$

$$\begin{cases} u_1^D(0, -\sqrt{2}L) = \varepsilon_{12}\sqrt{2}L \\ u_2^D(0, -\sqrt{2}L) = 0 \end{cases}$$

Fig. 4.32 A purely angular deformation may leave corner angles unchanged and vary the length of the square edges

how it deforms space and all objects embedded in it, is the same in all reference frames; that is why it is a tensorial magnitude.

We will now obtain general formulas for the changes of lengths, areas, volumes and angles caused by a homogeneous $\underline{\underline{\epsilon}}$.

1. Relative variation in the length L of a segment whose orientation is given by a unit vector $\underline{n}^A$. Placing one end of the segment at the origin,[12] its displacement is zero; the other end is at $L\underline{n}^A$ and its displacement (4.22):

$$\underline{\underline{\epsilon}} \cdot (L\underline{n}^A) = L\underline{\underline{\epsilon}} \cdot \underline{n}^A$$

so that the change in modulus is:

$$\begin{aligned} \Delta L &= \sqrt{(L\underline{n}^A + L\underline{\underline{\epsilon}} \cdot \underline{n}^A) \cdot (L\underline{n}^A + L\underline{\underline{\epsilon}} \cdot \underline{n}^A)} - L = \\ &= \sqrt{L^2 + 2L^2 (\underline{\underline{\epsilon}} \cdot \underline{n}^A) \cdot \underline{n}^A + O(\epsilon^2)} - L \simeq \\ &\simeq \sqrt{L^2 + 2L^2 \underline{\underline{\epsilon}} : \underline{n}^A \underline{n}^A} - L \simeq L\underline{\underline{\epsilon}} : \underline{n}^A \underline{n}^A \end{aligned} \qquad (4.25)$$

after using $(1+x)^n \simeq 1 + nx$ with $n = 1/2$ and $x = 2\underline{\underline{\epsilon}} : \underline{n}^A \underline{n}^A$, and neglecting $O(\epsilon^2)$ terms. The relative length variation is then:

$$\frac{\Delta L}{L} = \underline{\underline{\epsilon}} : \underline{n}^A \underline{n}^A \qquad (4.26)$$

---

[12]Which can always be done because the strain is homogeneous. All points are equivalent.

2. Relative variation of an area S of orientation defined[13] by two unit, mutually orthogonal vectors $\underline{n}^A$ and $\underline{n}^B$. Before the deformation, $\underline{n}^A$ and $\underline{n}^B$ define an area $|\underline{n}^A \times \underline{n}^B| = 1$. The strain changes both the lengths of these vectors and the angle they form, so that the increment of the area is:

$$\Delta S = |\underline{n}'^A||\underline{n}'^B|\sin(\pi/2 + \Delta\theta) - 1 =$$
$$= (1 + \underline{\underline{\epsilon}} : \underline{n}^A \underline{n}^A)(1 + \underline{\underline{\epsilon}} : \underline{n}^B \underline{n}^B)\underbrace{(\sin \pi/2 \cos \Delta\theta + \cos \pi/2 \sin \Delta\theta)}_{\simeq 1} - 1 =$$

$$= \underline{\underline{\epsilon}} : (\underline{n}^A \underline{n}^A + \underline{n}^B \underline{n}^B) \quad \Rightarrow \quad \frac{\Delta S}{S} = \underline{\underline{\epsilon}} : (\underline{n}^A \underline{n}^A + \underline{n}^B \underline{n}^B) \tag{4.27}$$

which is the desired result.

3. The relative variation of a volume V is given by (4.17):

$$\frac{\Delta V}{V} = \text{tr}(\underline{\underline{\epsilon}}) \tag{4.28}$$

which can also be calculated from any three unit, mutually orthogonal vectors $\underline{n}^A$, $\underline{n}^B$ and $\underline{n}^C$:

$$\frac{\Delta V}{V} = \underline{\underline{\epsilon}} : (\underline{n}^A \underline{n}^A + \underline{n}^B \underline{n}^B + \underline{n}^C \underline{n}^C) \tag{4.29}$$

Notice how similar the three formulas (4.26), (4.27), (4.29) are. They give the relative variations of a length (1D), an area (2D) or a volume (3D).

When the three vectors $\underline{n}^A$, $\underline{n}^B$ and $\underline{n}^C$ in (4.29) are chosen to be the base vectors, Eq. 4.29 reduces to Eq. 4.28.

4. Variation of an angle θ defined[14] by two unit vectors $\underline{n}^A$ y $\underline{n}^B$. Before the deformation, the angle defined by $\underline{n}^A$ y $\underline{n}^B$ is:

$$\cos\theta = \frac{\underline{n}^A \cdot \underline{n}^B}{|\underline{n}^A||\underline{n}^B|}$$

The strain $\underline{\underline{\epsilon}}$ modifies both the moduli of the vectors and the angle in this expression, so that:

$$\cos(\theta + \Delta\theta) = (\cos\theta \cos\Delta\theta - \sin\theta \sin\Delta\theta)$$
$$\simeq \cos\theta - \Delta\theta \sin\theta = \frac{\underline{n}'^A \cdot \underline{n}'^B}{|\underline{n}'^A||\underline{n}'^B|} \tag{4.30}$$

The moduli in the denominator denominator are (4.25):

---

[13]"Defined" means that the normal to S and $\underline{n}^A \times \underline{n}^B$ are parallel. S is flat but may have any size and shape.

[14]"Defined" means that, if their origins are brought to the same point, so that both point away from the common origin, the angle between the vectors is θ.

$$|\underline{n}'^A| = 1 + \underline{\underline{\epsilon}} : \underline{n}^A \underline{n}^A \qquad |\underline{n}'^B| = 1 + \underline{\underline{\epsilon}} : \underline{n}^B \underline{n}^B$$

and the reciprocal of their product:

$$\left[(1 + \underline{\underline{\epsilon}} : \underline{n}^A \underline{n}^A)(1 + \underline{\underline{\epsilon}} : \underline{n}^B \underline{n}^B) \right]^{-1} \simeq 1 - \underline{\underline{\epsilon}} : (\underline{n}^A \underline{n}^A + \underline{n}^B \underline{n}^B)$$

The numerator:

$$\underline{n}'^A \cdot \underline{n}'^B = (\underline{n}^A + \underline{\underline{\epsilon}} \cdot \underline{n}^A) \cdot (\underline{n}^B + \underline{\underline{\epsilon}} \cdot \underline{n}^B) \simeq \underline{n}^A \cdot \underline{n}^B + 2\underline{\underline{\epsilon}} : (\underline{n}^A \underline{n}^B)$$

substituting in (4.30) and neglecting terms quadratic in ϵ:

$$
\Delta\theta = \frac{-\dfrac{\underline{n}'^A \cdot \underline{n}'^B}{|\underline{n}'^A||\underline{n}'^B|} + \cos\theta}{\sin\theta} =
$$

$$
= \frac{-\left[\underline{n}^A \cdot \underline{n}^B + 2\underline{\underline{\epsilon}} : (\underline{n}^A \underline{n}^B)\right]\left[1 - \underline{\underline{\epsilon}} : (\underline{n}^A \underline{n}^A + \underline{n}^B \underline{n}^B)\right] + \cos\theta}{\sin\theta} =
$$

$$
= \frac{-2\underline{\underline{\epsilon}} : \underline{n}^A\underline{n}^B + (\underline{n}^A \cdot \underline{n}^B)\underline{\underline{\epsilon}} : (\underline{n}^A\underline{n}^A + \underline{n}^B\underline{n}^B)}{\sqrt{1 - (\underline{n}^A \cdot \underline{n}^B)^2}} \tag{4.31}
$$

which is the desired result.

Formulas (4.26) through (4.31) are written exclusively in terms of invariants. They are therefore valid in all frames of reference. The results obtained from the example of the square at the beginning of this subsection are particular cases of (4.26) and (4.31) for especially simple orientations of the unit vectors.

4.7 Elastic Compliance

The elastic compliance $\underline{\underline{s}}$, which relates stress and strain in Hooke's law $\underline{\underline{\epsilon}} = \underline{\underline{s}}\,\underline{\underline{\tau}}$, is very well suited to illustrate the projection of a fourth rank tensor. The projection of a generic fourth rank tensor along a direction specified by the unit vector $\underline{n}$ is:

$$T_{\underline{n}} = \underline{\underline{T}} \vdots \underline{n}\,\underline{n}\,\underline{n}\,\underline{n} = T_{ijkl}n_l n_k n_j n_i$$

In the case of the elastic compliance $\underline{\underline{s}}$ this projection can be related to a specific mechanical experiment, as it was done above for the electric resistivity.

A sample in the shape of a parallelepipedic bar of cross section A is cut out of a sample of a material (like the bar cut out of a layered composite in Fig. 4.33) so that the axis of the bar lies along an arbitrary direction given by $\underline{n}$. The sample is

Fig. 4.33 The lengths of all sides and the angles of all corners change when a normal stress $\frac{F}{A}\underline{n}\,\underline{n}$ is applied to a bar of a general anisotropic material

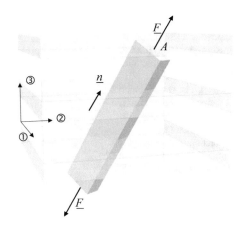

subjected to opposed forces $\underline{F}$ and $-\underline{F}$ so that a homogeneous normal stress:

$$\frac{F}{A}\underline{n}\,\underline{n} = \tau\underline{n}\,\underline{n}$$

where $\tau \equiv \dfrac{F}{A}$, is produced in the material. For a Hookean material $\underset{=}{\epsilon}$ is given by Hooke's law:

$$\underset{=}{\epsilon} = \underset{\equiv}{s} : \tau\underline{n}\,\underline{n}$$

with $\epsilon_{ij} \neq 0 \; \forall i, j$, i.e. the lengths of all sides and all angles change.

The deformation along the long dimension is given by the projection of the deformation along $\underline{n}$:

$$\epsilon_n = \epsilon : \underline{n}\,\underline{n} = (\underset{\equiv}{s} : \tau\underline{n}\,\underline{n}) : \underline{n}\,\underline{n}$$

Since the definition of Young's modulus E is the ratio of the normal stress to the longitudinal deformation, it turns out that:

$$E_n = \frac{1}{\underset{\equiv}{s} : \underline{n}\,\underline{n}\,\underline{n}\,\underline{n}} \tag{4.32}$$

where E_n indicates that the modulus is measured in the direction of $\underline{n}$. Thus, the projection of the elastic compliance $\underset{\equiv}{s}$ along a direction $\underline{n}$ is the reciprocal of Young's modulus measured in that direction.

The formal expression for the projection of a generic fourth rank tensor $\underset{\equiv}{T}$ in a direction:

Fig. 4.34 A shear stress applied to a sample of an anisotropic material. In general, the deformation of the sample will have all $\epsilon_{ij} \neq 0$. The shear modulus is defined as the ratio of the projection of the stress on $\underline{n}^A \underline{n}^B$ divided by the projection of $\underline{\epsilon}$ on $\underline{n}^A \underline{n}^B$. The two forces that act on the hidden sides of the slab are not shown

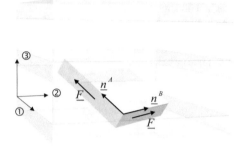

$$T_{\underline{n}} = \underline{\underline{T}} : \underline{n}\,\underline{n}\,\underline{n}\,\underline{n} = T_{ijkl}n_l n_k n_j n_i$$

has a simple and experimentally accessible interpretation in the case of the elastic compliance:

$$E_{\underline{n}} = \left(\underline{\underline{s}} : \underline{n}\,\underline{n}\,\underline{n}\,\underline{n}\right)^{-1} = \left(s_{ijkl}n_l n_k n_j n_i\right)^{-1} \tag{4.33}$$

It is now natural to try and project along more than one direction. Let's take two unit orthogonal vectors $\underline{n}^A$, $\underline{n}^B$ and cut a sample of the same material so that its sides point along $\underline{n}^A$ and $\underline{n}^B$.

The application of a homogeneous shear stress to the sample will produce a deformation $\underline{\epsilon}$ in which all components will be non-zero. The experimental measurement of the ratio between the shear stress applied in the plane defined by $\underline{n}^A$, $\underline{n}^B$, and the angular deformation of the angle defined by $\underline{n}^A$, $\underline{n}^B$ (independently of all other dimensional changes of the sample) is, by definition, the shear modulus the material (Fig. 4.34):

$$G_{\underline{n}^A \underline{n}^B} = \frac{1}{4}(\underline{\underline{s}} : \underline{n}^A\,\underline{n}^B\,\underline{n}^A\,\underline{n}^B)^{-1} = \frac{1}{4}\left(s_{ijkl}n_l^A n_k^B n_j^A n_i^B\right)^{-1}$$

which will depend on the orientations of $\underline{n}^A$ and $\underline{n}^B$.[15]

In the previous examples, the directions of the unit vectors were arbitrary. If they are made to coincide with the unit vectors of the conventional reference frame $\underline{\delta}_1$, $\underline{\delta}_2$, $\underline{\delta}_3$, expressions such as (4.32) reduce to:

---

[15]Not only if they are allowed to freely point in any direction, but even if they are rotated only within the same original plane keeping their relative angle of $90°$.

$$E_i = \frac{1}{\underset{\equiv}{s}\, :\underline{\delta}_i \underline{\delta}_i \underline{\delta}_i \underline{\delta}_i} = \frac{1}{s_{iiii}} \qquad i = 1, 2, 3 \quad \text{no sum over } i$$

i.e. the reciprocal of the projections of $\underset{\equiv}{s}$ along the base vectors are the three longitudinal moduli in the conventional axes.

Slightly more generally: the s_{ijkl} component of $\underset{\equiv}{s}$ relates, via Hooke's law $\epsilon_{ij} = s_{ijkl}\tau_{lk}$, the ijth component of $\underset{=}{\epsilon}$ with the lkth component of $\underset{=}{\tau}$. Thus, the meaning of a term like s_{36} is:

$$s_{36} = \frac{-\nu_{63}}{G_6} = \frac{1}{2}s_{3312} = \frac{1}{2}\underset{\equiv}{s}\, :\underline{\delta}_2\underline{\delta}_1\underline{\delta}_3\underline{\delta}_3 \tag{4.34}$$

i.e. the reciprocal of the projection along $\underline{\delta}_2, \underline{\delta}_1, \underline{\delta}_3, \underline{\delta}_3$ (or along the other index permutations allowed by the symmetries of $\underset{\equiv}{s}$).

In mechanical terms, a component like s_{36} is the proportionality constant between a shear stress in the plane ①② and a longitudinal strain ③③, due to Poisson's effect: it is a coupling term between shear stress and longitudinal strain, which is absent for all materials whose off-diagonal 3×3 subblocks of $\underset{\smile}{s}$ are empty.

$$
\begin{bmatrix}
\text{longitudinal} \\
\text{strains} \\
\\
\text{angular strains}
\end{bmatrix}
=
\begin{bmatrix}
\begin{array}{cc}
\text{normal-longitudinal} & \text{shear-longitudinal} \\
\text{coupling} & \text{coupling} \\
\\
\text{normal-angular} & \text{shear-angular} \\
\text{coupling} & \text{coupling}
\end{array}
\end{bmatrix}
\begin{bmatrix}
\text{normal stresses} \\
\\
\\
\text{shear stresses}
\end{bmatrix}
\tag{4.35}
$$

Experimentally, it corresponds to applying a shear stress in the plane ①② and measuring the longitudinal strain ③③. Although this may seem an unusual type of modulus, such coupling terms are not only non-zero but also important for many materials, such as fabric laminates used in aerospace applications.

To sum up, all engineering moduli (E_i, G_i) are just ratios of stresses and deformations for which the two indices of the stress and the two of the deformation coincide. They are reciprocals of projections of the elastic compliance $\underset{\equiv}{s}$ of the type:

$$\left(\underset{\equiv}{s}\, :\underline{n}^A\,\underline{n}^A\,\underline{n}^A\,\underline{n}^A\right)^{-1} \qquad \text{longitudinal, } E \text{ moduli.}$$

$$\frac{1}{4}\left(\underset{\equiv}{s}\, :\underline{n}^A\,\underline{n}^B\,\underline{n}^A\,\underline{n}^B\right)^{-1} \qquad \text{shear, } G \text{ moduli.}$$

and they appear on the diagonal of the condensed $\underset{\smile}{s}$. If the first and second pair of vectors in the projection do not coincide, such as in:

$$\left(\underline{\underline{s}} : \underline{n}^A \, \underline{n}^B \, \underline{n}^C \, \underline{n}^C \right)^{-1}$$

the (reciprocal of) the projection includes a Poisson coupling, and appears off-diagonally in $\underline{s}$.

References

1. Goldstein, H., Poole, C., Safko, J.: Classical Mechanics. Addison Wesley, Boston (2002)
2. Nalwa, H.S.: Ferroelectric Polymers: Chemistry, Physics, and Applications. CRC Press, Boca Raton (1995)
3. Asaka, K., Okuzaki, H.: Soft Actuators. Wiley, Hoboken (2013)
4. Salimi, A., Yousefi, A.A.: Analysis method: FTIR studies of β-phase crystal formation in stretched PVDF films. Polym. Test. **22**(6), 699–704 (2003)
5. Bird, B., Stewart, W.E., Lightfoot, E.N.: Transport Phenomena, 2nd edn. Wiley, Hoboken (2002)
6. Gurtin, M.E.: The Linear Theory of Elasticity. Springer, Berlin (1973)
7. Marsden, J.E., Hughes, T.J.R.: Mathematical Foundations of Elasticity. Prentice-Hall, Upper Saddle River (1983)
8. Dym, C.L., Shames, I.H.: Solid Mechanics. Springer, Berlin (1973)

Chapter 5
Scalar and Rank One Properties

In this chapter we meet the simplest RS. But before doing so, a brief review of some basic material and some general features of RSs are presented, as well as those specific of odd and even order properties which will be useful in later chapters.

5.1 Trigonometric Functions in Polar Coordinates

Representation surfaces give the orientation dependence of material properties according to the projection (5.2). The n_i factors in (5.2) are components of the unit vector that samples 3D space, and thus are sines or cosines of polar spherical angles, or Euler angles. Sines and cosines are thus a basic ingredient of representation surfaces, so it is convenient to briefly review their graphs in polar coordinates.

In the plots below the angle is generically called α. It stands for any of the spherical polar angles θ, φ, or for any of the Euler angles θ, φ, ψ.

A polar plot is constructed by drawing a point at a distance equal to the value of the function $r(\alpha)$ on a line inclined an angle α with respect to the origin of angles. As α sweeps the range $[0, 2\pi)$ the point at $r(\alpha)$ describes a curve; this is the polar plot.

As we saw in Chap. 2 the only axes of symmetry crystalline materials can have are of order 1, 2, 3, 4 and 6.

Figures 5.1, 5.2 and 5.3 show the polar $r(\alpha)$ representation of $\cos \alpha$ for $n = 1, 2, 3, 4, 6$.[1] Because of $\sin \alpha = \cos(\frac{\pi}{2} - \alpha)$, the corresponding plots for the sine function are identical, but rotated $\dfrac{\pi}{2n}$ counterclockwise.

Since the distance between the origin and the curve when moving along the radius vector is equal to the value of the function $r(\alpha)$ for a given value of α, a negative value of $r(\alpha) < 0$ corresponds to "advancing" a negative amount, i.e. moving backwards

---

[1] What do the polar plots of sine and cosine of the rotation angle look like for limit classes?

© Springer Nature Switzerland AG 2020
M. Laso and N. Jimeno, *Representation Surfaces for Physical Properties of Materials*, Engineering Materials, https://doi.org/10.1007/978-3-030-40870-1_5

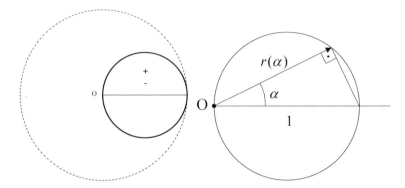

Fig. 5.1 The polar representation of $r(\alpha) = \cos \alpha$ is a circle of unit diameter. On the right, a right angled triangle inscribed in the circle. In this and the following two figures the angle α is measured counterclockwise from the horizontal line; symbols $+$ and $-$ denote positive and negative branches; the horizontal line is the origin of angles, the dashed circle has radius 1

in the radial direction defined by the argument α. As a consequence, although the function $r(\alpha)$ is single valued, the polar plot may consist of overlapping *branches* which we will call positive and negative, depending on the sign of $r(\alpha)$. Branches are marked with $+$ and $-$ signs in the figures below.

All plots for $\cos n\alpha$ have $2n$ branches: n positive and n negative. For n odd ($n = 1, 3$) positive and negative branches overlap; thus, only n double or overlapping branches are visible. For $n = 1$, the graph of $r(\alpha)$ is a simple circle (Fig. 5.1). For n even, all $2n$ branches are visible.

Higher powers of $\cos \alpha$ and $\sin \alpha$ also appear in the equation of the RS. An even value for the exponent m in $r(\alpha) = \cos^m n\alpha$ makes negative branches visible. The curves of even powers of $\cos n\alpha$ consist of $2n$ positive branches for all values of n. Overlapping positive and negative branches for $r(\alpha) = \cos^m n\alpha$ happen only if both m and n are odd.

The polar plots of $r(\alpha) = \cos^m n\alpha$ thus look qualitatively similar to those of Figs. 5.1, 5.2 and 5.3, except that for even m all branches are positive.

In addition, whether odd or even, larger values of m render the branches narrower, as shown in Fig. 5.4.

Products of these simpler trigonometric functions can easily be understood qualitatively by graphically "multiplying" the factor curves, i.e. multiplying, for every value of α, the value of $r(\alpha)$ for each the factors, including of course the sign of each branch. Overlapping, "hidden", branches may then appear, as in the example of Fig. 5.5.

Occasionally, it will be convenient to cast such products as functions of the multiple angle by means of well known identities as:

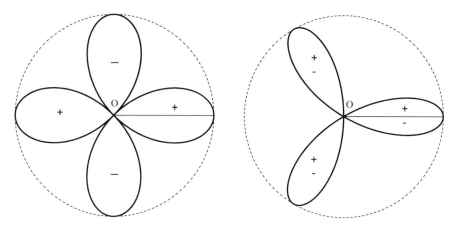

Fig. 5.2 Polar representation of $r(\alpha) = \cos 2\alpha$ and $r(\alpha) = \cos 3\alpha$

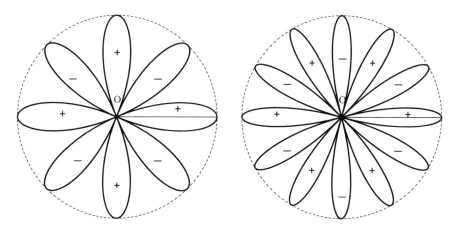

Fig. 5.3 Polar representation of $r(\alpha) = \cos 4\alpha$ and $r(\alpha) = \cos 6\alpha$

$$\begin{aligned}
\sin 2\alpha &= 2 \sin \alpha \cos \alpha \\
\cos 2\alpha &= \cos^2 \alpha - \sin^2 \alpha \\
\sin 3\alpha &= 3 \sin \alpha \cos^2 \alpha - \sin^3 \alpha \\
\cos 3\alpha &= -3 \sin^2 \alpha \cos \alpha + \cos^3 \alpha \\
\sin 4\alpha &= 4 \sin \alpha \cos^3 \alpha - 4 \sin^3 \alpha \cos \alpha \\
\cos 4\alpha &= \cos^4 \alpha - 6 \sin^2 \alpha \cos^2 \alpha + \sin^4 \alpha
\end{aligned} \tag{5.1}$$

Notice that the factor n in $\cos n\alpha$ or $\sin n\alpha$ is equal to the sum of the exponents in each of the terms of the RHSs.

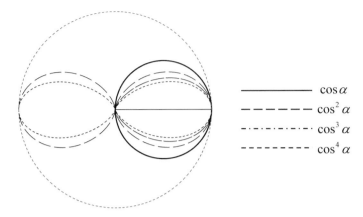

$$\text{\textemdash\textemdash\textemdash} \quad \cos \alpha$$
$$\text{-----} \quad \cos^2 \alpha$$
$$\text{-·-·-·-} \quad \cos^3 \alpha$$
$$\text{- - - - -} \quad \cos^4 \alpha$$

Fig. 5.4 Polar plots of the first four powers of $\cos \alpha$. The horizontal line is the origin of angles

Fig. 5.5 Polar plots of the functions $r(\theta) = \sin^3 \theta$, $r(\theta) = \cos \theta$ and their product $r(\theta) = \sin^3 \theta \cos \theta$. Multiplication of the two overlapping branches of opposite signs of each curve produce four branches

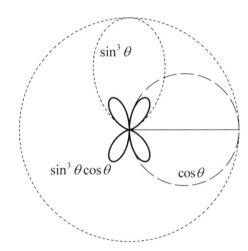

5.2 Some General Properties of the RS

For a general tensor $\underline{\underline{T}}$ of order p, the RS is given by projecting/contracting the property p times with the unit vector:

$$r(\varphi, \theta) = \underline{\underline{T}} \vdots \underbrace{\underline{n}(\varphi, \theta) \ldots \ldots \underline{n}(\varphi, \theta)}_{p \text{ times}} = T_{ij\ldots qr} \underbrace{n_r(\varphi, \theta) \ldots n_i(\varphi, \theta)}_{p \text{ factors}} \qquad (5.2)$$

Expression (5.2) is a sum of 3^p products, each of which consists of p factors, each of which is a component n_i of a unit vector $\underline{n}$ like (5.6). Thus $r(\varphi, \theta)$ is a polynomial in sine and cosine of the angles φ, θ, with real coefficients $T_{ij\ldots qr}$. The functions $\sin \varphi, \sin \theta, \cos \varphi, \cos \theta$ appear in (5.2) raised to, at most, the pth power.

This particularly simple functional form implies:

- since $|\sin(\ )| \le 1$ and $|\cos(\ )| \le 1$, the vector radius $r(\varphi, \theta)$ is also bounded. The RS will have neither asymptotic planes nor directions,
- since $\sin(\ )$ and $\cos(\ )$ are single-valued, the vector radius $r(\varphi, \theta)$, considered a function of n, will be single-valued as well. For all possible orientations of n, the vector radius either cuts the RS at a single point, or, if $r(\varphi, \theta) = 0$, the RS passes through the origin. This result prohibits the RS to be as sketched qualitatively in the left panel of Fig. 5.7,
- the RS $r(\varphi, \theta)$ is defined for all values of the angles and is infinitely differentiable at all points (i.e. for all orientations of n). As a consequence the RS is a smooth, closed surface.

The second item above needs clarification: that $r(\varphi, \theta)$ is single-valued means that to a given direction of n corresponds a unique value of $r(\varphi, \theta)$. This does not imply that the RS consists of simple lobes. It may and actually does happen, that the value of $r(\varphi, \theta)$ for n and for $-n$ have opposite signs, so that double lobes may occur, as we will see now.

5.2.1 Odd Rank Properties

Let's denote the value of a property in the direction of n by r_n and in the opposite direction $-n$ by r_{-n}. If the rank p of the property is odd, then:

$$r_{-n} = \underline{\underline{T}} \vdots \underbrace{(-n)......(-n)}_{p \text{ times}} = (-1)^p \, \underline{\underline{T}} \vdots \underbrace{n......n}_{p \text{ times}} = -r_n \qquad (5.3)$$

i.e. the values of the property in two opposite directions have opposite signs. In the direction n we advance r_n, in the opposite direction $-n$ we "advance" $-r_n$, so we end up at the same point. The RS is single valued for the argument n (for every n there is a single r_n), but lobes overlap so the RS for odd-rank properties must consist of overlapping positive and negative lobes. Furthermore, for each of the double $\pm$ lobes there is a correspondingly empty region opposite it: the empty space left by the negative lobe. This empty region is related to the lobe opposite it by a $\bar{1}$.

Let us know assume that the origin is strictly interior to the RS so that the RS must look as in Fig. 5.6. Let P and Q be the points where the vector radii r_n and r_{-n} intersect the RS (Fig. 5.6), and Γ an arbitrary curve contained in the RS with the condition that it joins P and Q. Let Γ be parametrized by the colatitude θ, i.e.:

$$\underline{\Gamma}(\varphi(\theta), \theta)$$

with $\varphi(\ )$ a C^0 function.

Fig. 5.6 Antipodal points P
and Q in an *incorrect* sketch
of a RS for an odd rank
property

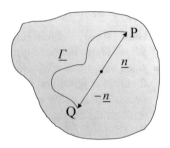

Let θ_1 and $\varphi(\theta_1)$ be the values of θ and φ at P. Since according to (5.3) Q is antipodal to P, the values θ_2 and $\varphi(\theta_2)$ of θ and φ at Q must be:

$$\theta_2 = \pi - \theta_1 \qquad \varphi_2 = \varphi_1 + \pi$$

so that

$$[\![\underline{n}(\theta_1)]\!] = \begin{bmatrix} \sin\theta_1 \cos\varphi(\theta_1) \\ \sin\theta_1 \sin\varphi(\theta_1) \\ \cos\theta_1 \end{bmatrix}$$

$$[\![\underline{n}(\theta_2)]\!] = \begin{bmatrix} \sin(\pi - \theta_1)\cos(\varphi(\theta_1)+\pi) \\ \sin(\pi - \theta_1)\sin(\varphi(\theta_1)+\pi) \\ \cos(\pi - \theta_1) \end{bmatrix} = -[\![\underline{n}(\theta_1)]\!]$$

i.e. $\varGamma$ starts at P and ends at the antipodal point Q following an arbitrary path. Then by (5.3):

$$r_P = r_{\underline{n}(\theta_1)} \qquad r_Q = r_{\underline{n}(\theta_2)} = -r_{\underline{n}(\theta_1)}$$

i.e. $r_{\underline{n}}(\theta)$ changes sign between P and Q. Since $\varGamma$ is continuous, there must be at least one value θ^* for which $r_{\underline{n}(\theta^*)} = 0$.

As a consequence, the RS for odd rank properties must pass through (contain) the origin. This result makes it impossible for the RS to look as in Fig. 5.6 or in the middle panel of Fig. 5.7.

The only possibility for the RS of an odd rank property is then to consist of pairs of overlapping lobes of opposite signs; it must qualitatively look as in the right panel of Fig. 5.7.

5.2.2   Even Rank Properties

For even rank properties, the value of the property in the direction of $\underline{n}$ and in the opposite direction, given by $-\underline{n}$, are related by:

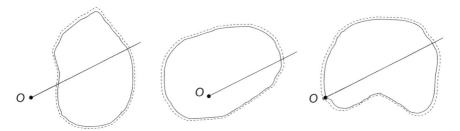

Fig. 5.7 The RS in the left panel is not radially single valued. The one in the middle panel does not pass through the origin. None of them is acceptable for odd rank properties. The one on the right is qualitatively correct. Solid and dashed lines represent positive and negative lobes respectively

$$r_{-\underline{n}} = \underline{\underline{T}} \vdots \underbrace{(-\underline{n}).....(-\underline{n})} = (-1)^p \, \underline{\underline{T}} \vdots \underbrace{\underline{n}.....\underline{n}} = r_{\underline{n}} \quad p \text{ even}$$
$$\phantom{r_{-\underline{n}} = \underline{\underline{T}}} p \text{ times} p \text{ times}$$

For every point at $r_{\underline{n}}$ on the RS, there is a corresponding antipodal point situated at $r_{-\underline{n}}$. Under the inversion operation these two points swap positions, so that the entire RS remains invariant under inversion.

Hence, the point group of the RS for all even rank properties must contain a $\bar{1}$ and belong to a centrosymmetric class, even for materials whose geometric structure does not belong to a centrosymmetric class.

For this reason, lobes of opposite signs of the RS for even rank properties do not overlap, which is compatible with the value of the property being positive, negative or zero. For second,[2] zero is attained only if the principal values are not all of the same sign or if at least one of them is zero.

5.3 Zero Rank Properties

The representation surface of scalar properties is the simplest one. Taking the mass density ρ as a representative example:

$$r = \rho \quad \text{or equivalently:} \quad r(\underline{n}) = r(\varphi, \theta) = \rho \tag{5.4}$$

The vector radius is constant. Since it does not depend on the direction, the representation surface is a sphere *centered at the origin* O, of radius equal to the value of the property, as shown in Fig. 5.8. The absence of orientational dependence is the signature of isotropy: for zero rank properties, all materials are perforce isotropic.

The centered sphere of unit radius is the only basic geometric object needed to construct the representation surface. If the value of the property is different from 1: $\rho \neq 1$, the representation surface is nothing but ρ times this basic object.

---

[2]And some fourth rank properties. See Sect. 8.4.

Fig. 5.8 The representation
surface for all zero rank
properties is a sphere
centered at the origin

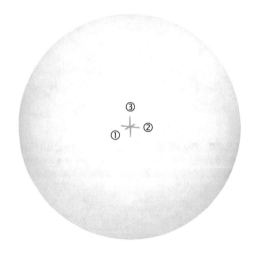

The representation surface itself, judged by its shape, belongs to the Curie class $\infty\infty m$, the class of isotropic materials, no matter what point group the material belongs to. It is no coincidence that the symbol conventionally used to represent the $\infty\infty m$ class is a sphere.

From now on, we will refer to the *class of the property* as the (crystallographic or limit) class to which its representation surface belongs, judged by its shape and *its position with respect to the origin*, but not by its orientation with respect to any reference frame.

5.4 First Rank Properties

The pyroelectric coefficient or modulus p appearing in the constitutive equation of pyroelectricity (4.7) serves to illustrate the shape of the RS for rank one tensors. Its equation is:

$$r(\underline{n}) = r(\varphi, \theta) = \underline{p} \cdot \underline{n} \quad \text{or equivalently:} \quad r(\varphi, \theta) = p_i n_i(\varphi, \theta) \tag{5.5}$$

where $\underline{p}$ is the known pyroelectric coefficient of the material, and the unit vector $\underline{n}$ samples all orientations in 3D space.

In the rest of the book we will always assume that all numerical values of the components of tensor properties are reported in the standard conventional axes [1].

In physical terms (5.5) can be interpreted as giving the value of the electric polarization $\underline{P} = \underline{p}\Delta T$ along the direction of the unit vector $\underline{n}$ when $\Delta T = 1$. Because of the linearity of the CE, the response of the material $(\underline{P})$ for any other value of ΔT is proportional.

Fig. 5.9 Sample slab of
pyroelectric material and
unit vector n

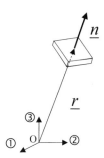

As illustrated in Fig. 4.10 (Sect. 4.3), in the general case all three components of
P will be non-zero, i.e. a surface electric charge density σ_{pol} will appear on all faces
of the sample upon a change in temperature.

If we choose the unit vector n to be colinear with r, n coincides with n^A of Fig. 4.3:

$$[\![n]\!] = \begin{bmatrix} \sin\theta\cos\varphi \\ \sin\theta\sin\varphi \\ \cos\theta \end{bmatrix} \tag{5.6}$$

with $\theta \in [0, \pi]$, $\varphi \in [0, 2\pi)$ (Fig. 5.9).

Let's now consider the PVDF sample of Fig. 4.10. The poled material belongs to
class $mm2$; according to the convention, the structure of the property is:

$$\mathrm{str}(p) = \begin{bmatrix} \cdot \\ \cdot \\ \bullet \end{bmatrix}$$

and the conventional axis ③ coincides with the twofold axis (see stereogram for class
$mm2$). The equation of the representation surface is then:

$$r(\theta, \varphi) = p \cdot n = p_3 \cos\theta \tag{5.7}$$

which is independent of φ and hence axisymmetrical about ③. The cross section of
this surface is a circumference of diameter p_3, that passes through the origin and
whose diameter lies on the ③ axis as represented in Fig. 5.10 (the generic angle α in
Fig. 5.1 is now the polar angle θ, its origin is the positive ③ semiaxis).

The $\pm$ symbol in Fig. 5.10 has been placed as a reminder that the surface consists
of two overlapping lobes.

This RS fulfills the conditions of Sect. 5.2.1: the oriented sphere contains the origin; opposite the sphere is an empty region. Both are separated by a plane through the
origin and perpendicular to p. In all directions contained in this plane, and therefore
perpendicular to p, the property is zero. First rank properties are transversally null,
in the sense that their value in the plane transversal to them is zero.

Fig. 5.10 The representation
surface $r(\theta, \varphi) = \cos\theta$

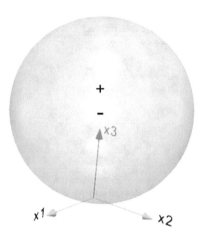

Unlike the centered sphere that represents zero rank properties, the surface for
rank one properties belongs to the Curie conical class ∞m. We will use the term
"oriented sphere" and include the origin to emphasize this fundamental difference.
The direction of the oriented sphere is that of the diameter that starts at the origin.

The oriented sphere belongs to ∞m because it possesses the same symmetry
elements as the cone, namely an ∞-fold axis and an infinite number of symmetry
mirror planes containing this axis. The directionality of a rank one tensor is more
immediately appreciated in the pointed shape of the cone.

We are actually even more used to expressing directionality with a simpler object:
an arrow.[3] The point of which should be a cone.

The oriented sphere, the cone and a three-dimensional, axisymmetric arrow are
equally valid representations of a rank one tensor. After all, they are usually called
"vectors", and vectors have traditionally been represented by arrows, not conical-
headed, but flat-headed arrows. A flat arrow is of course perfectly acceptable, so long
as we keep in mind that the flat arrow is a simplified sketch of an axisymmetrical,
3D arrow (Fig. 5.11).

Only materials belonging to the 12 polar classes can have rank one properties
(matter tensors). The RSs for the polar classes are displayed in Fig. 5.12. In this
figure, the straight line is the diameter starting at the origin.

For triclinic class 1, the orientation of the sphere is arbitrary, for monoclinic
class 2 the orientation is along the conventional ② axis, for monoclinic class m the
orientation direction lies in the conventional ①③ plane, and for all other classes
orientation is along ③.

The previous discussion is valid for any rank one matter or field tensor, like
velocity, force, etc., not just for the pyroelectric coefficient. For all rank one tensors,
the RS is a sphere through the origin. The only difference among the four panels in
Fig. 5.12 is the orientation of the sphere with respect to the conventional axes.

---

[3]In Knowles' words: *"When we are young, we learn that vectors are arrows"* [2].

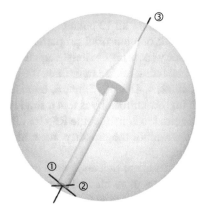

Fig. 5.11 An oriented sphere, and a 3D arrow

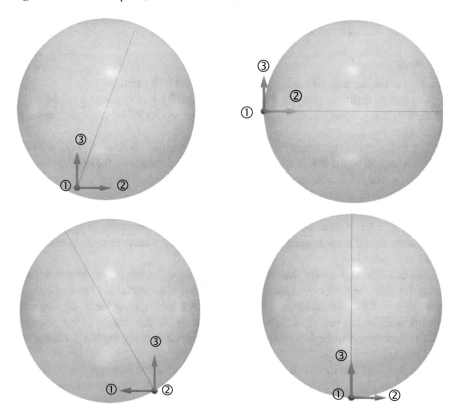

Fig. 5.12 Representation surfaces for the 12 polar classes: class 1 (upper left), class 2 (upper right), class *m* (lower left), classes *mm*2, 4, 4*mm*, 3, 3*mm*, 6, 6*mm*, ∞, ∞*m* (lower right)

Notice that the RS is the same for all 12 polar crystallographic and Curie classes, and as a consequence all rank one tensorial properties belong to class ∞m. The point symmetry of the property (∞m) is obviously equal to the point symmetry of a material of class ∞m; but includes, i.e. is larger than, the point symmetry groups of all other classes: 1, 2, m, $mm2$, 4, $4mm$, 3, $3mm$, 6, $6mm$, and ∞.

The immediate reason for this uniformity is that a single type of geometric object (the oriented sphere) is required to construct the RSs for all possible classes. Which is just a consequence of the rank of the property: for rank one, (5.2) contains only first powers of sine or cosine.

The functional form of (5.2) is not rich enough to describe any shape other than an oriented sphere. This was even more blatantly obvious for zero rank properties.

Thus, the point group of the property is as symmetrical as or more symmetrical than the material. We will return to this point later on.

Let us now try to do away with the conventional axes and consider a generic, rank one property $\underline{u}$ of *unit modulus*. Its representation surface is nothing but:

$$r = \underline{u} \cdot \underline{n} = \cos \alpha \tag{5.8}$$

where α is the angle between the property $\underline{u}$ and the unit vector $\underline{n}$.

The value of the property in the direction given by $\underline{n}$ is the cosine of the angle formed by the property $\underline{u}$ and $\underline{n}$, i.e. the usual projection of a vector, as shown in Fig. 5.13.

The left panel of this figure has been drawn on the plane defined by both $\underline{u}$ and $\underline{n}$, so it is a section of the RS. Since the value of r only depends on the angle α formed by $\underline{u}$ and $\underline{n}$, the RS is obtained by rotating the circumference about $\underline{u}$. The resulting RS is a sphere of diameter $|\underline{u}|$ (right panel of Fig. 5.13).

If the vector $\underline{u}$ is chosen according to the standard axes convention, all individual RSs of Fig. 5.12 are recovered; taking class $mm2$ as an example, if $\underline{u}$ is set to be the ③ axis, as the convention dictates, the RS of Fig. 5.8 turns into the bottom right panel of Fig. 5.12.

Just as a centered sphere was the single basic object for the RS of zero rank properties, the RSs for rank one properties of all classes are constructed from one single basic object, the oriented sphere consisting of two overlapping lobes, by correctly placing it in the conventional axes and scaling it by the magnitude of the property. This axes- or coordinate-free point of view is a useful one, applicable to matter and field tensors.

It is also possible to visualize the RS for the polarization that appears on faces B of the slab of Fig. 4.10 by projecting along the directions of the unit vector $\underline{n}^B$ of that figure:

$$[\![\underline{n}^B]\!] = \begin{bmatrix} \cos\theta \cos\varphi \\ \cos\theta \sin\varphi \\ -\sin\theta \end{bmatrix} \qquad r = \underline{u} \cdot \underline{n}^B = -\sin\theta$$

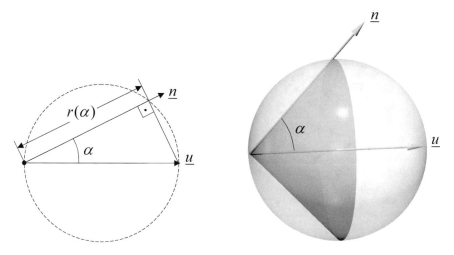

Fig. 5.13 Section of the RS for rank one properties in the plane defined by $\underline{u}$ and $\underline{n}$ (left), and the resulting RS

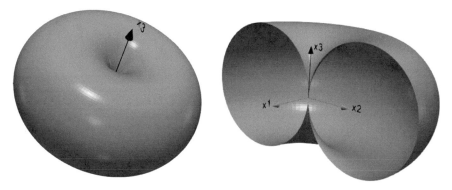

Fig. 5.14 Transversal RS for a rank one property. The RS consists of a positive and a negative overlapping toroidal lobes

The RS is again axisymmetric, the cross section being $\sin\theta$, which is the same circle as in Fig. 5.13, just advanced $\frac{\pi}{2}$ with respect to $\cos\theta$. Because of axisymmetry, a torus results, instead of a sphere (Figs. 5.15 and 5.14).

Reading the transversal RS requires understanding what is being depicted: the length of the vector radius from O to the transversal RS in the direction of $\underline{n}$ (the $\underline{n}^A$ of Fig. 4.10) is the value of $r = \underline{u} \cdot \underline{n}^B = -\sin\theta$, i.e. the value of the property along $\underline{n}^B$, although we read it off the transversal RS traveling in the direction of $\underline{n}^A$.

Let's distinguish the two surfaces by the "l" (longitudinal) and "t" (transversal) superindices:

$$r^l = \underline{u} \cdot \underline{n}^A = \cos\theta \qquad r^t = \underline{u} \cdot \underline{n}^B = -\sin\theta \qquad (5.9)$$

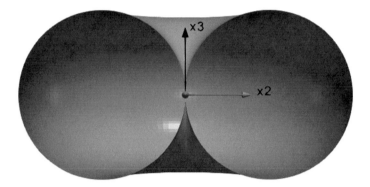

Fig. 5.15 Cross section of the transversal RS for a rank one property. The RS consists of a positive and a negative overlapping toroidal lobes

Fig. 5.16 Cross section of
the longitudinal and
transversal RSs for a rank
one property with the plane
defined by u and n. The
segments $\overline{\text{OA}}$ and $\overline{\text{OB}}$ are the
longitudinal and transversal
projections of u, respectively

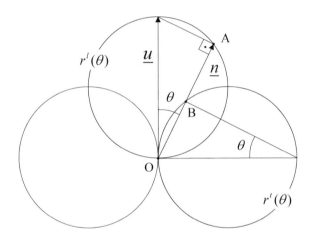

Figure 5.16 is a cross section of the longitudinal and transversal RSs through the plane defined by the generic property u and n. Traveling along n, the distances ($\overline{\text{OA}}$ and $\overline{\text{OB}}$) between O and the intersection points (A and B) with the longitudinal and transversal RSs fulfill:

$$\overline{\text{OA}} = |u| \cos \theta \qquad \overline{\text{OB}} = |u| \sin \theta$$

which are (5.9).[4] The intersection with the longitudinal RS gives the longitudinal projection, the intersection with the transversal RS gives the transversal component.

The results $u^l = \overline{\text{OA}} = |u| \cos \theta$, $u^t = \overline{\text{OB}} = |u| \sin \theta$ are nothing but the usual decomposition of a vector in two orthogonal components. For rank one properties, the RSs are a slightly different way of looking at a very familiar matter.

---

[4]What about the sign of r^t in the second equation of (5.9)? Draw n^B and remember there are two overlapping lobes in each of the RSs.

Fig. 5.17 Pyroelectricity causes a macroscopic polarization $\underline{P}$ to appear in the material and in any sample cut out of it with arbitrary orientation. Pairs of opposed faces develop surface charges proportional to the projection of $\underline{P}$ along the unit vectors normal to the faces

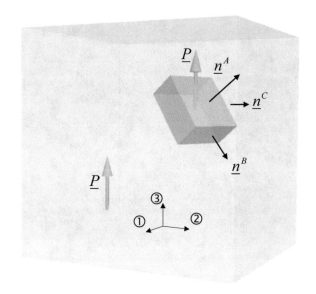

To come back to pyroelectricity: a change in temperature causes a homogeneous polarization $\underline{P} = p\Delta T$ to appear in the starting material (large block) and in any sample cut out of it with arbitrary orientation (smaller slab in Fig. 5.17). As a consequence, the three pairs of faces of the slab may develop surface density charges proportional to the projection of $\underline{P}$ along the unit vectors normal to the faces. For PVDF, class $mm2$, of Fig. 4.10, $\underline{P}$ points in the conventional ③ direction.

The physical meaning of (5.9) is that the charge densities on the faces perpendicular to the unit vectors $\underline{n}^A$ and $\underline{n}^B$ are given by the usual projections of $\underline{P}$ along the normal vectors $\underline{P} \cdot \underline{n}^A$ and $\underline{P} \cdot \underline{n}^B$, as in Fig. 4.7. The RSs show how the charge surface densities depend on the orientation of the faces.

Transversality is straightforward for rank one properties,[5] because projecting a vector is a very familiar operation, but it is a bit less simple for higher rank properties; we will return to it in Chap. 9.

References

1. Giacovazzo, C., Monaco, H.L., Artioli, G., Viterbo, D., Ferraris, G.: Fundamentals of Crystallography. Oxford University Press, Oxford (2002)
2. Knowles, J.K.: Linear Vector Spaces and Cartesian tensors. Oxford University Press, Oxford (1998)

---

[5] What about the faces perpendicular to $\underline{n}^C$? Why does $\underline{P} \cdot \underline{n}^C$ yield 0, despite $\underline{n}^C$ being transversal to $\underline{n}^A$? Hint: coplanarity.

Chapter 6
Symmetric Second Rank Properties

In the previous chapter RSs were first introduced for the polar classes in their conventional axes, and then in an axes-free setting. We saw that a single type of geometrical object, the oriented sphere, was enough to construct the RS for all crystallographic and limit classes. Their RSs exclusively differed in their orientation with respect to the standard conventional axes.

In this section we will start from the more general axes-free point of view, taking mechanical stress as a representative example of symmetric second rank tensors. Although a field tensor and not a matter tensor, mechanical stress is useful because it is quite familiar and general, in the sense that all its eigenvalues can be positive, negative or zero.

6.1 The Representation Surfaces of Dyadics

As for rank one properties, the value of the vector radius to the RS in direction $\underline{n}$ is given by the projection along this unit vector, or by the equivalent statement of "using up" the arguments of the bilinear map $\underline{\underline{\tau}}(\ ,\ )$:

$$r(\theta, \varphi) = \underline{\underline{\tau}}(\underline{n}, \underline{n}) = \underline{\underline{\tau}} : \underline{n}\,\underline{n} = \tau_{ij} n_j n_i \qquad (6.1)$$

The nine terms in the last expression of (6.1) are of two kinds: those containing $n_i n_i$ (no sum), and those of the form $n_i n_j$, $i \neq j$, where for the time being the indices refer to the conventional axes ①②③. Let us look at the terms with equal subindices. Terms like $\tau_{ii} n_i n_i$ in (6.1) come from the contraction of $\underline{n}$ with terms of the form:

$$\tau_{ii} \underline{\delta}_i \underline{\delta}_i \qquad \text{(no sum)}$$

© Springer Nature Switzerland AG 2020
M. Laso and N. Jimeno, *Representation Surfaces for Physical Properties
of Materials*, Engineering Materials, https://doi.org/10.1007/978-3-030-40870-1_6

Fig. 6.1 The RS for the
dyadic $\underline{u}\,\underline{u}$. It is obtained by
rotating the curve $r = \cos^2 \alpha$
about $\underline{u}$

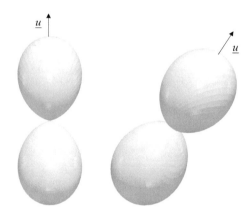

so it is natural to try and visualize the RS for objects like $\underline{u}\,\underline{u}$, where $\underline{u}$ is a unit vector,
as in Fig. 5.8 Its RS is:

$$r = \underline{u}\,\underline{u} : \underline{n}\,\underline{n} = (\underline{u} \cdot \underline{n})(\underline{u} \cdot \underline{n}) = \cos^2 \alpha$$

where α is the angle between $\underline{u}$ and $\underline{n}$.

By the same argument as was used in Fig. 5.8, the RS of $\underline{u}\,\underline{u}$ is a surface of
revolution about the vector $\underline{u}$. Its cross section is $r = \cos^2 \alpha$ (Fig. 6.1).

Since the vector radius r is proportional to the square of the cosine between $\underline{u}$
and $\underline{n}$, its value is always nonnegative. The two overlapping (positive and negative)
lobes of the RS for rank one properties are now positive and separately visible.

Let's extend this to a linear combination like $A\underline{u}\,\underline{u} + B\underline{v}\,\underline{v}$, where $\underline{u}$ and $\underline{v}$ are
unit vectors, orthogonal to each other but with otherwise arbitrary orientations. The
individual RSs for $\underline{u}\,\underline{u}$ and $\underline{v}\,\underline{v}$ are as represented in Fig. 6.2.

The RS of $A\underline{u}\,\underline{u} + B\underline{v}\,\underline{v}$ is just the linear combination of the individual RSs for
$\underline{u}\,\underline{u}$ and $\underline{v}\,\underline{v}$ of Fig. 6.2. Adding and subtracting the RSs for dyadics like $\underline{u}\,\underline{u}$, $\underline{v}\,\underline{v}$ as
in Fig. 6.2 is quite intuitive: graphically, the linear combination of two RSs, say $r_{\underline{u}\,\underline{u}}$
and $r_{\underline{v}\,\underline{v}}$, is nothing but the weighted sum of the values of the radius vector:

Fig. 6.2 The representation
surfaces of $\underline{u}\,\underline{u}$ (left) and $\underline{v}\,\underline{v}$
for orthogonal $\underline{u}$ and $\underline{v}$

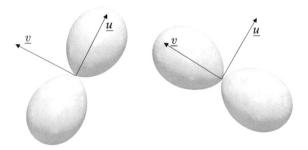

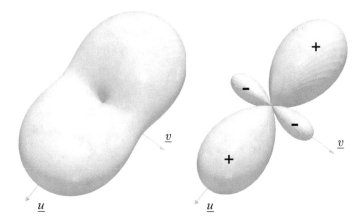

Fig. 6.3 The representation surface of $Au\underline{u} + Bv\underline{v}$ for $B = 0.5A$ (left) and $B = -0.5A$ with $A > 0$ (right)

$$r = Ar_{\underline{u}\,\underline{u}} + Br_{\underline{v}\,\underline{v}}$$

i.e. the radius vector of the sum RS, along a direction $\underline{n}$ is the sum[1] of the radius vectors of each individual RS along that same direction.

On the left panel of Fig. 6.3, both coefficients A and B have the same sign, on the right panel, opposite signs, so that positive and negative lobes appear.

The important particular cases of $A = B$ and $A = -B$ are shown in Fig. 6.4. While the way the difference $\underline{u}\,\underline{u} - \underline{v}\,\underline{v}$ looks is quite obvious, the fact that adding the two basic RSs of Fig. 6.2 in equal amounts results in the torus of the left panel of Fig. 6.4 is less intuitive.

The toroidal RS of $\underline{u}\,\underline{u} + \underline{v}\,\underline{v}$ has axial symmetry and belongs to the limit class ∞/mm. In the plane defined by $\underline{u}$ and $\underline{v}$, the value of the radius vector is the same in all directions, which is the signature of isotropicity (Sects. 6.3 and 6.4). The easiest way to see that $\underline{u}\,\underline{u} + \underline{v}\,\underline{v}$ is isotropic is to write the equation of its cross section in terms of the angles between $\underline{n}$, and $\underline{u}$ and $\underline{v}$, which are complementary:

$$r = (\underline{u}\,\underline{u} + \underline{v}\,\underline{v}) : \underline{n}\,\underline{n} = \cos^2\alpha + \cos^2\left(\frac{\pi}{2} - \alpha\right) = \cos^2\alpha + \sin^2\alpha = 1$$

The last step is to add the RS of $\underline{w}\,\underline{w}$, where $\underline{w}$ is a third unit vector orthogonal to both $\underline{u}$ and $\underline{v}$. Two particular cases of the RSs for the linear combination $Au\underline{u} + Bv\underline{v} + Cw\underline{w}$ are shown in Fig. 6.6.

In one of them (left panel of Fig. 6.6) all three coefficients have the same sign, which corresponds to either a positive definite or to a negative definite tensor property.

If one of the signs of the coefficients is different from the other two (an indefinite tensor), the RS looks like the right panel of Fig. 6.6, with lobes of different signs.

---

[1]The same is true for any other function of individual RSs.

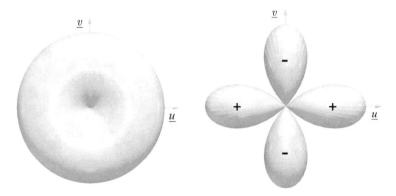

Fig. 6.4 The RSs of the sum $\underline{u}\,\underline{u} + \underline{v}\,\underline{v}$ (left) and the difference $\underline{u}\,\underline{u} - \underline{v}\,\underline{v}$ for orthogonal $\underline{u}$ and $\underline{v}$ of Fig. 6.2

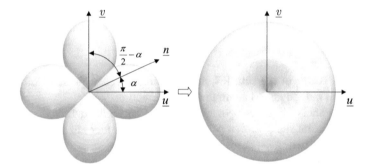

Fig. 6.5 The sum $\underline{u}\,\underline{u} + \underline{v}\,\underline{v}$ is plane isotropic (right panel). On the left panel, the RSs of $\underline{u}\,\underline{u}$ and $\underline{v}\,\underline{v}$ overlap because they are represented individually, without adding them

Qualitatively, there are only two visually different types of RSs for the three linear combinations of the dyadics $\underline{u}\,\underline{u}$, $\underline{v}\,\underline{v}$, $\underline{w}\,\underline{w}$. Depending on whether all three coefficients have the same sign, or only two of them have the same sign, the RS will look like either the left or the right panel of Fig. 6.6 respectively. Notice that these representations are independent of any reference frame; no axes whatsoever appear in Figs. 6.1, 6.2, 6.3, 6.4, 6.5 and 6.6.

How is this related to the RS of second rank properties? Since $\underline{u}$, $\underline{v}$ and $\underline{w}$ are of unit modulus and orthogonal, we are allowed to define our conventional axes ①②③ as coinciding with $\underline{u}$, $\underline{v}$ and $\underline{w}$. In addition, let's assign specific names to the constants A, B and C: $\tau_{11} = A$, $\tau_{22} = B$, $\tau_{33} = C$. In these axes, the stress tensor has the form:

$$\underline{\underline{\tau}} = \tau_{11}\underline{\delta}_1\underline{\delta}_1 + \tau_{22}\underline{\delta}_2\underline{\delta}_2 + \tau_{33}\underline{\delta}_3\underline{\delta}_3$$

or equivalently:

Fig. 6.6 The representation surface of $Au\,u + Bv\,v + Cw\,w$ for $C = 2B = 4A$ (positive definite if $A > 0$, negative definite if $A < 0$, left) and $C = -2B = -4A$ (indefinite, right). It has *mmm* symmetry

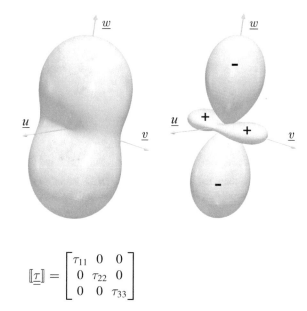

$$[\![\underline{\underline{\tau}}]\!] = \begin{bmatrix} \tau_{11} & 0 & 0 \\ 0 & \tau_{22} & 0 \\ 0 & 0 & \tau_{33} \end{bmatrix}$$

The stress is diagonal, in other words, the vectors $\underline{u}$, $\underline{v}$ and $\underline{w}$ (alias ①②③) are the eigenvectors of $\underline{\underline{\tau}}$, i.e. they lie along its principal directions. It is also particularly easy to write the equation of the RS in these axes. For the unit vector $\underline{n}$ from Eq. 5.6, the equation of the RS is:

$$r(n_1, n_2, n_3) = \underline{\underline{\tau}} : \underline{n}\,\underline{n} = \tau_{11}n_1^2 + \tau_{22}n_2^2 + \tau_{33}n_3^2$$
$$r(\varphi, \theta) = \tau_{11} \sin^2 \theta \sin^2 \varphi + \tau_{22} \sin^2 \theta \cos^2 \varphi + \tau_{33} \cos^2 \theta \qquad (6.2)$$

Surfaces represented by equations like (6.2) are called *ovaloids* [1]. Equation (6.2) is particularly simple because of the convenient orientation of the surface with respect to the axes. Needless to say, the equations of the RSs in Figs. 6.2 and 6.3 in the same axes are obtained by setting to zero the corresponding coefficients in (6.2).

It is very easy to check in (6.2):

- inversion symmetry: an inversion center changes the signs of all three components n_i simultaneously. Since all terms are quadratic in n_i, they do not change sign when $n_i \rightarrow -n_i$, $\forall i$. This argument also applies to the more general form (6.3).
- invariance with respect to reflection on the three planes ①②, ②③ and ③①: each of these reflections change the sign of one n_i; as before, (6.2) remains unchanged.
- invariance with respect to π rotation about binary axes along ①, ② and ③, each of which change the sign of two axes.

The previous invariances are characteristic of the *mmm* point group.

At this point we recognize that the RS of all second rank, symmetric properties is built up of just three basic objects, shown in Fig. 6.7.

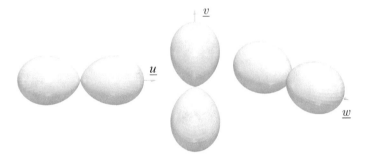

Fig. 6.7 The three basic objects from which the RS of all second rank, symmetric properties are built. The vectors $\underline{u}$, $\underline{v}$ and $\underline{w}$ are orthonormal and the three isomers are congruent

These three surfaces of revolution are congruent[2]; we will take the liberty to call them *isomers*. Their shapes are identical, they only differ in orientation. If we do not count isomers, there is really just one basic geometric object from which all second rank, symmetric properties are built up: the RS of Fig. 6.1.

Linear combinations of the three isomers give rise to all the RSs that appear in this chapter, including all crystallographic and limit classes of Figs. 6.9, 6.10 and 6.11.

The directionality of the property depends solely on the amounts in which the three isomers are combined. With the help of one simple shape, understanding anisotropy in second rank is a very straightforward matter.

In the following chapters we will also try and identify the basic geometric objects from which the RS is built up.

The RSs on the left panels of Figs. 6.3 and 6.6 have a "navel".[3] In the case of Fig. 6.6, the navel is a smooth dimple, while in Fig. 6.3 the dimple has shrunk to a point: the origin. On the right panel of Fig. 6.6 the dimple is negative, it protrudes through the other side. This is a consequence of the coefficient C being zero in Fig. 6.3, positive on the left panel of Fig. 6.6 and negative on the right panel of the same figure.

Figure 6.8 shows intermediate steps between the two panels of Fig. 6.6 as the coefficient C decreases and goes from positive to negative. When all coefficients are of the same sign (as in the first four panels of Fig. 6.8) the surface does not pass through the origin.[4]

The surface can contain the origin if (a) one of the signs is different from the other two, or (b) at least one of the coefficients is zero. In these cases, there is an infinite

---

[2]Two objects or figures or RSs are congruent if they can be transformed into each other by an isometry, i.e. by a combination of translations, rotations and reflections. An isometry does not change angles nor lengths.

[3]For $C > 0$, the navel is a finite region of elliptical points of the RS; for $C = 0$ the size of the elliptical region shrinks to zero and the RS becomes pinched at the origin.

[4]Neither the axes nor the signs of the lobes have been marked in this figure. Try to identify the axes and assign signs to positive and negative lobes.

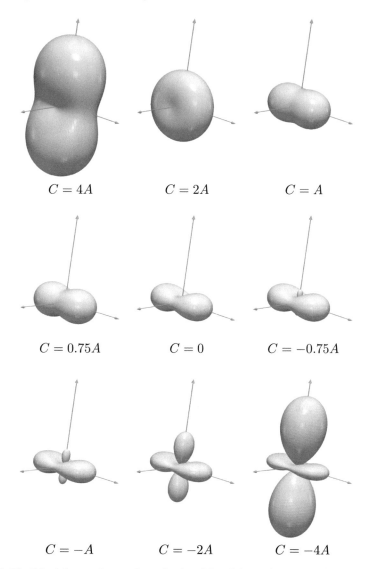

$C = 4A$ $C = 2A$ $C = A$

$C = 0.75A$ $C = 0$ $C = -0.75A$

$C = -A$ $C = -2A$ $C = -4A$

Fig. 6.8 The RS of $A\underline{u}\,\underline{u} + B\underline{v}\,\underline{v} + C\underline{w}\,\underline{w}$ for $B = 2A$ and decreasing values of C, with $\underline{u}$, $\underline{v}$ and $\underline{w}$ perpendicular to each other. The top left panel ($C = 4A$) corresponds to the left panel of Fig. 6.6; the bottom right one ($C = -4A$) to the right panel of the same figure

set of directions along which the property is zero: all the directions tangent to the RS at the origin (see also Sect. 6.8).

The middle panel of Fig. 6.8 corresponds to $C = 0$. This case is called a plane state (see Sect. 6.5).

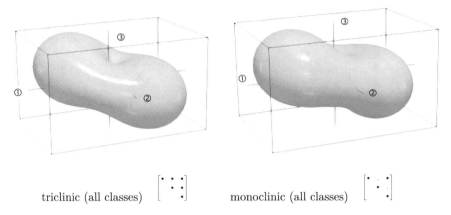

triclinic (all classes) $\begin{bmatrix} \cdot & \cdot & \cdot \\ & \cdot & \cdot \\ & & \cdot \end{bmatrix}$ monoclinic (all classes) $\begin{bmatrix} \cdot & \cdot \\ & \cdot & \\ & & \cdot \end{bmatrix}$

Fig. 6.9 The RSs for second rank symmetric properties for all triclinic and monoclinic classes, and their structures in their conventional axes. The shapes correspond to the numerical values of the thermal conductivity of triclinic 4-bromobenzophenone [2] (left), and of β-Ga_2O_3 [3] (right)

While stress, strain and some properties like the thermal expansion coefficient are indefinite, i.e can have positive and negative eigenvalues, properties like generalized conductivities and resistivities can only have positive eigenvalues: their representation matrices are positive definite and their RS will, qualitatively look like the left panel of Fig. 6.6.

Figures 6.9, 6.10 and 6.11 show the RSs for all second rank symmetric properties for all crystallographic and Curie classes in their conventional axes. In some cases, the numerical values of the components are those of specific materials and properties.

The important point of these figures is that all RSs are particular cases of the general form shown in Fig. 6.6, with different numerical values for A, B and C. The appearance of the RS is qualitatively different when the signs of the principal values are the same, or when only two are the same. But only the basic shapes shown in Fig. 6.6 are necessary to represent second rank symmetric properties.

Apart from the values of the parameters, their RSs in Figs. 6.9, 6.10 and 6.11 differ in the orientation with respect to the conventional axes. When the conventional axes coincide with the principal directions of the property, the RS is placed in particularly simple, i.e. aligned, position. Otherwise, it appears rotated. For monoclinic classes, the conventional axis ② coincides with a principal direction, so the RS is aligned with this axis and arbitrarily rotated in the the ①③ plane.

The great simplicity of the RS is obscured in its analytical expression in spherical coordinates, which for the most general case (triclinic materials) and using $\underline{n}$ from (4.2) is:

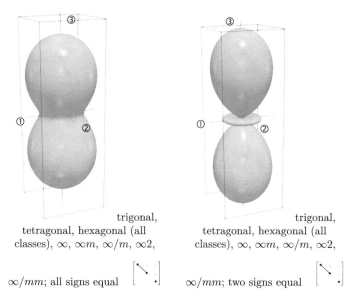

trigonal,
tetragonal, hexagonal (all
classes), ∞, ∞m, ∞/m, $\infty 2$,
∞/mm; all signs equal

trigonal,
tetragonal, hexagonal (all
classes), ∞, ∞m, ∞/m, $\infty 2$,
∞/mm; two signs equal

Fig. 6.10 The RSs for second rank symmetric properties for all trigonal, tetragonal, hexagonal crystallographic classes, and conical and cylindrical Curie classes, and their structures in their conventional axes. The shapes correspond to the numerical values of the properties of the following representative materials: thermal conductivity of HCP iron [4] (left), thermal expansion coefficient of calcite [5] (right)

$$r(\varphi, \theta) = \underline{\underline{\tau}} : \underline{n}\,\underline{n} = \tau_{11} \sin^2 \theta \sin^2 \varphi + \tau_{22} \sin^2 \theta \cos^2 \varphi + \tau_{33} \cos^2 \theta +$$
$$+ 2\tau_{12} \sin^2 \theta \sin \varphi \cos \varphi$$
$$- 2\tau_{23} \sin \theta \cos \theta \cos \varphi$$
$$- 2\tau_{13} \sin \theta \cos \theta \sin \varphi \tag{6.3}$$

Out of the six coefficients in the general τ_{ij}, only three are relevant for the shape of the RS, and therefore for its point group symmetry. The other three are a consequence of the reference frame begin "misaligned" with respect to the natural symmetry axis of the material. Referring the property to unit vectors along the principal directions makes the off-diagonal terms τ_{ij}, $i \neq j$ vanish.

In the general case, i.e. for arbitrary values of the components, the RS belongs to the orthorhombic holoedric class mmm, with three orthogonal symmetry planes and three orthogonal twofold axes, which coincide with $\underline{u}$, $\underline{v}$ and $\underline{w}$. Some of the RSs in Figs. 6.9, 6.10 and 6.11 have higher symmetries as a consequence of two or three of the eigenvalues being equal. If two of them are equal, like in panels $C = 2A(= B)$ and $C = A$ in Fig. 6.10, the RS belongs to the Curie class ∞/mm.

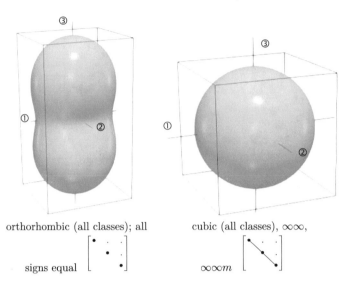

orthorhombic (all classes); all

signs equal $\begin{bmatrix} \bullet & \cdot & \cdot \\ & \bullet & \cdot \\ & & \bullet \end{bmatrix}$

cubic (all classes), $\infty\infty$,

$\infty\infty m$ $\begin{bmatrix} \bullet & \cdot & \cdot \\ & \bullet & \cdot \\ & & \bullet \end{bmatrix}$

Fig. 6.11 The RSs for second rank symmetric properties for all orthorhombic and cubic crystallographic classes, and spherical Curie classes, and their structures in their conventional axes. The orthorhombic case for semidefinite properties is shown on the right panel of Fig. 6.6. The shapes correspond to the numerical values of the properties of the following representative materials: thermal expansion coefficient of untwinned $YBa_2Cu_3O_{7-\delta}$ [6] (left), electric resistivity of isotropic materials (right)

Again, as a consequence of being necessarily centrosymmetric, the symmetry of the RS and hence of the property will always be equal or greater than that of the structure of the material. We have already encountered a similar situation for rank one properties. For rank one, the degree of the RS considered a polynomial in the director cosines n_i was so low that the RSs for all classes was the same, except for orientation. For second rank however, the RS is quadratic in n_i, which is already rich enough to produce discrimination between spherically symmetric, simply axisymmetric and non axisymmetric classes, as an inspection of Figs. 6.9, 6.10 and 6.11 confirms.

In general, the higher the tensorial rank of the property, the higher the degree of the RS considered as polynomial in n_i and the more oscillatory its behavior can be.

The statement that "the symmetry of the RS and hence of the property will always be equal or greater than that of the structure of the material" is known as Neuman's Principle.

For second rank properties, the requirement of the RS being centrosymmetric is one of the reasons for the property being equally or more symmetric than the structure of the material. The angular (spatial) frequency of the oscillatory behavior of the polynomial expression of the RS is another one.

6.2 Neuman's Principle

We follow Newnham's example in quoting Nye [1] verbatim:

> The symmetry elements of any physical property of a crystal must include the symmetry elements of the point group of the crystal.

This postulate [1] is called Neumann's Principle. It applies to all materials and properties that depend on the point group of the material.[5] It is a general statement that includes the two particular examples we have encountered in previous sections.

In the context of RS, Neuman's Principle means that the set of symmetry operations (the point group) of the crystallographic or limit class to which the RS of a property of a material belongs must include or be equal to the point group of the material.

Taking pyroelectricity as an example, while the point group of a class 1 triclinic material consists only of the identity E, $L_E = \begin{bmatrix} 1 & 0 & 0 \\ 0 & 1 & 0 \\ 0 & 0 & 1 \end{bmatrix}$, its pyroelectricity (the RS of p) belongs to class ∞m, whose point group consists, apart from L_E, of the uncountably infinite symmetry operations generated by the infinite-fold axis and by the infinite planes that contain this axis.

As another example, in Fig. 6.12 eumorphic crystals of two materials, azurite (point group $2/m$) and aragonite (point group mmm) are shown together with the RSs of two properties: pyroelectric coefficient and thermal conductivity, respectively. In the first case, the property has greater symmetry ($\infty m \supset 2/m$), in the second the same symmetry (mmm) as the material.

In determining the class, both the eumorphic crystal and the RS of the property have to be judged according to their shape only (remember the remarks on eumorphic crystals on Sect. 2.4.2). The orientation of any kind of axes, crystallographic or conventional with respect to an external reference frame, plays no role whatsoever in this respect. Just as the possible shapes an eumorphic crystal can adopt are intrinsic characteristics of a crystalline material, so is the shape of the RS of a property.

The existence or absence of axes and planes is important for the determination of the point group of an eumorphic crystal, but their orientation with respect to an external reference frame is not. The external frame, if any, is defined a posteriori, once the class of the material has been determined, e.g. according to the crystallographic convention or to other criteria.

The same is true for the determination of the class to which a property (an RS) belongs: polar axes for rank one properties, principal directions of second rank properties, etc do of course contain important physical information, but their orientation is irrelevant for the determination of the point group.

---

[5]But see [1] for a distinction between properties depending on the point group and those that depend on the space group.

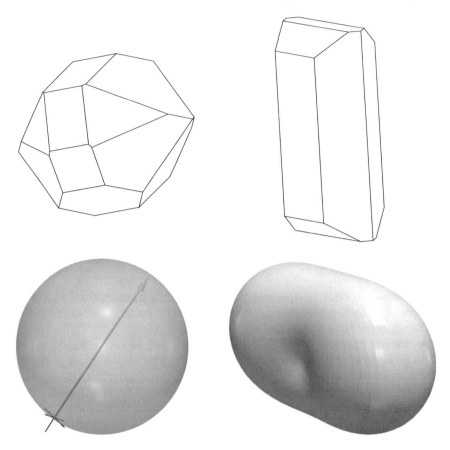

Fig. 6.12 Eumorphic monocrystal of azurite (top left), point group $2/m$, and the RS of its pyro-electric coefficient, point group ∞m (bottom left); eumorphic monocrystal of aragonite (top right), point group mmm, and the RS of its thermal conductivity, point group mmm (bottom right)

Neumann's Principle is however not constructive: it does not predict why or how much more symmetric than the material the RS can be.

We have already encountered two reasons for the property (the RS) to be more symmetric than the material: even rank tensors are centrosymmetric, hence their RSs will automatically contain a $\bar{1}$ in addition to all the geometric symmetry elements of the material. Thus, the RS of all second rank (among them, all generalized resistivities and conductivities) and fourth rank (foremost among them the elastic flexibility $\underline{\underline{s}}$ and stiffness $\underline{\underline{c}}$) properties will necessarily belong to one of the 13 centrosymmetric (11 crystallographic and two Curie) classes. The RSs, and thus the properties, of the remaining 26 classes will automatically be more symmetric than the material.

The second, slightly less obvious cause is a consequence of the definition of the RS. As we have seen (5.2) for a generic tensor $\underline{T}$ of rank p the RS is given by:

$$r(\varphi, \theta) = \underline{T} \vdots \underbrace{n......n}_{p \text{ times}} = T_{ij....qr} \underbrace{n_r n_q ....n_j n_i}_{p \text{factors}}$$

where each of the n_i is one of the components of a unit vector like (5.6). Thus, each of the terms in (5.2) is a trigonometric expression of degree at most p in the sine and cosine of each angle involved, and (5.2) is the sum of 3^p such terms.

If we now recall the equality:

$$\cos p\alpha + i \sin p\alpha = e^{p\alpha} = (e^{\alpha})^p = (\cos \alpha + i \sin \alpha)^p =$$

$$= \sum_{k=1}^{p} \binom{p}{k} i^{p-k} (\cos \alpha)^k (\sin \alpha)^{p-k} \qquad (6.4)$$

and identify real and imaginary parts of the first and last terms for particular values of p, we recover expressions such as (5.1). From (6.4) it follows that a sum of terms of order p in sine and cosine of an angle α is related to a sine or cosine of the mulitple angle $p\alpha$.

As a consequence, the RS for a second rank property (Eq. 6.3) consists of terms of at most second degree in sine and cosine of φ and θ, and cannot be rewritten in terms of sine and cosine of $p\varphi$ and $p\theta$ for $p > 2$.

Geometrically, the RS cannot have more than twofold symmetry in any of the angles (polar or Euler) alone. If the material has a twofold 2 axis or a plane $m \equiv \bar{2}$, the RS will have this symmetry as well (apart from the compulsory $\bar{1}$ for being of even rank).

But if the material has threefold, fourfold or sixfold axes, the second degree polynomial in trigonometric functions of the RS equation will still have at most multiplicity two in the angles φ or θ, and will be unable to reproduce the oscillations of higher angular frequency.

Since Neumann's principle demands that the RS be of equal or higher symmetry than the property, how then can the RS consist of second degree polynomials, i.e. have binary multiplicity, and yet include threefold, fourfold or sixfold axes? The (partial) answer is that a single second degree polynomial of a single angle cannot produce threefold, fourfold or sixfold multiplicity in that angle.

But there are ways out of this apparent contradiction for the RS to be built up of *several* low order (second, third) symmetry elements of the material and yet display higher order symmetry (cubic or hexagonal):

- the RS may become axisymmetric, i.e. independent of the angle. The ③ axis becomes an infinite-fold axis, thus including both the binary symmetry of the property and the higher order axis of the material. The RS belongs then to a Curie class.

This is clearly observable in the two panels of Fig. 6.10: the RS for second rank properties of classes having threefold, fourfold or sixfold symmetry (trigonal, tetragonal and hexagonal) systems all belong to Curie class ∞/mm; for cubic classes (Fig. 6.11, right panel) to $\infty\infty m$. Only the RSs for triclinic, monoclinic and orthorhombic classes do still retain the binary symmetry elements of point group mmm.

Similar non-crystallographic, axisymmetric RSs for materials belonging to crystallographic classes will also appear for third and fourth rank properties.

- another possibility is for an axis of low order, say third order, to combine with a mirror plane (remember that $m \equiv \bar{2}$) perpendicular to it to produce sixfold multiplicity.

  We also know that $\bar{6} \Leftrightarrow 3 \perp m$, so a third degree polynomial in the rotation angle about a given axis, plus a mirror plane normal to it will generate an RS having a $\bar{6}$ axis (see Fig. 7.9).

- it is also possible for more than two axes of lower order to combine and produce an RS of higher symmetry. Four threefold axes along the diagonals of a cube may produce fourfold symmetry, as in Fig. 7.12.

We will be encountering examples of these ways of evading insufficient angular sampling in the following chapters.

Centrosymmetry (for even rank properties) and combination of symmetry elements explain the equal or higher symmetry of the RS with respect to the material, as demanded by Neumann's principle.

6.3 Isotropy

As we have seen (Fig. 6.6), the general RS for symmetric second rank properties can be written as:

$$\underline{T} = A\underline{u}\,\underline{u} + B\underline{v}\,\underline{v} + C\underline{w}\,\underline{w}$$

and belongs to the orthorhombic point group mmm. Yet some of the RSs in Figs. 6.9, 6.10 and 6.11 look qualitatively different and indeed belong to classes of higher symmetry as a consequence of Neumann's principle (all classes with at least one axis of order higher than two will be at least ∞/mm), or to follow from two or more of the coefficients A, B, C having specific numerical values.

The simplest case, in the sense of most symmetrical, is $A = B = C$. Since:

$$\underline{u}\,\underline{u} + \underline{v}\,\underline{v} + \underline{w}\,\underline{w} = \underline{\underline{\delta}}$$

the case $A = B = C$ is 3D isotropy:

$$\underline{\underline{T}} = T\underline{\underline{\delta}} \quad \Rightarrow \quad [\![\underline{\underline{T}}]\!] = \begin{bmatrix} T & 0 & 0 \\ 0 & T & 0 \\ 0 & 0 & T \end{bmatrix} \tag{6.5}$$

The RS is a centered sphere (Fig. 6.11, right) and, not surprisingly, belongs to the spherical Curie class $\infty\infty m$. Second rank properties for cubic materials have all components T_{ii} (no sum) equal and are isotropic. Similarly, pressure, being an isotropic stress: $\underline{\underline{\tau}} = p\underline{\underline{\delta}}$ has a centered sphere as RS.

The only building block required for second rank isotropy is thus $\underline{\underline{\delta}}$.

6.4 Transversal Isotropy

If only two of the components are numerically equal, e.g. $A = B \neq C$, the RS loses some symmetry: now it is only axially symmetric but still retains a mirror plane perpendicular to the symmetry axis (Figs. 6.4). The RS is a surface of revolution about the vector whose coefficient is unequal, C in this case (Fig. 6.10), and belongs to the Curie class ∞/mm. The property is called *transversally* isotropic:

- the property is the same for all directions contained in the plane defined by $\underline{u}$ and $\underline{v}$ (in this example), which is called the *plane of isotropy* or the *transversal* plane. Very often, the literal subindex "t" is used to denote any one of these infinite transversal directions,
- the direction of $\underline{w}$ (in this example) is unique, often called *longitudinal*, and the literal subindex "l" is used to denote this direction. A reference-free way to write a 3D transversally symmetric tensor is:

$$\underline{\underline{T}} = T_t(\underline{u}\,\underline{u} + \underline{v}\,\underline{v}) + T_l\underline{w}\,\underline{w} \tag{6.6}$$

where the transversal plane is that defined by the two vectors $\underline{v}$ and $\underline{v}$.

If the axes of the reference frame ①②③ are made to coincide with $\underline{u}$, $\underline{v}$, $\underline{w}$, the matrix representation of $\underline{\underline{T}}$ is especially simple:

$$[\![\underline{\underline{T}}]\!] = \begin{bmatrix} T_t & 0 & 0 \\ 0 & T_t & 0 \\ 0 & 0 & T_l \end{bmatrix}$$

Since there are infinite transversal directions, an expression such as (6.6) seems to single out two of them ($\underline{u}$ and $\underline{v}$), as if they had some distinguished position.

In transversal isotropy, the only distinguished direction is the normal to the plane of isotropy, i.e $\underline{w}$ in (6.6). All transversal directions stand on an equal footing. It is therefore slightly clearer to construct transversal isotropy from the normal $\underline{w}$ to the plane by using the transversal $\underline{\underline{\delta}} - \underline{w}\,\underline{w}$ and the longitudinal $\underline{w}\,\underline{w}$ projectors:

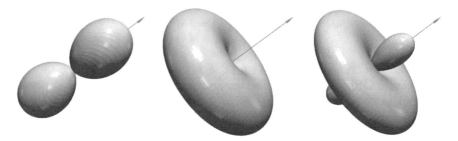

Fig. 6.13 The RS for the longitudinal part $\underline{w}\,\underline{w}$ (left), for the transversal part $\underline{\underline{\delta}} - \underline{w}\,\underline{w}$ (middle), and for the complete transversally isotropic property $C_1\underline{w}\,\underline{w} + C_2(\underline{\underline{\delta}} - \underline{w}\,\underline{w})$ for $C_1 = -0.8C_2$ (right). The arrow points in the direction of $\underline{w}$

$$\underline{\underline{T}} = C_1\underline{w}\,\underline{w} + C_2(\underline{\underline{\delta}} - \underline{w}\,\underline{w}) \tag{6.7}$$

where the two constants C_1, C_2 specify how large the longitudinal and the transversal contributions are. If $C_1 = C_2$ complete isotropy is recovered.

Let's visualize the contributions $\underline{\underline{\delta}}$ and $\underline{\underline{\delta}} - \underline{w}\,\underline{w}$ to the RS:

- the RS for $\underline{\underline{\delta}}$ is given by

$$\underline{\underline{\delta}} : \underline{n}\,\underline{n} = \delta_{ij}n_i n_j = n_i n_i = \underline{n} \cdot \underline{n} = 1$$

  a sphere of unit radius centered at the origin.
- for $\underline{\underline{\delta}} - \underline{w}\,\underline{w}$:

$$(\underline{\underline{\delta}} - \underline{w}\,\underline{w}) : \underline{n}\,\underline{n} = 1 - (\underline{n} \cdot \underline{w})^2 = 1 - \cos^2\alpha = \sin^2\alpha$$

which is the surface of revolution of cross section $r = \sin^2\alpha$ about the direction of $\underline{w}$, shown in the middle of Fig. 6.13. The RS for a transversally isotropic property is a linear combination of these two basic objects, as shown in the right panel of Fig. 6.13.

The expression (6.7) is equivalent to:

$$\underline{\underline{T}} = A\underline{\underline{\delta}} + B\underline{w}\,\underline{w} \tag{6.8}$$

with $A \equiv C_2$ and $B \equiv C_1 - C_2$. Which suggests that the simplest[6] basic objects necessary to construct transversal isotropy are $\underline{\underline{\delta}}$ and $\underline{w}\,\underline{w}$.

The general criteria for the selection of such basic building blocks are simple; any expression is valid as a building block for transversal isotropy if:

---

[6]In the sense of having the shortest names.

- it is of the correct tensorial rank (2 in this case),
- it has the correct index symmetry ($T_{ij} = T_{ji}$ in this case),
- it is either either isotropic, i.e. directionally neutral (as $\underset{=}{\delta}$ in this case) or is built up from the only directional information available: $\underline{w}$, that defines the longitudinal direction, perpendicular to the plane of isotropy.

For second rank, $\underset{=}{\delta}$ and $\underline{w}\,\underline{w}$ obviously fulfill these criteria. Other combinations, like $\underset{=}{\delta} \cdot \underline{w}\,\underline{w}$ or $\underset{=}{\delta} \cdot \underline{w}\,\underline{w} \cdot \underset{=}{\delta}$ reduce to one of them:

$$\underset{=}{\delta} \cdot \underline{w}\,\underline{w} = (\underset{=}{\delta} \cdot \underline{w})\underline{w} = \underline{w}\,\underline{w}$$
$$\underset{=}{\delta} \cdot \underline{w}\,\underline{w} \cdot \underset{=}{\delta} = (\underset{=}{\delta} \cdot \underline{w})(\underline{w} \cdot \underset{=}{\delta}) = \underline{w}\,\underline{w}$$

so that (6.8) is the most general form.

The structure str($\underline{\underline{T}}$) of the transversal property is obtained by filling up the elements of $\underline{\underline{T}}$ using (6.8). In an arbitrary frame of reference:

$$[\![\underline{\underline{T}}]\!] = \begin{bmatrix} A + Bw_1^2 & Bw_1w_2 & Bw_1w_3 \\ & A + Bw_2^2 & Bw_2w_3 \\ & & A + Bw_3^2 \end{bmatrix} \tag{6.9}$$

which is apparently triclinic: $\begin{bmatrix} \bullet & \bullet & \bullet \\ & \bullet & \bullet \\ & & \bullet \end{bmatrix}$. The reason is that the orientation of $\underline{w}$ in (6.8) is arbitrary with respect to the reference frame. If, according to the crystallographic standard, the normal $\underline{w}$ to the plane of isotropy is chosen as conventional axis ③, then, in this conventional reference frame, $[\![\underline{w}]\!] = \begin{bmatrix} 0 \\ 0 \\ 1 \end{bmatrix}$. If these values of w_i are inserted in $[\![\underline{\underline{T}}]\!]$ (6.9), then:

$$[\![\underline{\underline{T}}]\!] = \begin{bmatrix} A & 0 & 0 \\ & A & 0 \\ & & A + B \end{bmatrix}$$

whose str($\underline{\underline{T}}$) $= \begin{bmatrix} \ddots & \bullet \\ & \bullet \end{bmatrix}$ is the correct one for transversally isotropic materials. As expected, the number of independent components (two), hence of independent building blocks, is also the correct one.

A similar approach will be used to construct transversally isotropic fourth rank properties (8.6).

6.5 Plane States

Another particular case is $A \neq B$, $C = 0$. If the setting is still 3D, this situation is usually called a *plane* state. Thus

$$\underline{\underline{\tau}} = \tau_{uu}\underline{u}\,\underline{u} + \tau_{vv}\underline{v}\,\underline{v} \qquad \underline{\underline{\epsilon}} = \epsilon_{uu}\underline{u}\,\underline{u} + \epsilon_{vv}\underline{v}\,\underline{v}$$

are plane stress and plane strain states, in the plane defined by $\underline{u}$ and $\underline{v}$. From the shape of the RS (Fig. 6.14), it is clear that a 3D state of plane stress or strain does *not* imply that $\underline{\underline{\tau}} : \underline{n}\,\underline{n}$ or $\underline{\underline{\epsilon}} : \underline{n}\,\underline{n}$ are zero outside the plane. Although the stress looks like:

$$[\![\underline{\underline{\tau}}]\!] = \begin{bmatrix} \tau_{11} & 0 & 0 \\ 0 & \tau_{22} & 0 \\ 0 & 0 & 0 \end{bmatrix}$$

if ①②③ are made to coincide with $\underline{u}$, $\underline{v}$, $\underline{w}$, its projection in directions outside the ①② plane are *not* zero, except in the direction of $\underline{w}$. In plane states, the RS has a navel that is pinched to a point, as in Fig. 6.3.

Three dimensional plane states are not to be confused with 2D states. A 2D stress expressed in its principal directions is:

$$[\![\underline{\underline{\tau}}]\!] = \begin{bmatrix} \tau_{11} & 0 \\ 0 & \tau_{22} \end{bmatrix}$$

In 2D the words "outside the plane" do not even have a meaning. The perspective view on the right panel of Fig. 6.14 is somewhat misleading in this respect, as it makes the RS appear as if it were defined in 3D space.

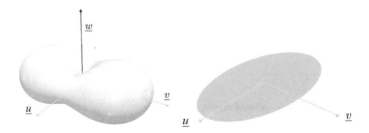

Fig. 6.14 The RS for a plane property, e.g. a 3D plane stress (left) and a 2D stress

6.6 The RS Under a Change of Reference Frame

Let's see now how the RS gives a visual understanding of the meaning of a change of reference frame. Focusing on stress once more, let's assume it has the following form:

$$\underline{\underline{\tau}} = \tau(\underline{u}\,\underline{u} - \underline{v}\,\underline{v})$$

If we take ①②③ to be $\underline{u}, \underline{v}, \underline{w}$ as in Fig. 6.15, its matrix representation is:

$$[\![\underline{\underline{\tau}}]\!] = \begin{bmatrix} \tau & 0 & 0 \\ 0 & -\tau & 0 \\ 0 & 0 & 0 \end{bmatrix} \tag{6.10}$$

This state consists of two normal stresses of the same magnitude, one positive and one negative. Its RS, $r(\underline{n}) = r(\varphi, \theta) = \underline{\underline{\tau}} : \underline{n}\,\underline{n}$, using:

$$[\![\underline{n}]\!] = \begin{bmatrix} \sin\theta \sin\varphi \\ -\sin\theta \cos\varphi \\ \cos\theta \end{bmatrix}$$

results in:

$$r(\underline{n}) = \tau \sin^2\theta(\sin^2\varphi - \cos^2\varphi) \tag{6.11}$$

as shown on the left panel of Fig. 6.15. We have made the dependence on $\underline{n}$ (i.e. φ, θ) explicit.

We now want to express the same stress in a reference frame ①'②'③' rotated anticlockwise by 45about ③ (Fig. 6.15, right):

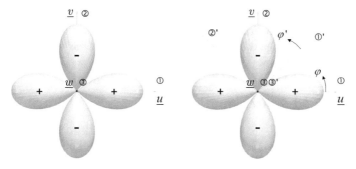

Fig. 6.15 The RS of a shear stress, and two reference frames related by a rotation of 45about the ③ axis. View is along the ③ axis

$$[\![\underline{\underline{\tau}}']\!] = \begin{bmatrix} \sqrt{2}/2 & \sqrt{2}/2 & 0 \\ -\sqrt{2}/2 & \sqrt{2}/2 & 0 \\ 0 & 0 & 1 \end{bmatrix} \begin{bmatrix} \tau & 0 & 0 \\ 0 & -\tau & 0 \\ 0 & 0 & 0 \end{bmatrix} \begin{bmatrix} \sqrt{2}/2 & -\sqrt{2}/2 & 0 \\ \sqrt{2}/2 & \sqrt{2}/2 & 0 \\ 0 & 0 & 1 \end{bmatrix} = \begin{bmatrix} 0 & \tau & 0 \\ \tau & 0 & 0 \\ 0 & 0 & 0 \end{bmatrix} \quad (6.12)$$

Written in these new axes, we immediately recognize $\underline{\underline{\tau}}'$ in (6.12) as a (plane) shear stress. Both forms of the stress (6.11) and (6.12) represent the same physical situation, but their matrix expressions look different because they are written in two different frames.

In yet a third frame, of axes ①"②"③", which were rotated arbitrarily with respect to ①②③, the matrix $[\![\underline{\underline{\tau}}]\!]$ would be full. We would still recognize it as representing a shear stress because $\mathrm{tr}(\underline{\underline{\tau}}) = 0$, but it would not be possible to know in which plane without carrying out an eigenvalue/eigenvector calculation.

In the ①'②'③' axes, the equation of the RS is:

$$r(\underline{n}') = r(\varphi', \theta') = \underline{\underline{\tau}}' : \underline{n}'\underline{n}'$$

where $\underline{n}'$ is now referred to the new axes, i.e. the angles φ', θ' are now measured with respect to the news axes[7]:

$$[\![\underline{n}']\!] = \begin{bmatrix} \sin\theta' \sin\varphi' \\ -\sin\theta' \cos\varphi' \\ \cos\theta' \end{bmatrix}$$

so that:

$$r(\underline{n}') = 2\tau \sin^2\theta' \sin\varphi' \cos\varphi' = \tau \sin^2\theta' \sin 2\varphi' \quad (6.13)$$

where the dependence is now on $\underline{n}'$ (i.e. φ', θ').

Although the matrix representations of the stress (Eqs. 6.10 and 6.12), and the equations of the RS (Eqs. 6.11 and 6.13) look quite different, the RS itself is the same. This cannot be otherwise: the term $2'\sin\varphi'\cos\varphi'$ is just $\sin 2\varphi'$ (Fig. 5.2), a version of $(\sin^2\varphi - \cos^2\varphi)$ rotated by 45.

A change of reference frame does not change the physical situation (the stress) nor its RS. Quite literally, the RS is an invariant picture of the property or physical phenomenon it represents.

6.7   Simplified Representations

As we saw in Sect. 5.11, an oriented sphere or an axisymmetric arrow are correct simplified symbols for a rank one property. What about a simplified representation for symmetric, second rank tensors? Fig. 6.6 immediately suggests the use of double tipped arrows, as in Fig. 6.16.

---

[7] Since ③ =③', $\theta = \theta'$.

Fig. 6.16 A schematic representation of a second rank symmetric property with three positive (left), two positive and one negative (middle), and two zero and one positive eigenvalues (right)

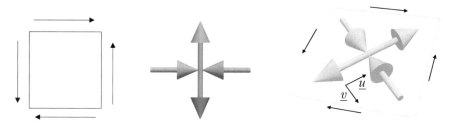

Fig. 6.17 A very usual representation of a shear stress (left); arrows as simplified RSs for shear (middle), and the relation between both (right)

The combination of double tipped axisymmetric arrows preserve the correct symmetry, *mmm* in the general case.

It is not unusual in the literature to find other types of simplified representations. Referring again to stress, the right panel of Fig. 6.16 is a very usual representation of a normal stress, class ∞/mm, whereas a shear stress is often represented as in Fig. 6.17.

Unlike for rank one properties, *flat* arrows (Fig. 6.18) are also a correct simplified representation of a general second rank RS of point group *mmm*.

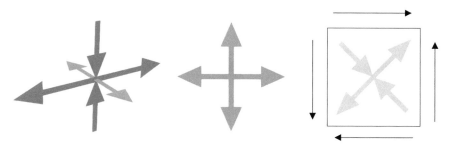

Fig. 6.18 Schematic representations of a general symmetric second rank, indefinite tensor (left); a 2D isotropic state; and a 2D shear stress

Fig. 6.19 The usual
schematic drawing of a shear
stress has all symmetry
elements (three orthogonal
planes, three orthogonal
binary axes) of point group
mmm. Vectors $\underline{u}$ y $\underline{v}$
correspond to those of
Fig. 6.16

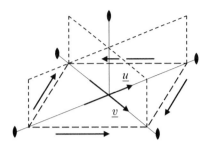

In the representation of a shear stress in the right panel of Fig. 6.17 the orientations
of the arrows and of the surface normals directly correspond to the definition of stress
in Fig. 4.14. In the other two panels, the arrows point along the diagonals of the square
in the right panel.

Only in special cases of higher symmetry would it be justified to use 3D axisymmetric arrows. As for rank one properties, the symbol for a normal stress (Fig. 6.16,
right) should be a 3D, axisymmetric arrow of the proper symmetry, although a flat
arrow does equally well, as long as it is interpreted correctly.

Figure 6.19 shows how the standard representation of a shear stress correctly has
all symmetry elements of point group *mmm*.

6.8 The Representation Quadric

A different kind of representation surface, the representation quadric or RQ from now
on, has been used for a long time as an aid to visualize the orientational dependence
of second rank properties. It is frequently used in optics. We will mention it only
briefly, since very extensive and more specialized literature on the subject is available
[1, 7, 8].

Taking the electric resistivity $\underline{\underline{\rho}}$ as an example, let's consider the surface S of
equation:

$$\underline{\underline{\rho}} : \underline{r}\,\underline{r} = \rho_{ij} x_i x_j = 1 \tag{6.14}$$

Equation 6.14 is the equation of the representation quadric of the property. It is
quadratic in the spatial coordinates and thus a quadric: an ellipsoid, hyperboloid, etc
[9].

The vector radius from the origin O to a point P of coordinates x_1, x_2, x_3 on S is
given by:

$$r = |\underline{r}| = \sqrt{x_1^2 + x_2^2 + x_3^2}$$

whereas the components of a unit vector in this direction OP are:

$$n_i = x_i/r \text{ so that } x_i = r\, n_i$$

Substituting $x_i = r\, n_i$ in Eq. 6.14:

$$\rho_{ij} x_i x_j = \rho_{ij} n_i n_j r^2 = 1 \;\Rightarrow\; \rho_{ij} n_i n_j = \underline{\rho} : \underline{\underline{nn}} = \rho_{\underline{n}} = \frac{1}{r^2} \qquad (6.15)$$

Since $\rho_{ij} n_i n_j = \rho_{\underline{n}}$ is the value of the property $\underline{\underline{\rho}}$ in the direction of the radius vector $\underline{r}$, it turns out that:

$$r = \frac{1}{\sqrt{\rho_{ij} n_i n_j}} = \frac{1}{\sqrt{\rho_{\underline{n}}}} \qquad (6.16)$$

that is, the length of the radius vector $\underline{r}$ of the RQ is proportional to the *reciprocal of the square root* of the property in that direction; whereas the radius vector of the RS is proportional to the property in the direction of $\underline{r}$.

In directions in which the property is large, the RS will be far away from the origin, whereas the RQ will be close to it. In directions where the property is zero, the RQ will have an asymptotic direction.

Although at first sight it seems awkward to use a surface with the property $r = \frac{1}{\sqrt{\rho_{\underline{n}}}}$ instead of $r = \rho_{\underline{n}}$, there are several reasons for the use of the RQ. The first and most trivial is that a great deal is known about quadrics; dealing with them is unproblematic.

A weightier argument is that an important optical property, the refraction index n_R, is related to the electrical impermeability of a material (7.1) by an expression of precisely the same type as (6.16), hence, the radius vector of the RQ of the electrical impermeability is directly proportional to the refraction index. The RQ of the electrical impermeability is the RS of the refractive index.

Finally, there is a neat graphical construction that visually illustrates the cause-effect relationship for second rank properties. It is called the *radius-normal* construction. If we again refer to the electrical resistivity, $\underline{\underline{\rho}}$ relates the electric field $\underline{E}$ (effect) to the electric current density $\underline{J}$ (cause) through Ohm's law:

$$\underline{E} = \underline{\underline{\rho}} \cdot \underline{J} \qquad E_i = \rho_{ij} J_j$$

Let's take a radius vector of the RQ of components $\underline{r} = x_i \underline{\delta}_i$ in the direction of $\underline{J}$. The point of coordinates x_i will be on the RQ and will be proportional to the components of $\underline{J}$: $x_i \propto J_i$.

The vector normal to the RQ at that point is given by the gradient of Eq. 6.14:

$$(\underline{\nabla} S)_i = \underline{\delta}_i \frac{\partial}{\partial x_i}(\rho_{kj} x_k x_j - 1) = \underline{\delta}_i \rho_{kj}\left(\frac{\partial x_k}{\partial x_i} x_j + x_k \frac{\partial x_j}{\partial x_i}\right)$$

$$= \underline{\delta}_i \rho_{kj}\left(\delta_{ki} x_j + x_k \delta_{ji}\right) = \underline{\delta}_i(\rho_{ij} x_j + \rho_{ki} x_k) = \underline{\delta}_i(2\rho_{ij} x_j) = 2\underline{\underline{\rho}} \cdot \underline{r} \qquad (6.17)$$

Fig. 6.20 Radius-normal construction. If the radius points along the cause ($\underline{J}$), the effect ($\underline{E}$) points along the normal to the tangent plane (although this sketch is two dimensional, the construction is 3D, i.e. valid for any point on the surface of the RQ)

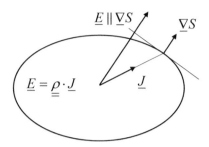

where the symmetry of $\underset{=}{\rho}$ has been used. It turns out that the gradient of the RQ is twice $\underset{=}{\rho} \cdot \underline{r}$, i.e. proportional to the value of the electric field $\underline{E}$, as predicted by Ohm's law:

$$E_i = \rho_{ij} J_j \propto \rho_{ij} x_j \propto (\nabla S)_i$$

so that the direction of $\underline{E}$ is that of the normal to the RQ. When the radius vector points in the direction of the cause ($\underline{J}$ in this case), the effect ($\underline{E}$ in this case) points along the normal to the RQ at the point where the radius vector touches the RQ.

In addition, the modulus of the effect is equal to the reciprocal of the square root of the length of the radius vector.

This graphical construction (Fig. 6.20) was useful when numerical computations had to be carried out by hand. It is still useful as a means to visualize how an anisotropic material deviates the effect with respect to the cause, for symmetric second rank properties.

When the construction of Fig. 6.20 is carried out for the particular case of the cause ($\underline{J}$) pointing along one of the three axes of the RQ, the normal (the effect, $\underline{E}$) also points in the same direction. This is the condition fulfilled by the principal directions of a property: the axes of the RQ point along the principal directions. In the frame of the eigenvectors the equation of the RQ is particularly simple because all cross terms $x_i x_j$ vanish.

The axes of the RQ are then aligned with the axes of the RS (Fig. 6.21). The maximum value of the property (its largest eigenvalue) happens along the largest axis of the RS, which is the shortest axis of the RQ. The lengths of the semiaxes of the RQ are $\dfrac{1}{\sqrt{\rho_{ii}}}$ (no sum), whereas the semiaxes of the RS are ρ_{ii} (no sum).

Depending on the signs of the eigenvalues, the RQ can be an ellipsoid (three positive eigenvalues), a hyperboloid of one sheet (two positive, one negative), a hyperboloid of two sheets (one positive, two negative), an imaginary ellipsoid (all three negative) (Fig. 6.22).[8]

---

[8]Special cases are possible for particular numerical values of the eigenvalues. The RQ is e.g an elliptic or hyperbolic cylinder if one of them is zero, the other two unequal. If two eigenvalues are zero, the quadric degenerates into a pair of parallel of real or imaginary planes, etc.

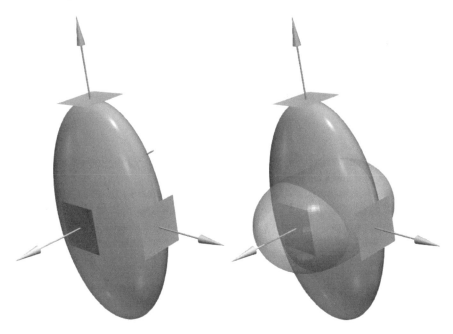

Fig. 6.21 The axes of the RQ are the principal directions of the property. The axes of the RQ and of the RS (right) are aligned

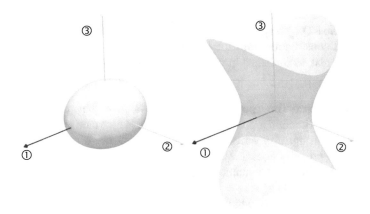

Fig. 6.22 RQ for the thermal expansion coefficient of an orthorhombic material, point group 222, with $\alpha_{11} : \alpha_{22} : \alpha_{33} = 2 : 3 : 4$ (ellipsoid on the left) and for $\alpha_{11} : \alpha_{22} : \alpha_{33} = 2 : 3 : -4$ (hyperboloid of one sheet on the right)

When there is at least one negative eigenvalue, there are directions in which the vector radius does not intersect the RQ. These are the directions in which the property is negative, so that $r = \frac{1}{\sqrt{\rho_n}}$ is imaginary. The extreme case corresponds to three negative eigenvalues: the property is negative in all directions and the RQ is an imaginary ellipsoid. The simple way out of this is to use the RQ:

$$\underset{=}{\rho} : \underline{r}\,\underline{r} = \rho_{ij}x_ix_j = -1$$

which is equivalent to reversing the signs of the eigenvalues. Imaginary sheets of the RQ now turn real and visible (Fig. 6.23).

The radius-normal construction is also valid for any other symmetric, second rank tensor, such as strain $\underset{=}{\epsilon}$. In this, case, the radius-normal construction yields the value of $\underline{u} = \underset{=}{\epsilon} \cdot \underline{r}$, i.e. the displacement of the tip of a vector $\underline{r}$ emanating from the origin.

The radius-normal construction can also be carried out when the RQ form (6.14) is indefinite. Let's take the deformation tensor $\underset{=}{\epsilon}$ as an illustrative example, and assume $\underset{=}{\epsilon}$ is diagonal with $\epsilon_{11}, \epsilon_{33} > 0$, $\epsilon_{22} < 0$, in which case $\epsilon_{ij}x_ix_j = 1$ is a one sheet hyperboloid. A section of this hyperboloid is shown on the left of Fig. 6.24.

In a direction such as $\underline{r}$ on the left of the figure, the radius vector intersects the RQ and the displacement $\underline{u} = \underset{=}{\epsilon} \cdot \underline{r}$ is given by the radius-normal construction with the hyperboloid $\epsilon_{ij}x_ix_j = 1$. The value of the property, in this case the displacement parallel to $\underline{r}$ is given by $\dfrac{1}{\sqrt{\epsilon_r}}$ (Fig. 6.25, left).

However, for a direction such as $\underline{r}$ on right of Fig. 6.24, the radius vector does not intersect the RQ $\epsilon_{ij}x_ix_j = 1$. In this case, the displacement is obtained from the radius-normal construction with the hyperboloid $\epsilon_{ij}x_ix_j = -1$, and changing the sign of the normal, as indicated in the figure. The displacement parallel to $\underline{r}$ is now negative; its modulus is $\dfrac{1}{\sqrt{-\epsilon_r}}$ (Fig. 6.25, right).

For indefinite tensors both positive and negative values of the property are possible depending on the direction $\underline{n}$ points to. Since the RS is a surface without discontinuities, the change in sign as $\underline{n}$ varies implies that there must be directions along which the value of the property goes through zero.

These are the directions contained in the cone shown in Fig. 6.26, called the *zero cone*, whose equation is $\epsilon_{ij}x_ix_j = 0$. This is also the cone common to both RQs $\epsilon_{ij}x_ix_j = 1$ y $\epsilon_{ij}x_ix_j = -1$, which they approach as the radius vector gets larger. For this reason it is also called the *asymptotic* cone.

The physical meaning of the zero cone is clear. An indefinite matter tensor, for example the thermal expansion coefficient of some materials, will have eigenvalues of different signs. A thin bar cut along any of the directions of the zero cone will not change its length when the temperature changes.[9] Figure 6.26 is a schematic cross section of the RQ of Fig. 6.26, with $\alpha_{11}, \alpha_{33} > 0$, $\alpha_{22} < 0$.

---

[9] So long as the linear relationship $\underset{=}{\epsilon} = \Delta T \underset{=}{\alpha}$ is valid. For a large enough temperature variation, the linear law may not be valid.

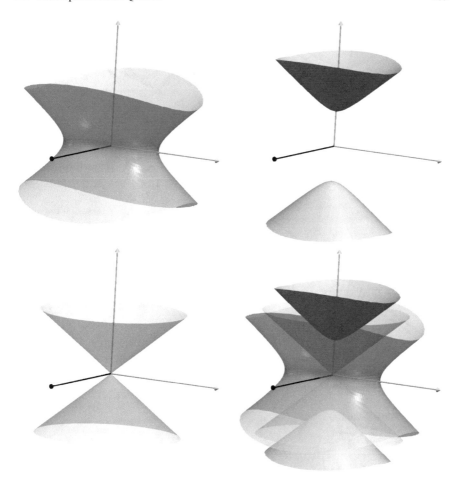

Fig. 6.23 The RQ $\rho_{11}x_1^2 + \rho_{22}x_2^2 + \rho_{33}x_3^2 = 1$ (top left); the RQ $\rho_{11}x_1^2 + \rho_{22}x_2^2 + \rho_{33}x_3^2 = -1$ (top right); the common asymptotic cone $\rho_{11}x_1^2 + \rho_{22}x_2^2 + \rho_{33}x_3^2 = 0$ (bottom left); all three surfaces (bottom right). In all cases $\rho_{11} : \rho_{22} : \rho_{33} = 2 : 4 : -3$. Axes ① ② ③

A slight advantage of the RS when a non-homogeneous field of indefinite second rank tensors has to be represented (as in medical imaging [10], complex fluid flow visualization [11], etc.), is that it is bounded, so that a field of symbols of limited size can be drawn (Fig. 6.27).

Unless truncated, neighboring hyperboloidal RQs will overlap and yield a confusing field representation. RSs on the contrary, are always of finite size, and display a negative lobe for the orientations in which the radius vector has imaginary intersection with the RQ. In visualization it is usual to assign different colors to positive and negative lobes to emphasize contrast.

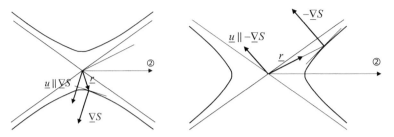

Fig. 6.24 Section of the RQ $\epsilon_{11}x_1^2 + \epsilon_{22}x_2^2 + \epsilon_{33}x_3^2 = 1$ for $\epsilon_{11}, \epsilon_{33} > 0, \epsilon_{22} < 0$ (hyperboloid of one sheet, left), and of the RQ $\epsilon_{11}x_1^2 + \epsilon_{22}x_2^2 + \epsilon_{33}x_3^2 = -1$ with $\epsilon_{11}, \epsilon_{33} > 0, \epsilon_{22} < 0$ (hyperboloid of two sheets, right)

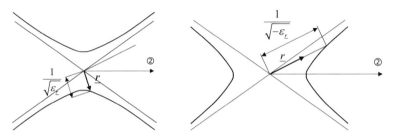

Fig. 6.25 Value of the displacement $\underline{u}$ from the radius-normal construction of Fig. 6.24 for two different directions: on the left, the direction of $\underline{r}$ intersects the RQ $\epsilon_{ij}x_ix_j = 1$; on the right, it does not, but it intersects the RQ $\epsilon_{ij}x_ix_j = -1$

Fig. 6.26 The RS, the RQ and the zero cone for the thermal expansion coefficient with $\alpha_{11} : \alpha_{22} : \alpha_{33} = 2 : -3 : 4$.

Axes ① ②

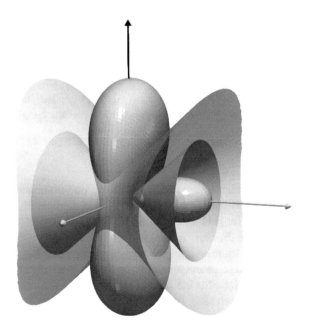

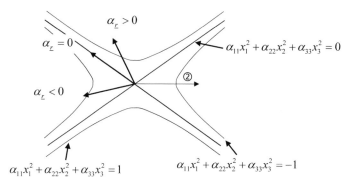

Fig. 6.27 Along any direction $\underline{r}$ contained in the zero cone, thermal expansion vanishes. A rod of this material cut along any asymptotic direction would neither shrink nor expand on a temperature change because $\alpha_r = 0$

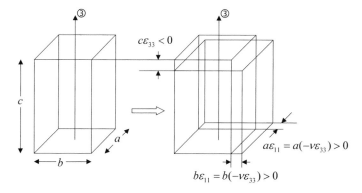

Fig. 6.28 Block of material with $\nu > 0$ subjected to homogeneous deformation. A homogeneous stress makes the material shrink in the ③ direction and expand in the ① and ② directions

The zero cone and the radius-normal construction can also be applied to field tensors: a block of an isotropic material of Poisson ratio $\nu > 0$ is subjected to a homogeneous normal compressive stress $\tau_{33} < 0$ (Fig. 6.28) so that it undergoes a homogeneous deformation $\underline{\underline{\epsilon}}$ given by:

$$\llbracket \underline{\underline{\epsilon}} \rrbracket = \begin{bmatrix} -\nu\epsilon_{33} > 0 & 0 & 0 \\ 0 & -\nu\epsilon_{33} > 0 & 0 \\ 0 & 0 & \epsilon_{33} = \dfrac{\tau_{33}}{E} < 0 \end{bmatrix}$$

in the axes of Fig. 6.28. Since $\underline{\underline{\epsilon}}$ is indefinite, there will be directions along which $\epsilon_n = \underline{\underline{\epsilon}} : \underline{n}\,\underline{n}$ will be zero. These are the directions contained in the cone:

$$\epsilon_{11}x_1^2 + \epsilon_{22}x_2^2 + \epsilon_{33}x_3^2 = \epsilon_{33}(-\nu x_1^2 - \nu x_2^2 + x_3^2) = 0$$

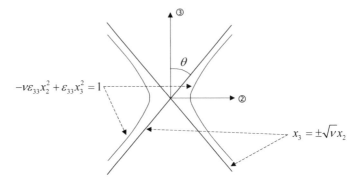

Fig. 6.29 Section of the RQ through $x_1 = 0$. The semiangle of the cone is $\theta = \arctan \pm \dfrac{1}{\sqrt{\nu}}$

where $\epsilon_{33} < 0$ y $\nu > 0$. A section of the cone through the plane $x_1 = 0$ is shown in Fig. 6.29

In this case ($\epsilon_{11} = \epsilon_{22}$) the cone is circular and its semiangle can be obtained from the equation of the cross section:

$$x_1 = 0 \quad \Rightarrow \quad -\nu x_2^2 + x_3^2 = 0 \quad \Rightarrow \quad x_3 = \pm\sqrt{\nu}$$

Thus, if a deformable, straight piece of thread of length L were embedded in the material in an asymptotic direction before deformation takes place, there would be no change in its length when the stress were applied.

Although not immediately obvious, this result is natural: if the thread were originally oriented parallel to the ③ axis, it would shorten. If parallel to the ① or ② axis, it would lengthen. Therefore, there must necessarily be directions along which it neither contracts nor expands. These are the infinite directions of the generatrices of the zero cone.[10]

---

[10]If, on the one hand, the length of the thread does not change, and, on the other hand, every point of the material suffers some displacement, then what kind of change would an observer tied to the thread experience?

References

1. Nye, J.F.: Physical Properties of Crystals. Oxford University Press, Oxford (1985)
2. Strzhemechny, M.A., Krivchikov, A.I., Jeżowski, A., Zloba, D.I., Buravtseva, L.M., Churiukova, O., Horbatenko, Y.V.: New thermal conductivity mechanism in triclinic 4-bromobenzophenone crystal. Chem. Phys. Lett. **647**, 55–58 (2016)
3. Zhang, C., Li, Z., Cong, H., Wang, J., Boughton, R.I.: Crystal growth and thermal properties of single crystal monoclinic NdCOB(NdCa$_4$O(BO$_3$)$_3$). J. Alloy. Compd. **507**(2), 335–340 (2010)
4. Wohlfarth, E.P.: Iron, cobalt and nickel. Handbook of Ferromagnetic Materials. Elsevier, Amsterdam (1980)
5. Rao, K.V.K., Naidu, S.V.N., Murthy, K.S.: Precision lattice parameters and thermal expansion of calcite. J. Phys. Chem. Solids **29**(2), 245–248 (1968)
6. Meingast, C., Kraut, O., Wolf, T., Wühl, H., Erb, A., Müller-Vogt, G.: Large a-b anisotropy of the expansivity anomaly at T_c in untwinned YBa$_2$Cu$_3$O$_7$-δ. Phys. Rev. Lett. **67**(12), 1634 (1991)
7. Born, M., Wolf, E.: Principles of Optics: Electromagnetic Theory of Propagation, Interference and Diffraction of Light. Elsevier, Amsterdam (2013)
8. Hecht, E.: Optics. Pearson, London (2015)
9. Audin, M.: Geometry and Topology. Springer, Berlin (2003)
10. Flower, M.A.: Webb's Physics of Medical Imaging. CRC Press, Boca Raton (2012)
11. Globus, C., Levit, A., Lasinski, T.: A tool for visualizing the topology of three-dimensional vector fields. In: Nielson, G., Rosenblum, L. (eds.) IEEE Visualization. Institute of Electrical and Electronics Engineers (1991)

Chapter 7
Third Rank Properties

Odd rank properties are, in Newnham's words, *null* properties, meaning that they may vanish for certain point groups (like all centrosymmetric ones, see Sect. 3.6). As a result, not all materials will display third rank properties. Also as a consequence of being of odd rank, the RS will consist of overlapping positive and negative lobes, as shown in Fig. 7.1.

The left part of Fig. 7.1 shows the RS for the longitudinal piezoelectric modulus $r = \underline{\underline{d}} \vdots \underline{n}\,\underline{n}\,\underline{n}$ for point group 6*mm*, which consists of overlapping positive and negative lobes.

In Fig. 7.1 positive and negative lobes are shown separately in the middle and right panels by plotting the RS of the absolute value $|r| = \left| \underline{\underline{d}} \vdots \underline{n}\,\underline{n}\,\underline{n} \right|$.

All RSs in this chapter[1] are to be interpreted as consisting of pairs of positive and negative overlapping lobes.

All odd rank RSs contain the origin (Figs. 7.4, 7.5, 7.6, 7.7, 7.8, 7.9, 7.10, 7.11 and 7.12), meaning that there always exist directions along which the property, e.g. the longitudinal piezoelectric modulus, is zero. These are all directions tangent to the surface at the origin. In Fig. 7.2, the null directions lie along the generatrices of a cone and on the plane transversal to the symmetry axis of the RS.

7.1 Piezoelectricity and Linear Electro-Optic Effect

Among third rank properties, the piezoelectric modulus $\underline{\underline{d}}$ and the linear electro-optic effect (Pöckels coefficient) $\underline{\underline{r}}$ are two of the most practically important ones. They are symmetric in the second, and first pair of indices respectively:

---

[1] And all directed spheres of Chap. 5.

© Springer Nature Switzerland AG 2020
M. Laso and N. Jimeno, *Representation Surfaces for Physical Properties of Materials*, Engineering Materials, https://doi.org/10.1007/978-3-030-40870-1_7

Fig. 7.1 RS for the longitudinal piezoelectric modulus of point group $6mm$. On the left, the RS $r = \underline{\underline{d}} \vdots \underline{n}\,\underline{n}\,\underline{n}$ consisting of overlapping positive and negative lobes. In the middle, the RS of the absolute value $|r| = \left| \underline{\underline{d}} \vdots \underline{n}\,\underline{n}\,\underline{n} \right|$; right, a cross section of the latter. In the middle and right panels, lobes corresponding to positive values of r have been shaded light grey, negative values dark grey

Fig. 7.2 For odd rank properties space is partitioned in regions of positive and negative values of the property. The straight lines are the cross sections of the null cone and the null transversal plane

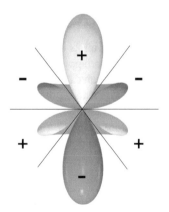

$$\underline{\underline{d}}^T = \underline{\underline{d}} \qquad d_{ijk} = d_{ikj}$$

$$^T\underline{\underline{r}} = \underline{\underline{r}} \qquad r_{ijk} = r_{jik}$$

They appear in the CEs for:

$$\text{direct piezoelectricity:} \quad \underline{P} = \underline{\underline{d}} : \underline{\underline{\tau}}$$
$$\text{inverse piezoelectricity:} \quad \underline{\underline{\epsilon}} = \underline{\underline{E}} \cdot \underline{\underline{d}}$$
$$\text{Pöckels effect:} \quad \Delta\underline{\underline{B}} = \underline{\underline{r}} \cdot \underline{E}$$

where the relative dielectric impermeability[2] $\underline{\underline{B}}$ is defined as $B_{ij} = \kappa_0 \frac{\partial E_i}{\partial D_j}$ and is very closely related to the refractive index of the material. In the linear electro-optic effect, small changes in an electric field applied to a material produce small changes in its refractive index.

Notice how similar the CEs for inverse piezoelectricity and for the linear electro-optic effect are. Furthermore, an inspection of Tables 3.8 and 3.9 shows that

$$\text{str}(\underline{r}) = \text{str}(\underline{d})^T.$$

The equality $\text{str}(\underline{r}) = \text{str}(\underline{d})^T$ plus the full permutation symmetry of $\underline{n}\,\underline{n}\,\underline{n}$ makes the RSs for $\underline{\underline{d}}$ and $\underline{\underline{r}}$ the same, irrespective of the left ($\underline{\underline{r}}$) or right ($\underline{\underline{d}}$) index symmetry.

The RS, $\underline{r} = \underline{\underline{d}} : \underline{n}\,\underline{n}\,\underline{n}$, can have several interpretations, depending on the application. If we take the direct piezoelectric effect, and replace $\underline{\underline{\tau}}$ by two of the $\underline{n}$'s, i.e. we use two of the slots of the tensor $\underline{\underline{d}}$ considered as a linear mapping:

$$\underline{P} = \underline{\underline{d}} : \underline{n}\,\underline{n}$$

we would obtain the polarization $\underline{P}$ caused by a unit longitudinal stress ($\underline{n}\,\underline{n}$). One further contraction with $\underline{n}$ yields the component of the polarization $\underline{P}$ along $\underline{n}$. The RS:

$$r = \underline{\underline{d}} : \underline{n}\,\underline{n}\,\underline{n} = (\underline{\underline{d}} : \underline{n}\,\underline{n}) \cdot \underline{n} = \underline{P} \cdot \underline{n} = P_{\underline{n}}$$

can be interpretated as the RS of the longitudinal piezoelectric effect.

A second possibility is to consider inverse piezoelectricity. If we replace $\underline{E}$ by one of the $\underline{n}$'s:

$$\underline{\underline{\epsilon}} = \underline{n} \cdot \underline{\underline{d}}$$

we obtain the strain caused in the material by a unit electric field in the direction of $\underline{n}$. Two further contractions project the resulting strain $\underline{\underline{\epsilon}}$ along $\underline{n}$, i.e. the radius vector r gives the longitudinal strain in the direction of $\underline{n}$ caused by a unit electric field in the same direction.

$$r = \underline{\underline{d}} : \underline{n}\,\underline{n}\,\underline{n} = (\underline{n} \cdot \underline{\underline{d}}) : \underline{n}\,\underline{n} = \underline{\underline{\epsilon}} : \underline{n}\,\underline{n} = \epsilon_{\underline{n}}$$

---

[2] The RQ for $\underline{\underline{B}}$ is the indicatrix for the refractive index [1].

Fig. 7.3 Piezoelectricity
causes a macroscopic
polarization $\underline{P}$ to appear in
the material. Surface charges
appear on pairs of opposed
faces. The charges are
proportional to the
projections of $\underline{P}$ along the
unit vectors normal to the
faces

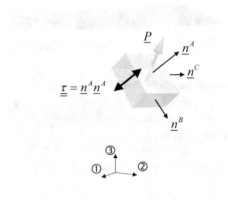

This second interpretation applies to the Pöckels effect as well: its RS gives the longitudinal change in relative dielectric impermeability $\underline{\underline{B}}$ in the direction $\underline{n}$ of the radius vector, when a unit electric field in the same direction is applied to the material[3]:

$$r = \underline{\underline{r}} \vdots \underline{n}\,\underline{n}\,\underline{n} = (\underline{r} \cdot \underline{n}) : \underline{n}\,\underline{n} = \Delta\underline{\underline{B}} : \underline{n}\,\underline{n} = \Delta B_{\underline{n}}$$

The interpretation of the RS for other third rank properties is similar and depends on the specific field or matter tensor under consideration.

It is also clear that longitudinal projection is not the only possibility: recall Fig. 5.14 for the transversal RS of a rank one property. The higher the rank of the property, i.e. the greater the number of $\underline{n}$'s that appear in the projection, the more possibilities for constructing RS that give transversal information.

It is therefore legitimate to ask for the polarization that appears transversally to the direction of the applied stress. Figure 7.3 illustrates the situation. It is intended to be almost identical to Fig. 5.17 in Sect. 5.4, the only difference being the cause of the polarization: a change in temperature in pyroelectricity $\underline{P} = \underline{p}\Delta T$, an applied stress in direct piezoelectricity $\underline{P} = \underline{\underline{d}} : \underline{\underline{\tau}}$.

The RS for the transversal polarization on the faces perpendicular to $\underline{n}^B$ is given by the projection of the polarization $\underline{P}$ caused by the unit normal stress $\underline{n}^A\underline{n}^A$ along the face normal $\underline{n}^B$:

$$r = \underbrace{(\underline{\underline{d}} : \underline{n}^A\,\underline{n}^A)}_{\underline{P}} \cdot \underline{n}^B = \underline{\underline{d}} \vdots \underline{n}^A\,\underline{n}^A\,\underline{n}^B$$

---

[3]No difficulty should arise from the use of r and $\underline{\underline{r}}$ in the same equation. Although the literal symbol is the same, their tensorial rank makes it impossible to confuse them.

Written in subindex notation, the RSs for the longitudinal and transversal projections with the unit vectors $\underline{n}^A$ and $\underline{n}^B$ are:

$$r = d_{ijk} n_k^A n_j^A n_i^A \qquad r = d_{ijk} n_k^A n_j^A n_i^B$$

Clearly, the transversal piezoelectric effect given by the projection $\underline{\underline{d}} : \underline{n}^A \, \underline{n}^A \, \underline{n}^B$ is not the only possibility. An equally transversal possibility is:

$$r = \underline{\underline{d}} : \underline{n}^A \, \underline{n}^A \, \underline{n}^C$$

It is also possible to represent the polarization along a given direction, say $\underline{n}^A$ caused by a *shear* stress, say in the plane defined by $\underline{n}^B$ and $\underline{n}^C$,[4] in which case the RS would be:

$$r = \underline{\underline{d}} : \underline{n}^B \, \underline{n}^C \, \underline{n}^A$$

Obviously, the number of possible combinations of projecting directions grows with the tensorial rank of the property. For each of them a separate RS can be drawn. While some of these RSs have very direct applicability (i.e. the elastic modulus of Eq. 4.33), others, like the modulus (4.34) may be useful in more specialized situations only.

To sum up: while for first and second rank properties a single RS is enough for a visual understanding of the orientational dependence of a property, no single RS can convey all the information on directional dependence of higher rank properties.

In addition, transversal projection also rises a question about the sampling of orientations by the unit vectors used for projection. We will come back to this point in Chap. 9. The rest of this chapter focuses on the longitudinal RS.

7.2 Representation Surfaces for Crystallographic and Limit Classes

The RS for generic third rank properties like the longitudinal piezoelectric modulus or the Pöckels coefficient for all classes are shown in Figs. 7.4–7.5, 7.6, 7.7, 7.8, 7.9, 7.10, 7.11 and 7.12. Where possible, numerical values of the components for specific materials have been used.

Properties such as $\underline{\underline{d}}$ or $\underline{\underline{r}}$ may have a large number of nonzero components. Except for a few point groups, whose RS depends on a single component,[5] the shape RS will depend on the numerical values of several parameters. For this reason, the RSs

---

[4]How would you write this shear stress without resorting to a specific reference frame?.
[5]For example, trigonal class 32.

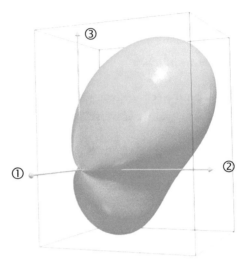

Fig. 7.4 RS for the piezoelectric modulus $\underline{\underline{d}}$ for triclinic class 1. RS shape belongs to the same point group 1

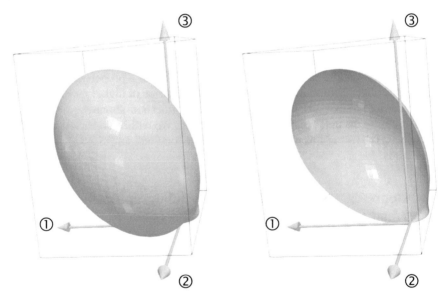

Fig. 7.5 RS for the piezoelectric modulus $\underline{\underline{d}}$ for monoclinic class m. RS shape belongs to the same point group m. The section is through the symmetry plane $m \perp ②$

$$n_1 \rightarrow -n_1 \quad n_2 \rightarrow n_2 \quad n_3 \rightarrow -n_3$$

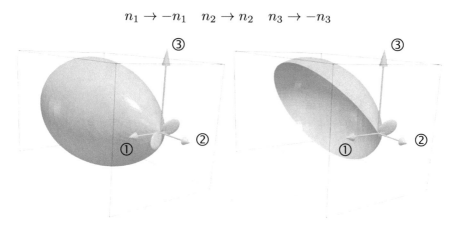

Fig. 7.6 RS for the piezoelectric modulus $\underset{=}{d}$ for monoclinic class 2; lithium sulfate monohydrate $Li_2SO_4 \cdot H_2O$. RS shape belongs to the orthorhombic point group $mm2$. The binary axis points along the ② axis. The section is through one of its two orthogonal symmetry planes. The vectors normal to the symmetry planes are given by (7.3)

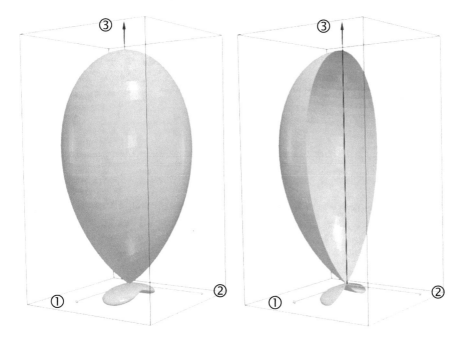

Fig. 7.7 RS for the piezoelectric modulus $\underset{=}{d}$ for orthorhombic class $mm2$ Barium magnesium fluoride, $BaMgF_4$. RS shape belongs to the orthorhombic point group $mm2$. The binary axis points along the ② axis. The section is through the symmetry plane $m \perp ②$, one of the two perpendicular symmetry planes

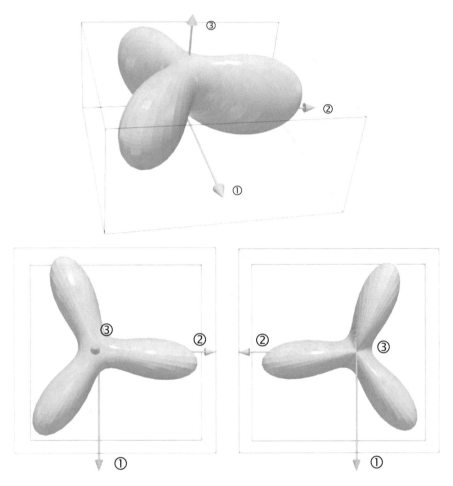

Fig. 7.8 RS for the piezoelectric modulus $\underset{=}{d}$ for a material of trigonal class 3. RS shape belongs to the trigonal point group $3m$. Views in the bottom panels are along $-$③ and $+$③

for two materials belonging to the same class may look quite different, although they will always share the symmetry elements of the same point group.

In the following we will try and show how all the RSs of Figs. 7.4, 7.5, 7.6, 7.7, 7.8, 7.9, 7.10, 7.11 and 7.12 are actually built up from a limited number of basic shapes.

First of all, some non-centrosymmetric classes do not appear in the captions of Figs. 7.4, 7.5, 7.6, 7.7, 7.8, 7.9, 7.10, 7.11 and 7.12. There are two reasons for their absence:

- the cubic class 432 and the Curie class $\infty\infty$, although not centrosymmetric, have other elements of symmetry that cancel all components of their $\underset{\sim}{d}$.

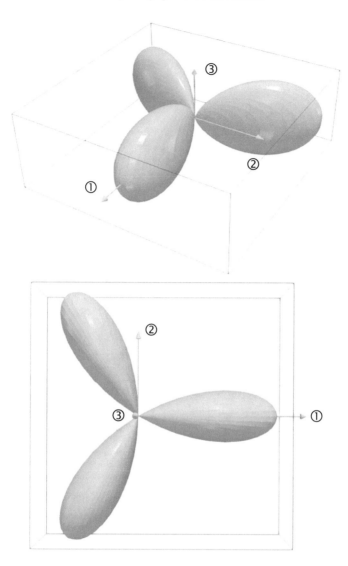

Fig. 7.9 RS for the piezoelectric modulus $\underline{\underline{d}}$ for materials of trigonal class 32, α-SiO_2, and of hexagonal classes $\bar{6}$, $\bar{6}m2$. RS shape belongs to the hexagonal point group $\bar{6}m2$. View of second panel is along ③

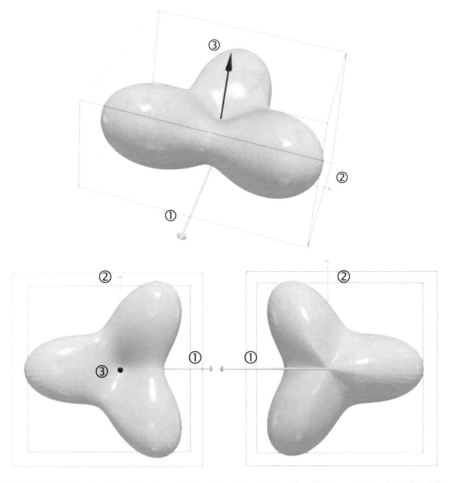

Fig. 7.10 RS for the piezoelectric modulus $\underline{\underline{d}}$ for trigonal class $3m$ lithium niobate, LiNbO$_3$. RS shape belongs to the same point group $3m$. Views in the bottom panels are along $-$③ and $+$③

- classes 422, 622 and $\infty2$ do have some non-zero components in $\underline{d}$ but they are also absent in the figures. In this case the reason is the full index permutation symmetry of the projector $\underline{n}\,\underline{n}\,\underline{n}$, which groups together terms in $d_{ijk}n_k n_j n_i$ and leads to cancellations for the longitudinal RS. These three classes have the same structure:

$$\mathrm{str}(\underline{d}) = \begin{bmatrix} \cdot & \cdot & \cdot & \bullet & \cdot & \cdot \\ \cdot & \cdot & \cdot & \cdot & \searrow & \cdot \\ \cdot & \cdot & \cdot & \cdot & \circ & \cdot \\ \cdot & \cdot & \cdot & \cdot & \cdot & \cdot \end{bmatrix} \quad \text{so that} \quad d_{25} = -d_{14}.$$

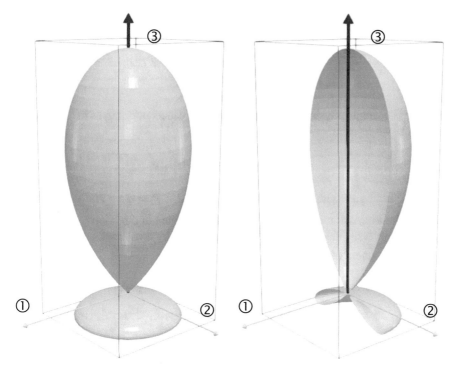

Fig. 7.11 RS for the piezoelectric modulus $\underline{\underline{d}}$ for barium nitrite monohydrate $Ba(NO_2)_2 \cdot H_2O$ (hexagonal class 6), for $6mm$; for tetragonal classes 4, $4mm$, and Curie classes ∞, ∞m. RS shape belongs to Curie class ∞m

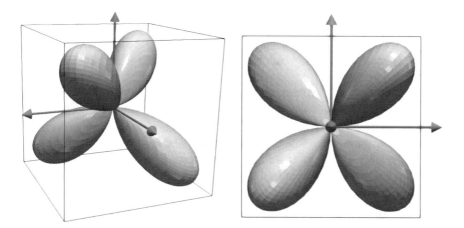

Fig. 7.12 RS for the piezoelectric modulus for cubic classes 23 and $\bar{4}3m$, for orthorhombic class 222 (barium formate $Ba(COOH)_2$), and for tetragonal class $\bar{4}2m$ (ammonium dihydrogen arsenate $NH_4H_2AsO_4$). The panel on the right is a view along the conventional ① axis of the cubic system. Views along ② and ③ are identical

The resulting RS is:

$$r = \underline{\underline{d}} : \underline{n}\,\underline{n}\,\underline{n} = d_{123}n_3n_2n_1 + d_{132}n_2n_3n_1 + d_{213}n_3n_1n_2 + d_{231}n_1n_3n_2$$
$$= (d_{14} + d_{25})n_1n_2n_3 = 0$$

There is a simple physical interpretation for this cancellation; if, as in the previous section, we rewrite:

$$r = \underline{\underline{d}} : \underline{n}\,\underline{n}\,\underline{n} = (\underline{\underline{d}} : \underline{n}\,\underline{n}) \cdot \underline{n}$$

Identifying terms with the direct piezoelectric law: $\underline{P} = \underline{\underline{d}} : \underline{\underline{\tau}}$ and recalling that $\underline{n}\,\underline{n}$ is the expression of a (unit) normal stress applied in direction $\underline{n}$,[6] the term in parenthesis is the electric polarization $\underline{P}$ produced by that stress, and $(\underline{\underline{d}} : \underline{n}\,\underline{n}) \cdot \underline{n}$ is the component of $\underline{P}$ along $\underline{n}$:

$$\underline{P} = \underline{\underline{d}} : \underline{n}\,\underline{n} = \underline{\delta}_1(d_{123}n_3n_2 + d_{132}n_2n_3) + \underline{\delta}_2(d_{213}n_3n_1 + d_{231}n_1n_3)$$
$$= \underline{\delta}_1 d_{14}n_2n_3 + \underline{\delta}_2 d_{25}n_1n_3 = d_{14}(\underline{\delta}_1 n_2n_3 - \underline{\delta}_2 n_1n_3)$$

The polarization $\underline{P}$ caused by the stress is not zero, but cancels when projected onto $\underline{n}$. The RS for these three classes vanishes because the resulting polarization is perpendicular to $\underline{n}$ for all possible orientations of $\underline{n}$.

We notice in passing that the left 3×3 sub-block of $\mathrm{str}(\underline{d})$ for classes 422, 622, $\infty 2$ together with 222, $\bar{4}2m$, $\bar{4}3m$, 23 is empty. When carrying out the multiplication $\underline{d}\,\vec{\tau}$, the first three elements of $\vec{\tau}$ are multiplied by zero, while a nonzero result would be obtained if some of the three last components of $\vec{\tau}$, (τ_4, τ_5, τ_6), were nonzero.

In Voigt notation the components τ_4, τ_5 and τ_6 are shear stresses: these classes exclusively react to stresses which have shear components when expressed *in the conventional axes*.

Except for the highly symmetric classes 432 and $\infty\infty$, for which all d_{ij} are zero, there are no piezoelectric classes with empty right 3×3 sub-block of $\mathrm{str}(\underline{d})$, i.e. materials belonging to all these classes will react to one or more shear stresses.

We go now through the RSs of the remaining point groups. Those that have the same RS have been gathered in a single figure. In some cases, repeated factors (i.e. $n_1n_1n_2$) in equations have been kept separate instead of using exponents (i.e. $n_1^2n_2$) in order to make explicit the contributions of the different basic objects to the RS.

• triclinic: the RS of the only monoclinic piezoelectric class 1 (Fig. 7.4) has the same symmetry as the material.
  The expression for the RS of a triclinic material is, in terms of the components of $\underline{n}$ and $\underline{d}$:

---

[6]Which is *not* a normal stress in the conventional axes to which $\mathrm{str}(\underline{d})$ is referred.

$$r = d_{11}n_1n_1n_1 + d_{22}n_2n_2n_2 + d_{33}n_3n_3n_3 +$$
$$+ (d_{12} + d_{26})n_2n_2n_1 + (d_{21} + d_{16})n_1n_1n_2 +$$
$$+ (d_{23} + d_{34})n_3n_3n_2 + (d_{32} + d_{24})n_2n_2n_3 +$$
$$+ (d_{31} + d_{15})n_1n_1n_3 + (d_{13} + d_{35})n_3n_3n_1 +$$
$$+ (d_{14} + d_{25} + d_{36})n_1n_2n_3$$

which has no symmetry apart from the unit E.

This is the first time we encounter a maximally asymmetric (triclinic class 1) RS. The least symmetric RSs for first and second rank properties were still quite symmetric: they belonged to classes ∞m and mmm respectively. In general, only odd-rank properties of rank ≥ 3 can have RSs of minimum, triclinic point group 1 symmetry.

- monoclinic: the shape of the RS of the monoclinic piezoelectric class m (Fig. 7.5) has the same m symmetry, with the symmetry plane normal to the ② axis. Its RS is:

$$r = d_{11}n_1n_1n_1 + d_{33}n_3n_3n_3 +$$
$$+ (d_{12} + d_{26})n_2n_2n_1 + (d_{32} + d_{24})n_2n_2n_3 + (d_{31} + d_{15})n_1n_1n_3 +$$
$$+ (d_{13} + d_{35})n_3n_3n_1$$

The $m \perp ②$ plane of symmetry transforms the n_i as:

$$n_1 \rightarrow n_1 \quad n_2 \rightarrow -n_2 \quad n_3 \rightarrow n_3 \tag{7.1}$$

The terms that contain n_2 in (7.1) are quadratic and thus invariant under (7.1), which confirms the invariance of the RS under reflection on $m \perp ②$.

- however, the RS of the monoclinic piezoelectric class 2 (Fig. 7.6) has higher, orthorhombic $mm2$ symmetry. Its RS is:

$$r = d_{22}n_2n_2n_2 + (d_{21} + d_{16})n_1n_1n_2 + (d_{23} + d_{34})n_3n_3n_2 +$$
$$+ (d_{14} + d_{25} + d_{36})n_1n_2n_3 \tag{7.2}$$

To check that the RS has a twofold axis, a 2 along ② transforms the n_i as:

$$n_1 \rightarrow -n_1 \quad n_2 \rightarrow n_2 \quad n_3 \rightarrow -n_3$$

Substitution in (7.2) confirms the invariance under an axis 2 along ②.

Checking that (7.2) is also invariant under two orthogonal reflection planes which contain the ② (see the stereogram for class $mm2$), is slightly more involved because the symmetry planes are not necessarily oriented perpendicular to ① and ③ (see Fig. 7.6, where a section along one of the two symmetry planes is shown), so we cannot proceed as with the 2 axis. But we can factor out n_2 in (7.2):

$$r = d_{22}n_2n_2n_2 +$$
$$n_2 \left[(d_{21} + d_{16})n_1n_1 + (d_{23} + d_{34})n_3n_3 + (d_{14} + d_{25} + d_{36})n_1n_3 \right]$$

Reflections on planes that contain the ② axis do not alter n_2, so we only have to check the invariance of the expression in brackets under two orthogonal reflections. Since the d_{ij} are constants, the bracket is just a quadratic form in n_1 and n_3, whose RQ in the ①③ plane has two orthogonal symmetry planes. The eigenvectors of this RQ give the vectors normal to the symmetry planes:

$$[\![v_+]\!], [\![v_-]\!] = \begin{bmatrix} \dfrac{1}{\sqrt{1 + (K_\pm)^2}} \\[3mm] \dfrac{K_\pm}{\sqrt{1 + (K_\pm)^2}} \end{bmatrix} \tag{7.3}$$

where

$$K_\pm \equiv \frac{(d_{21} + d_{16} - d_{23} - d_{34}) \pm \sqrt{(d_{21} + d_{16} - d_{23} - d_{34})^2 + (d_{14} + d_{25} + d_{36})^2}}{(d_{14} + d_{25} + d_{36})}$$

Reflections on the planes with normal vectors (7.3) leave (7.2) unchanged. Factoring out n_2 in (7.2) implies fixing the value of n_2, i.e. cutting the RS by planes perpendicular to the ② axis. The result (7.3) shows that each of these sections has the same symmetry planes, so that the whole RS fulfills the symmetry criteria of point group $mm2$.

There is a point here which needs clarification; the structure (atomic, molecular, macroscopic) of a material of class 2 does not have mirror planes, only a twofold axis; its stereogram only defines the position of this twofold axis.

The numerical values of the components (e.g. the d_{ij} in this case) are reported referred to a Cartesian system the ② axis of which coincides with the twofold axis, but whose ① and ③ axes are chosen (and also reported!) at will by the researcher(s) that carry out the measurements.

The RS does have mirror planes but only as a consequence of the projection $n\,\underline{n}\,n$, it is therefore not surprising that they appear rotated with respect to the ① and ③ axes chosen in the experimental setup.

The same applies to materials of other classes lacking mirror planes. If their RS does have mirror planes, they will appear at positions dictated by the numerical values of the components, not by the geometry of the material.

If the material does have mirror planes, then the RS will have at least the same planes, and in the same position. Class m above (Fig. 7.5) is actually the first such case.

- orthorhombic: the RS for class $mm2$ (Fig. 7.6) has the same orthorhombic $mm2$ symmetry as the material, with the twofold axis along the ③ axis. The equation of the RS is:

$$r = (d_{31} + d_{15})n_3n_1n_1 + (d_{32} + d_{24})n_3n_2n_2 + d_{33}n_3n_3n_3 \qquad (7.4)$$

Mirror planes that contain the ③ axis transform the components n_i as:

$$\begin{array}{ccc} n_1 \to -n_1 & n_2 \to n_2 & n_3 \to n_3 \ (m \perp ①) \\ n_1 \to n_1 & n_2 \to -n_2 & n_3 \to n_3 \ (m \perp ②) \end{array}$$

Carrying out these substitutions in (7.4) leaves it unchanged because terms in n_1 or n_3 are quadratic.

Under the action of the twofold axes, the n_i transform as:

$$n_1 \to -n_1 \qquad n_2 \to -n_2 \qquad n_3 \to n_3$$

which also leave (7.4) unchanged.

- trigonal: the RS for class 3 (Fig. 7.8) belongs to the more symmetric but still trigonal point group $3m$, with the threefold axis along the ③ axis. The equation of its RS is:

$$\begin{aligned} r = & d_{33}n_3n_3n_3 + (d_{15} + d_{31})n_3(n_1n_1 + n_2n_2) + \\ & + d_{11}(n_1n_1n_1 - 3n_1n_2n_2) + d_{22}(n_2n_2n_2 - 3n_1n_1n_2) \qquad (7.5) \end{aligned}$$

Proper rotations about the ③ axis leave n_3 unchanged, while the combination $n_1n_1 + n_2n_2$ is isotropic in the ①② plane, i.e. also invariant under rotations about the ③ axis.

While twofold or fourfold symmetry is best checked using the n_i's, polar spherical angles are better suited to check for threefold symmetry. Using the components of $\underline{n}$ from (4.3), the RS (7.5) can be re-written in terms of the triple azimuth 3φ:

$$\begin{aligned} r = & d_{33}\cos^3\theta + (d_{15} + d_{31})\cos\theta\sin^2\theta + \\ & d_{11}\sin^3\theta(\cos^3\varphi - 3\sin^2\varphi\cos\varphi) + d_{22}\sin^3\theta(\sin^3\varphi - 3\cos^2\varphi\sin\varphi) = \\ = & d_{33}\cos^3\theta + (d_{15} + d_{31})\cos\theta\sin^2\theta + \sin^3\theta(d_{11}\cos 3\varphi - d_{22}\sin 3\varphi) \quad (7.6) \end{aligned}$$

which has the symmetry elements of point group $3m$.[7] Similarly to monoclinic class 2, the material does not have symmetry planes, but the projection along $\underline{n}\,\underline{n}\,\underline{n}$ leads to the RS having three mirror planes staggered at $\dfrac{2\pi}{3}$ that contain the ③ axis. The last term in parentheses in (7.6) causes an angular shift so that the position of these planes does not conform with their placement in the stereogram of class $3m$. The higher symmetry of the RS is not an artifact of the projection: experimental measurements of the longitudinal piezoelectric modulus do display $3m$ symmetry.

- a similar increase in symmetry is found for the RS of trigonal class 32 (Fig. 7.9). Its RS:

---

[7]Check that (7.6) has three mirror planes at $\dfrac{2\pi}{3}$ to each other; find their position (their φ values).

$$r = d_{11}(n_1 n_1 n_1 - 3n_1 n_2 n_2) = d_{11} \sin^3 \theta \cos 3\varphi \qquad (7.7)$$

belongs to hexagonal point group $\bar{6}m2$. The threefold axis (term $\cos 3\varphi$) combines with the mirror $m \perp ③$ symmetry of the $\sin^3 \theta$ factor to yield a sixfold rotation-inversion axis (remember $\bar{6} \Leftrightarrow m \perp ③$). Since the material has mirror planes, so does the RS; furthermore, they are necessarily located at the same angular positions: 0, $\frac{2\pi}{3}$ and $\frac{4\pi}{3}$.

The equation of the RS depends on a single parameter (d_{11}). Apart from this scale factor, the RS for all materials belonging to this class (and those listed in the caption of Fig. 7.9) is the same.

• for trigonal class $3m$ the RS (Fig. 7.10) belongs to the same $3m$ point group. Its RS:

$$r = d_{33} n_3 n_3 n_3 + (d_{31} + d_{24}) n_3 (n_1 n_1 + n_2 n_2) + d_{22}(n_2 n_2 n_2 - 3n_1 n_1 n_2) =$$
$$= d_{33} \cos^3 \theta + (d_{31} + d_{24}) \cos \theta \sin^2 \theta - d_{22} \sin^3 \theta \sin 3\varphi \qquad (7.8)$$

is of the same form as (7.5) except for the missing term $d_{11} \sin^3 \theta \cos 3\varphi$, which was responsible for the angular shift of the RS lobes with respect to the standard position of the stereogram. As in the case of materials with point group symmetry 32, the planes of the RS now coincide with those of the material.

• for hexagonal class 6 materials, the rank of the property, hence the degree of the polynomial RS in trigonometric functions of the polar angles is too low to reproduce the sixfold angular periodicity demanded by a 6 axis. Besides, there is now no combination analogous to $\bar{6} \Leftrightarrow m \perp ③$ which successfully produced a sixfold inversion-rotation axis for point group 32.

As already mentioned, the way out of the conflict between third degree polynomial and sixfold multiplicity is for the RS to have continuous, i.e. infinite, rotational symmetry about the sixfold ③ axis of the material.

The RS:

$$r = d_{33} n_3 n_3 n_3 + (d_{31} + d_{24}) n_3 (n_1 n_1 + n_2 n_2) =$$
$$= d_{33} \cos^3 \theta + (d_{31} + d_{24}) \cos \theta \sin^2 \theta \qquad (7.9)$$

is independent of the azimuth φ and thus axisymmetric. It lacks symmetry under reflection on a plane perpendicular to the ③ axis ($n_3 \rightarrow -n_3$), hence belongs to the conical point group ∞m.

The same incompatibility between the rank of the property and the multiplicity of the axis exists for tetragonal classes 4, $4mm$, so that their RSs fall in the same point group.

• cubic classes 23 and $\bar{4}3m$, orthorhombic class 222, and tetragonal class $\bar{4}2m$ have a particularly simple $\mathrm{str}(d)$, which only contains diagonal terms in its right sub-block. All three subindex pairs 14, 25, 36 in Voigt notation are collected in a single coefficient $d_{14} + d_{25} + d_{36}$ by the indices of the projector, so that:

$$r = (d_{14} + d_{25} + d_{36})n_1 n_2 n_3 \qquad (7.10)$$

where d_{14}, d_{25} and d_{36} may all be different (orthorhombic class 222), two equal (tetragonal $\bar{4}2m$), or all equal (cubic 23 and $\bar{4}3m$). Apart from this scale factor, the RS is the same for all materials belonging to these classes, as was the case for all trigonal 32, hexagonal $\bar{6}$ and $\bar{6}m2$ materials.

The expression $n_1 n_2 n_3$ of (7.10) is highly symmetric. It remains invariant under threefold rotation about the four body diagonals of the cube. These rotations transform the n_i's, changing the signs of none or of two of them as:

$$
\begin{array}{lll}
n_1 \rightarrow n_2 & n_2 \rightarrow n_3 & n_3 \rightarrow n_1 \\
n_2 \rightarrow -n_1 & n_1 \rightarrow -n_3 & n_3 \rightarrow n_2 \\
n_3 \rightarrow -n_1 & n_1 \rightarrow n_2 & n_2 \rightarrow -n_3 \\
n_1 \rightarrow n_3 & n_3 \rightarrow -n_2 & n_2 \rightarrow -n_1
\end{array}
$$

Reflections on mirror planes along the face diagonals of the cube swap the identities of two axes with and without changing their signs:

$$
\begin{array}{llll}
\text{plane diagonal between } ①② : & n_1 \rightarrow n_2 & n_2 \rightarrow n_1 & n_3 \rightarrow n_3 \\
\text{plane diagonal between -}①② : & n_1 \rightarrow -n_2 & n_2 \rightarrow -n_1 & n_3 \rightarrow n_3
\end{array}
$$

etc. The resulting RS has cubic $\bar{4}3m$ point group symmetry.[8]

We again encounter the combination of several elements of symmetry resulting in a higher order element of symmetry, thus bypassing the limitation in the degree of the polynomial.

In this case a combination of a third degree monomial in θ and a second degree one in φ produce an RS with fourfold rotation-inversion axes, which is not exactly obvious in the polar spherical form of the RS:

$$r = (d_{14} + d_{25} + d_{36}) \cos\theta \sin^2\theta \sin\varphi \cos\varphi \qquad (7.11)$$

- tetragonal class $\bar{4}$ also has a rather simple str(d), very similar to those of 222, $\bar{4}2m$, 23 and $\bar{4}3m$ plus two additional terms of opposite signs. Judged by its shape, its RS:

$$r = (2d_{14} + d_{36})n_1 n_2 n_3 + d_{31}(n_3 n_1 n_1 - n_3 n_2 n_2) \qquad (7.12)$$

also belongs to the cubic point group $\bar{4}3m$.

There is a small difference with respect to the RS of Fig. 7.10: it is not completely aligned with the sides of the cube, but rotated about the ③ axis (bottom left panel of Fig. 7.10).

---

[8]Try and identify the remaining geometric elements (axes, planes, etc.) of symmetry that leave (7.10) invariant. How many *elements* of the symmetry group (individual geometric transformations, each represented by its L) do these geometric elements of symmetry generate?

The reason has already been discussed previously for materials of monoclinic class 2 and trigonal class 3. Unlike $\bar{4}3m$ materials, whose three fourfold rotation-inversion axes are aligned with the conventional ①, ② and ③ axes, the $\bar{4}$ material only has a $\bar{4}$ axis, which is inherited by the RS and therefore correctly aligned with the ③ axis.

There is nothing in the symmetry of the material to define where the other two conventional axes should be located. This freedom is also inherited by the RS, hence the rotation of the ① and ② axes of the RS with respect to those of the material. Otherwise, the RS is qualitatively the same as in Fig. 7.10.

The term $d_{31}(n_3n_1n_1 - n_3n_2n_2)$ in the RS (7.12) does not alter the shape of the RS with respect to (7.10), but this term is responsible for the rotation of the RS we just discussed (Fig. 7.13).

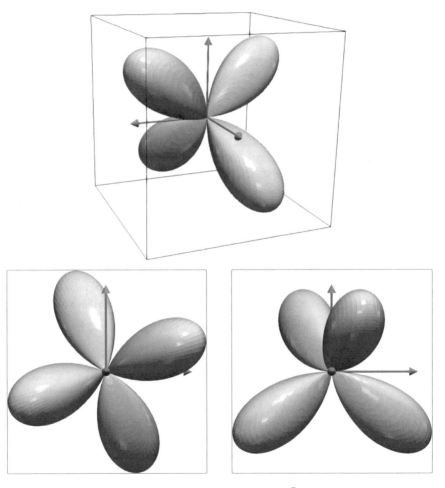

Fig. 7.13 RS for the piezoelectric modulus $\underline{\underline{d}}$ for tetragonal class $\bar{4}$. The bottom left panel is a view along the conventional ③ axis of the tetragonal system

7.3   The Basic RSs for Third Rank Properties

We saw in Sects. 5.4 and 6.1 that only one basic object (Fig. 5.13) and three basic objects (Fig. 6.7) (actually, three orientational isomers of a single object) are required to construct the RS of a rank one and of a symmetric second rank tensor, respectively.

A similar question can be posed about the RS of third rank tensors with the index symmetry of the piezoelectric modulus or the Pöckels effect coefficient.[9]

We have seen in the previous section how the full index symmetry of the projector $\vdots\, \underline{n}\,\underline{n}\,\underline{n}$ introduces further simplifications in $\mathrm{str}(\underset{\smile}{d})$ by gathering together some of its components. In the expression:

$$r = \underline{\underline{d}} \vdots \underline{n}\,\underline{n}\,\underline{n} = d_{ijk}n_k n_j n_i$$

all components of $\underline{\underline{d}}$ obtained by permutation of the indices i, j, k of d_{ijk} will con-

tribute equally to the factor containing $n_k n_j n_i$. E.g. both $d_{112} = d_{121} = \dfrac{1}{2}d_{16}$ and $d_{211} = d_{21}$ will all appear as coefficients of the term containing $n_1 n_1 n_2$. Thus, d_{16} and d_{21} will not contribute separate basic objects to the RS. The number of basic objects required to construct the RS is much lower than the apparent maximum of 18 independent components of $\underset{\smile}{d}$ or $\underset{\smile}{r}$ for triclinic point group 1.

Table 7.1 shows the elements of $\underset{\smile}{d}$ that are gathered in a single coefficient when evaluating the RS in the general, least symmetric case.

Figure 7.14 presents this information in graphical form. Each symbol represents one of the three types of $n_i n_j n_k$ terms from Table 7.1.

In Table 7.1, each line in the list contains the indices of the coefficients that multiply the common $n_i n_j n_k$ factor listed on the right. The full index permutation symmetry of $\underline{n}\,\underline{n}\,\underline{n}$ reduces the 18 components of $\underset{\smile}{d}$ to only ten, naturally arranged in three groups of three, six and one congruent objects. Within each of the first two groups the isomers only differ in their orientation in space, thus the name orientational isomers.

- the three isomers with three equal indices are shown in Fig. 7.15. The three panels of this figure show the terms $n_i n_i n_i$, $i = 1, 2, 3$ (no sum). They are surfaces of revolution about the ①, ②, ③ axes, of cross section $\cos^3 \alpha$, with α the angle between the radius vector and the axis. The shape belongs to limit point group ∞/m.
- the six isomers with two equal indices are shown in Fig. 7.16. They are again congruent shapes belonging to orthorhombic point group $mm2$.
- the only isomer with all indices different is the already familiar four-lobed surface shown in Fig. 7.17. It belongs to cubic point group $\bar{4}3m$.

The RS for the piezoelectric modulus for any material can be written as a linear combination of the ten geometrical objects of Figs. 7.15, 7.16 and 7.17. They are shown in Fig. 7.18 at the same scale for comparison. Their sizes are bounded by (7.13).

---

[9]For rigorous treatments of tensor decomposition see [2–5].

Tabla 7.1 Elements of a generic third rank tensor $\underset{\equiv}{T}$ and $\underset{\sim}{T}$ (left) that are gathered in a single coefficient of $n_i n_j n_k$ (middle) in the RS. The type of basic geometric object that corresponds to each group in shown on the right

All indices equal	$\begin{cases} 111(11) \\ 222(22) \\ 333(33) \end{cases}$	$\begin{matrix} n_1 n_1 n_1 \\ n_2 n_2 n_2 \\ n_3 n_3 n_3 \end{matrix}$	
Two indices equal	$\begin{cases} 122(12),\ 212,\ 221(26) \\ 233(23),\ 323,\ 332(34) \\ 311(31),\ 131,\ 113(15) \\ 211(21),\ 112,\ 121(16) \\ 322(32),\ 223,\ 232(24) \\ 133(13),\ 313,\ 331(35) \end{cases}$	$\begin{matrix} n_1 n_2 n_2 \\ n_2 n_3 n_3 \\ n_3 n_1 n_1 \\ n_2 n_1 n_1 \\ n_3 n_2 n_2 \\ n_1 n_3 n_3 \end{matrix}$	
No indices equal	$123(14),\ 231(25),\ 312(36)$	$n_1 n_2 n_3$	

$$\begin{bmatrix} \bullet & \triangle & \triangle & \blacksquare & \square & \square \\ \triangle & \bullet & \triangle & \square & \blacksquare & \square \\ \triangle & \triangle & \bullet & \square & \square & \blacksquare \end{bmatrix}$$

Fig. 7.14 The three types of $n_i n_j n_k$ terms from Table 7.1. $\bullet\ n_i n_i n_i$ all indices equal (no sum), $\triangle$ $n_i n_i n_j$ two indices equal (no sum), $\blacksquare\ n_i n_j n_k$, all indices different

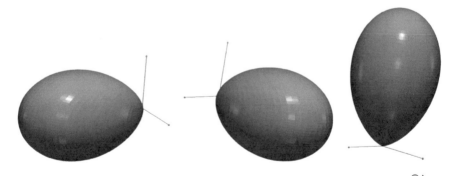

Fig. 7.15 RS for the three isomers of the form $n_i n_i n_i$, all indices equal (no sum). Axes ① ② ③

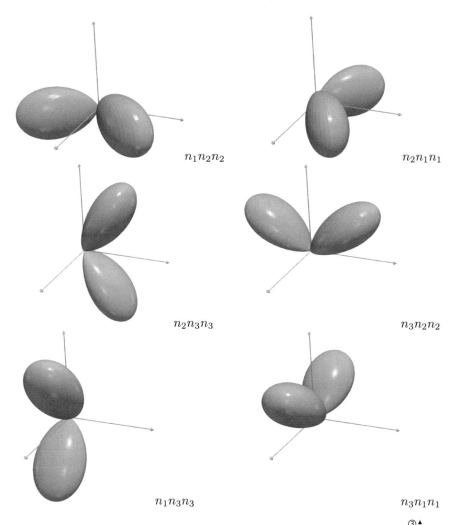

Fig. 7.16 RS for the six isomers of the form $n_i n_j n_j$, two indices equal (no sum). Axes ①②③

A few combinations of these basic objects deserve a short discussion:

- the qualitative aspect of the RS for combinations of terms of the type $n_i n_i n_i$ (all indices equal, no sum) is not difficult to predict.
  The RS of Fig. 7.19 is the linear combination $a n_1 n_1 n_1 + b n_2 n_2 n_2 + c n_3 n_3 n_3$ with $a : b : c = 1 : 2 : 3$. The combination of three unequal lobes belongs to point group 1 and offers a simple proof that the RS for third properties can be max-imally asymmetric. The contributions $n_i n_i n_i$ (no sum), $n_i n_i n_j$ (no sum), $n_i n_j n_k$ are bounded by:

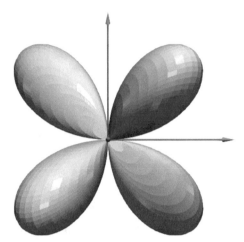

Fig. 7.17 RS for the only isomer of the form $n_1 n_2 n_3$, all indices different. The projections along the three axes are identical

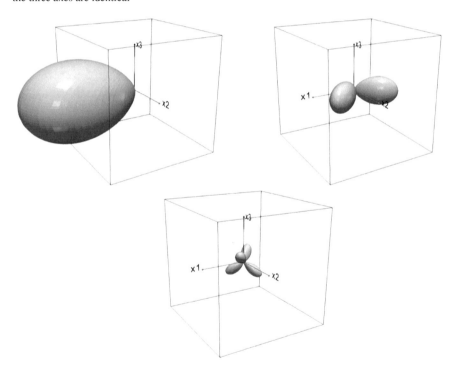

Fig. 7.18 Size comparison of the three types of geometric contributions (isomers corresponding to $n_1 n_1 n_1$, $n_2 n_1 n_1$ and $n_1 n_2 n_3$, left to right, top to bottom) to the RS of the longitudinal piezolectric modulus $\underline{\underline{d}}$ represented at the same scale. The edge of the enclosing boxes has unit length

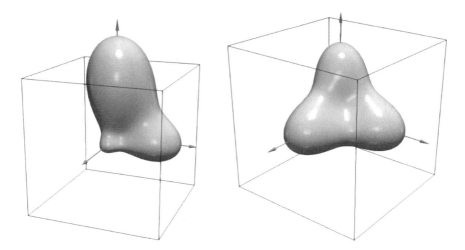

Fig. 7.19 Linear combination $an_1n_1n_1 + bn_2n_2n_2 + cn_3n_3n_3$ with $a : b : c = 1 : 2 : 3$ (left), and with $a : b : c = 1 : 1 : 1$ (right). Axes ① ③ ②

$$\max_{\varphi,\theta}(n_i n_i n_i) = 1 \quad \max_{\varphi,\theta}(n_i n_i n_j) = \frac{2\sqrt{3}}{9} \quad \max_{\varphi,\theta}(n_i n_j n_k) = \frac{\sqrt{3}}{9} \qquad (7.13)$$

and let $d_{ijk} \neq 0 \ \forall i, j, k$ be the components of $\underset{\sim}{d}$ for point group 1. Choose $d_{111} > \sum_i \sum_j \sum_k {}' d_{ijk}$, where the prime indicates that the term d_{111} is excluded from the sum. Furthermore, let $d_{ijk} \neq d_{mnp}$ if $ijk \neq mnp$. Since all bounds above are ≤ 1 and all components are unequal, the symmetry of the RS shape is then determined by the $d_{111}n_{111}$ term. The RS then belongs to point group 1.

The index permutation symmetrical combination $n_i n_i n_i$:

$$r = n_1 n_1 n_1 + n_2 n_2 n_2 + n_3 n_3 n_3 \qquad (7.14)$$

belongs to trigonal point group $3m$ (Fig. 7.19, right). The similarity between the RS of (7.14) (Fig. 7.19, right) and that for point group $m3$ (Fig. 7.10) is remarkable. This is no coincidence: let's transform (7.14) to a set of axes, the third of which points along the $[\bar{1}\,\bar{1}\,\bar{1}]$ cubic crystallographic direction (the negative body diagonal of the cube):

$$
\underset{\sim}{L} = \begin{bmatrix} -\dfrac{\sqrt{2}}{2} & \dfrac{\sqrt{2}}{2} & 0 \\[2mm] -\dfrac{\sqrt{6}}{6} & -\dfrac{\sqrt{6}}{6} & \dfrac{\sqrt{6}}{3} \\[2mm] -\dfrac{\sqrt{3}}{3} & \dfrac{\sqrt{3}}{3} & \dfrac{\sqrt{3}}{3} \end{bmatrix} \tag{7.15}
$$

The transformation of (7.14) to this new reference changes $\mathrm{str}(\underset{\sim}{d})$ into:

which is identical to $\mathrm{str}(\underset{\sim}{d})$ for point group $3m$ except for the equality of d'_{33} to d'_{31} and d'_{32}, whereas $d'_{33} \neq d'_{31} = d'_{32}$ for class $3m$. The reason is that class $3m$ has an extra degree of freedom of the form $n_3n_3n_3$ when written in the conventional axes for class $3m$.

The apparent complication of $\mathrm{str}(\underset{\sim}{d})$ for point group $3m$ is a consequence of being referred to the conventional trigonal axes, and not to the frame in which the basic objects have a simple form.

In the conventional trigonal frame, the RS of Fig. 7.10 is built up of a relatively complicated combination of several basic shapes with three and two equal indices (Eq. 7.8) in the precise amounts to produce threefold symmetry.

When referred to the natural axes of the $n_in_in_i$ isomers, the RS is much simplified and threefold symmetry is attained by construction: the RS consists of equal amounts of the three $n_in_in_i$ isomers plus two additional axisymmetric contributions (of the type $\cos^3 \alpha$ and $\cos \alpha \sin^2 \alpha$) along the cube body diagonal (Fig. 7.20). The RS itself is the same in both cases because its definition is an invariant expression. When the frame of reference changes, only the values of the components of $\underset{\sim}{d}$ and which associated geometrical building blocks are used to construct the RS change, not the RS.

• the six isomers of the form $n_in_in_j$ (two indices equal, no sum, Fig. 7.16) offer more possibilities for combination.

The addition of equal amounts of two of these V-shaped RSs objects leads to three different possibilities, depending on their relative orientations.

The sums $n_1n_2n_2 + n_2n_1n_1$ and $n_1n_2n_2 + n_2n_3n_3$ (Fig. 7.21), differ in whether only two or three different indices appear in the sum.

In the first case, RS of the left panel of Fig. 7.21, the two V's are coplanar. This RS looks remarkably similar to one of those shown in the first part of this section, except for a $\frac{\pi}{4}$ rotation. It is left as an exercise to the reader to transform the RS of the left panel of Fig. 7.21 into the most similar of the RSs shown in Figs. 7.4, 7.5, 7.6, 7.7, 7.8, 7.9, 7.10, 7.11 and 7.12 (or viceversa) through a suitable rotation of the frame of reference, as it was done for class $3m$ above.

The sum $n_1n_2n_2 + n_1n_3n_3$ (Fig. 7.22) displays axial symmetry as a consequence of the transversal isotropy of the term in parenthesis when n_1 is factored out:

$$r = n_1n_2n_2 + n_1n_3n_3 = n_1(n_2n_2 + n_3n_3)$$

It is not difficult to recognize this RS as the "foot" of the RS in Fig. 7.11, or as the term $n_3(n_1n_1 + n_2n_2)$ in (7.9). Furthermore, the second term in (7.9) $d_{33}n_3n_3n_3$ is one of the three simple RSs of $n_in_in_i$ isomers. The RS of Fig. 7.11 is thus a visually intuitive linear combination of RSs like those of Figs. 7.15 and 7.22 with weights d_{33} and $d_{31} + d_{24}$ respectively.

Differences of the $n_in_in_j$ isomers do not lead to any new RSs for combinations like $n_in_jn_j - n_jn_in_i$ (no sum) or $n_in_jn_j - n_jn_kn_k$ (no sum). However, the RSs of differences of the form $n_in_jn_j - n_in_kn_k$ like:

$$n_1n_2n_2 - n_1n_3n_3 = n_1(n_2n_2 - n_3n_3)$$

in (Fig. 7.23) bring us back to an old acquaintance we will be meeting again soon. Notice the similarity between the RSs for $n_1n_2n_2 + n_1n_3n_3$ (Fig. 7.22) and $n_1n_2n_2 - n_1n_3n_3$ (Fig. 7.23), and those for the second rank sum $\underline{u}\,\underline{u} + \underline{v}\,\underline{v}$ and difference $\underline{u}\,\underline{u} - \underline{v}\,\underline{v}$ of Fig. 6.4. The extra factor n_3 for the third rank products adds "three-dimensionality" to the RSs of the second rank products.

Clearly, combinations with arbitrary coefficients have minimal, point group 1 symmetry (Fig. 7.24).

• the final case $n_1n_2n_3$ is the only isomer in which all indices are different. It is shown in Fig. 7.25. It is not just similar to the RS of the difference $n_1n_2n_2 - n_1n_3n_3$, it is, except for a scale factor of $\dfrac{1}{2}$, congruent with it. In a set ①②③ of axes the term $n_1n_2n_3$ is the same geometric object as $\frac{1}{2}(n_1'n_1'n_3' - n_2'n_2'n_3')$ in a rotated set ①'②'③' of axes.

$$n_1n_2n_3 \quad \Leftrightarrow \quad \frac{1}{2}(n_1'n_1'n_3' - n_2'n_2'n_3') \text{ or}$$

$$n_1n_2n_3 \quad \Leftrightarrow \quad \frac{1}{2}(n_1'n_1' - n_2'n_2')n_3' \tag{7.16}$$

Notice how the part $n_1'n_2'$ is transformed as when we went from (6.10) to (6.12) in Sect. 6.6.

The coefficients of d_{ij} are grouped due both to relations between components in $\mathrm{str}(\underset{\sim}{d})$ and to the index symmetry of the projector. This grouping reduces the number of effectively independent parameters.

Since the absolute size of the RS is not relevant for its shape, one of the d_{ij} can be taken as the unit, which reduces by a further one the number of parameters

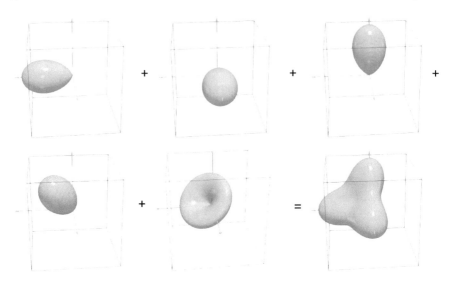

Fig. 7.20 The RS of trigonal class $m3$ (bottom left) can also be built up from equal amounts of the three $n_i n_i n_i$ isomers (top three images) plus two axisymmetric contributions $\cos^3 \alpha$ and $\cos \alpha \sin^2 \alpha$ along the cube diagonal (bottom left and center). In this case, the threefold axes points along the body diagonal of the cube. ① ③ ②

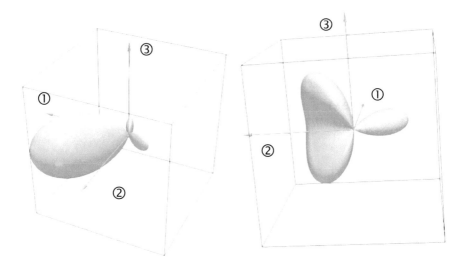

Fig. 7.21 The RS for the sums $n_1 n_2 n_2 + n_2 n_1 n_1$ (left) and $n_1 n_2 n_2 + n_2 n_3 n_3$ (right)

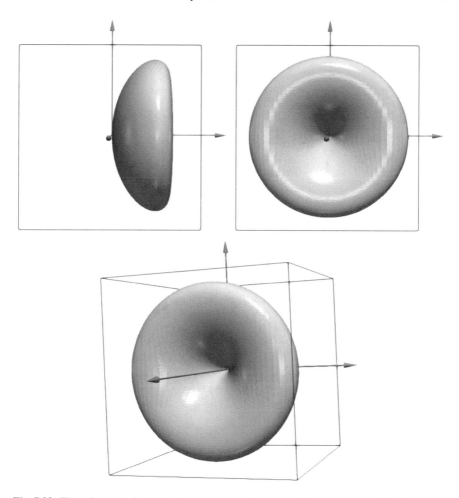

Fig. 7.22 The axisymmetric RS for the sum $n_1n_2n_2 + n_1n_3n_3$. The symmetry axis is ①

that actually determines the RS shape, irrespective of its orientation in space and its absolute size.

Table 7.2 lists the number of effective degrees of freedom, together with the reasons for the reduction.

The number of effective degrees of freedom in RS shape is considerably less than $\mathrm{str}(d)$ suggests. In several cases the shape of the RS contains, apart from a scale factor, no free parameters. The complication of $\mathrm{str}(d)$ is in many cases only apparent.

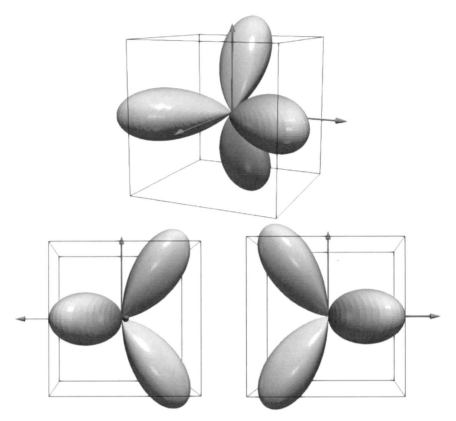

Fig. 7.23 The RS for the difference $n_1n_2n_2 - n_1n_3n_3$ belongs to point group $\bar{4}3m$. The bottom left view is along the ③ axis, the bottom right view along the ①. The enclosing box has the same size as in Fig. 7.25

7.4 Independence and Minimum Set of Basic Objects

In view of such equivalences, it is natural to ask if all RSs of the previous catalog of shapes for the $n_in_jn_k$ terms are necessary to construct the most general RS for third rank properties like $\underline{\underline{d}}$ or $\underline{\underline{r}}$.

Let's briefly recall:

- for rank one properties $\underline{T}$, the RS is an oriented sphere for all materials (all point groups). The RS of the different classes differ in its orientation with respect to the conventional axes. Two[10] of the three degrees of freedom in str($\underline{T}$) for the most general (triclinic) material define the orientation of the RS but have no influence on its size (its shape is fixed anyway); the remaining degree of freedom is the material specific property, which determines the size (diameter) of the RS.

---

[10]Because only one axis has to be oriented.

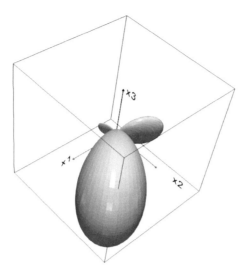

Fig. 7.24 RS of the sum of terms $an_1n_2n_2 + bn_2n_1n_1$ with $a : b = 2 : 3$. The RS belongs to triclinic point group 1

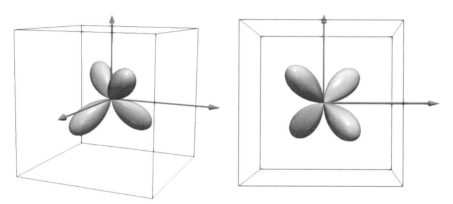

Fig. 7.25 The RS for $n_1n_2n_3$. The right view is identical along all three ①, ②, ③ axes. The enclosing box has the same size as in Fig. 7.23

A single basic geometric object of the form n_i is required to define the property. This is especially clear for those classes in which the orientation of the RS coincides with the ③ axis: the str($\underline{T}$) of the property contains a single non-zero component. The simplest str($\underline{T}$) then is:

Tabla 7.2 Number of effective degrees of freedom that determine the shape of the RS. Only some (third column) of the original number of parameters (second column) from $str(\underline{d})$ may appear in the functional form of the RS; the number is further reduced because absolute size does not influence shape (fourth column), because components appear grouped (fifth column), and by elimination of arbitrary rotations (sixth column). The net number of effective degrees of freedom is listed in the last column. Minus signs indicate loss of degrees of freedom

System/class	From $str(\underline{d})$	In RS appear	Scale factor	Grouping of d_{ij}	Arbitrary rotation	Eff. d.o.f.
1	18	18	−1	−8	−3	6
m	10	10	−1	−4	−1	4
2	8	8	−1	−4	−1	2
$mm2$	5	5	−1	−2	0	2
3	6	5	−1	−1	−1	2
$3m$	4	4	−1	−1	0	2
$32, \bar{6}$	2	1	−1	0	0	0
$\bar{6}m2$	1	1	−1	0	0	0
$6, 4, \infty$	4	3	−1	−1	0	1
$6mm, 4mm, \infty m$	3	3	−1	−1	0	1
$23, \bar{4}3m$	1	1	−1	0	0	0
222	3	3	−1	−2	0	0
$\bar{4}2m$	2	2	−1	−1	0	0
$\bar{4}$	4	3	−1	−1	−1	0

1 independent component

or one the other two equivalent alternatives, i.e. $p_1 = p_3 = 0$; $p_2 \neq 0$, or $p_2 = p_3 = 0$; $p_1 \neq 0$.

- for symmetric second rank properties $\underline{\underline{T}}$ the most general RS is a linear combination of only three basic objects of the form $n_i n_i$, in spite of the fact that the $\mathrm{str}(\underline{\underline{T}})$ for triclinic classes contains six independent coefficients.

Cross-terms of the form $n_i n_j$, $i \neq j$ can be made to vanish by a special choice of reference frame, namely, that of the eigenvectors. The three geometric objects corresponding to $n_1 n_2$, $n_2 n_3$ and $n_3 n_1$ are not necessary to construct the full RS for the most general material because terms of the form $n_i n_j$ can be written as linear combinations of the $n_i n_i$ terms.

The cross-shaped RS of the right panel of Fig. 6.15 that represents (6.12) is thus not essential, it can be built from the RSs of $n_i n_i$, $i = 1, 2, 3$. The linear combination that achieves this is just the usual transformation rule of second rank tensors. The key is finding the angles of the transformation, which is what an eigenvalue-eigenvector analysis does.

The simplest $\mathrm{str}(\underline{\underline{T}})$ then is:

3 independent components

The triclinic material contains three degrees of freedom which define the orientation of the RS in space, but have no influence on its shape; the remaining three define the shape and the symmetry of the RS.

Through a suitable change of reference it is possible to cancel three elements in the 3×3 matrix $\mathrm{str}(\underline{\underline{T}})$ of the property.

But not just *any* three elements. Would it be possible to cancel the three diagonal elements of

$$\mathrm{str}(\underline{\underline{T}}) = \begin{bmatrix} \bullet & \bullet & \bullet \\ & \bullet & \bullet \\ & & \bullet \end{bmatrix}$$

through a change in reference in order to obtain a matrix with empty diagonal:

$$\mathrm{str}(\underset{\smile}{T'}) = \begin{bmatrix} \bullet & \bullet \\ & \bullet \\ & \end{bmatrix} ?$$

The answer depends on whether the matrix $\mathrm{str}(T)$ is positive definite or not. If it is positive definite, it will have three positive eigenvalues, so $\mathrm{tr}(\underset{\smile}{T}) > 0$. Since an empty diagonal implies $\mathrm{tr}(\underset{\smile}{T}) = 0$ and the trace is an invariant under orthogonal transformations, no change of reference frame will void the diagonal.

Another hallmark of positive definiteness is that the determinants of all upper-left submatrices (i.e. principal submatrices, of sizes 1×1 through $d \times d$) must be positive. Any attempt to cancel elements of $\mathrm{str}(T)$ that renders at least one of these determinants non-positive will necessarily fail.

The more general answer is that any attempt to cancel some of its elements will fail if this cancellation does not preserve its three invariants (3.10). In the example above, cancelling all three diagonal elements would violate the invariance of $\mathrm{str}(\underset{\smile}{T})$.

The situation is similar for properties of higher rank. We can ask whether a change of reference frame will make the full $\mathrm{str}(\underset{\smile}{d})$ look "less full", just as a diagonal $\mathrm{str}(\underset{\smile}{T})$ is emptier that a full $\mathrm{str}(\underset{\smile}{T})$ for second rank properties, or a $\underline{T}$ with $T_1 = T_2 = 0$; $T_3 \neq 0$ for rank one properties.

The full $\mathrm{str}(\underset{\smile}{d})$ of a material can be simplified by eliminating the three degrees of freedom that define the orientation of the material (and of the RS). This helps us clarify the question of the basic geometric objects (Table 7.1) needed to construct the RS.

Let's take a concrete example and consider $\mathrm{str}(\underset{\smile}{d})$ for point group $\bar{4}3m$:

$$\begin{bmatrix} \cdot & \cdot & \cdot & \bullet & \cdot & \cdot \\ \cdot & \cdot & \cdot & \cdot & \bullet & \cdot \\ \cdot & \cdot & \cdot & \cdot & \cdot & \bullet \end{bmatrix}$$

$$\tag{7.17}$$

For this $\underset{\smile}{d}$ in the conventional axes, the RS of the longitudinal piezoelectric modulus is:

$$r = \underset{\equiv}{d} : \underline{n}\,\underline{n}\,\underline{n} = 3d_{14}n_1n_2n_3$$

which is the RS of Fig. 7.23 scaled up by a factor of three, because there are three identical non-zero elements in $\underset{\smile}{d}$.

Is it possible to make the elements d_{14}, d_{25} and d_{36} vanish through a change of reference frame? As a matter of fact a simple rotation of $\frac{\pi}{4}$ about the ③ axis will do the trick.[11] Using:

$$\varphi = \frac{\pi}{4} \quad \theta = 0 \quad \psi = 0 \quad \Rightarrow \quad \underset{\smile}{L} = \begin{bmatrix} \frac{\sqrt{2}}{2} & \frac{\sqrt{2}}{2} & 0 \\ -\frac{\sqrt{2}}{2} & \frac{\sqrt{2}}{2} & 0 \\ 0 & 0 & 1 \end{bmatrix} \tag{7.18}$$

we transform each element of $\underset{\smile}{d}$ to the rotated system ①'②'③' and obtain (assuming $d_{14} = d_{25} = d_{36} = 1$ without loss of generality; d_{14} is just a scale factor):

$$\underset{\smile}{d'} = \begin{bmatrix} 0 & 0 & 0 & 0 & 1 & 0 \\ 0 & 0 & 0 & -1 & 0 & 0 \\ \frac{1}{2} & -\frac{1}{2} & 0 & 0 & 0 & 0 \end{bmatrix} \tag{7.19}$$

in which d'_{14}, d'_{25} and d'_{36} have indeed vanished.

In the rotated frame (using 7.19), the RS is:

$$r' = \underset{=}{d'} : \underline{n}' \, \underline{n}' \, \underline{n}' = d'_{31} n'_1 n'_1 n'_3 + d'_{32} n'_2 n'_2 n'_3 + d'_{15} n'_3 n'_1 n'_1 + d'_{24} n'_3 n'_2 n'_2 =$$

$$= \frac{3}{2} (n'_1 n'_1 n'_3 - n'_2 n'_2 n'_3)$$

which is exactly the same geometric object, but referred to the rotated set of axes ①'②'③'.

Since both $\underset{\smile}{d}$ and $\underset{\smile}{d'}$ are the same RS, except for the $\frac{\pi}{4}$ rotation, the RS can be built either from objects with three equal indices, $n_1 n_2 n_3$, or alternatively from a linear combination of objects with two equal indices, $n'_1 n'_1 n'_3$ and $n'_2 n'_2 n'_3$ in this example.[12]

The set of three types of geometric objects of Table 7.1 seems to contain some redundancy. In the example above we did away with the $n_1 n_2 n_3$ object in a special $\mathrm{str}(\underset{\smile}{d})$ which was empty but for the identical d_{14}, d_{25} and d_{36} elements. Is it possible to make d_{14} d_{25} and d_{36} vanish for the most general (full) $\underset{\smile}{d}$?, i.e. is there an orthogonal transformation that accomplishes this change in $\underset{\smile}{d}$:

---

[11] This is not the only possibility. Can you find another one?.

[12] Can you find another two similar ways of constructing $n_1 n_2 n_3$?

$$
\begin{bmatrix} \bullet & \bullet & \bullet & \bullet & \bullet \\ \bullet & \bullet & \bullet & \bullet & \bullet \\ \bullet & \bullet & \bullet & \bullet & \bullet \end{bmatrix} \Rightarrow \begin{bmatrix} \bullet & \bullet & \cdot & \bullet & \bullet \\ \bullet & \bullet & \bullet & \cdot & \bullet \\ \bullet & \bullet & \bullet & \bullet & \cdot \end{bmatrix}
$$

from 18 to 15 independent components

$$(7.20)$$

The answer is affirmative: the number of components in the most general (triclinic) d can be brought to the form above right with $d_{14} = d_{25} = d_{36} = 0$ by an orthogonal transformation.

Let's try to find this transformation. A direct and rather brute force way to find this transformation is to solve the following three equations in the three Euler angles:

$$
\begin{cases} 0 = l_{i1}l_{2j}l_{3k}d_{ijk} \\ 0 = l_{i2}l_{3j}l_{1k}d_{ijk} \\ 0 = l_{i3}l_{1j}l_{2k}d_{ijk} \end{cases}
$$

where the unknowns φ, θ and ψ appear in the components l_{ij} of Euler's matrix. Each of the three equations above consists of 27 terms, of up to third degree in sine and cosine of φ, θ and ψ. Solution is feasible but requires a numerical approach.

We can get some insight for a more direct method from lower rank tensors:

- in rank one, the reduction of the number of independent components from three to one implies eliminating transversal components, and leaving the tensor aligned with one of the axes.
- the off-diagonal components of a second rank tensor, like a σ_{12} conductivity, also represent a transversal phenomenon: σ_{12} relates field E_2 and current density J_1 at right angles. Removing components that cause mixed coupling leads to a diagonal $\underset{=}{\sigma}$ with $6 - 3 = 3$ components.
- components such as d_{ijk} give the i-th component of the polarization $\underline{P}$ caused by a stress σ_{kj}. Although the elements of d we aim to remove, d_{14}, d_{25} and d_{36}, lie on the diagonal of the right 3×3 sub-block of d, they actually describe a coupling between a shear stress e.g. τ_{12} and components of $\underline{P}$, like P_3 that are transversal to the stress plane (Fig. 7.26).[13]

A transformation that casts shear stresses as a combination of normal stresses, as in (7.16) is all we need. The only difference is that now we have to make *three* shear stresses "look like" normal stresses simultaneously, and not just one. If we define the symmetric traceless matrix $\underset{\smile}{M}$ as:

---

[13] You can think of $\underline{P}$ as longitudinal, of $\underset{\smile}{d}$ as the upper half of a matrix like $\underset{\smile}{s}$ and then recall the block coupling (4.35).

Fig. 7.26 A term like d_{321} couples a component of the polarization (P_3) that is transversal to the plane where the cause (shear stress τ_{12}) acts

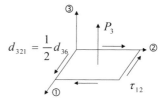

$$M \equiv \begin{bmatrix} 0 & d_{14} & d_{25} \\ d_{14} & 0 & d_{36} \\ d_{25} & d_{25} & 0 \end{bmatrix}.$$

(or any cyclic permutation of the d_{ij}) its eigenvectors $\vec{v}^{*①}, \vec{v}^{*②}, \vec{v}^{*③}$ form a reference frame in which the three shear stresses appear as normal stresses.

The desired transformation matrix L is:

$$L = \begin{bmatrix} v_1^{*①} & v_2^{*①} & v_3^{*①} \\ v_1^{*②} & v_2^{*②} & v_3^{*②} \\ v_1^{*③} & v_2^{*③} & v_3^{*③} \end{bmatrix} \text{ where } \vec{v}^{*①}, \vec{v}^{*②}, \vec{v}^{*③} \text{ are the eigenvectors of } M \quad (7.21)$$

Matrix L performs the elimination of three components of d (7.20).[14]

In a sense, the reference frame (7.21) is particularly simple, not just because it empties three elements of d, but because it has a clear physical meaning as well: for direct piezoelectricity, $d_{14} = d_{25} = d_{36} = 0$ implies that shear stresses do not produce polarizations P_3, P_1, P_2 perpendicular to the plane in which these stresses act.

In the frame $① \equiv \vec{v}^{*①}$, $② \equiv \vec{v}^{*②}$, $③ \equiv \vec{v}^{*③}$ only 15 independent coefficients are enough to describe the most general material. As in the case of first and second rank properties, the transformation (7.20) can be interpreted as eliminating the three degrees of freedom that describe the orientation in space of the RS without changing its invariant shape.

These $15 = 18 - 3$ components are the equivalent of the $3 = 6 - 3$ components of a symmetric second rank tensor expressed in the frame of its eigenvectors, or of the $1 = 3 - 2$ component of a rank one tensor in a frame in which one axis points along the tensor's direction.

Although we have managed to reduce d or r to a simpler form, we still have to address the more general question of which elements of r or d can be cancelled

---

[14]Check that for the example (7.17) the eigenvectors of: $M = \begin{bmatrix} 0 & 1 & 1 \\ 1 & 0 & 1 \\ 1 & 1 & 0 \end{bmatrix}$ form a basis in which $d_{14} = d_{25} = d_{36} = 0$. Is there anything special about this reference frame? What are the eigenvalues of M?

by a suitable transformation of the reference frame. The answer is the same as for second rank properties: at most three elements of $\underset{\smile}{d}$ can be cancelled by an orthogonal transformation, so long as the cancellation does not alter the invariants.

Unlike for second rank tensors, the determination of a complete set of independent linear and nonlinear invariants for third and fourth rank tensors of different symmetries is a work still in progress [6–8].

The only linear invariant for a third rank tensor with the symmetry of the piezoelectric modulus or the Pöckels coefficients is:

$$\underset{\equiv}{T} : \underset{\equiv}{\epsilon} = T_{ijk}\epsilon_{kji}$$

By pairing the terms of this expression and using $T_{ijk} = T_{ikj}$, for every term of the form $T_{ijk}\epsilon_{kji}$ there is another term which cancels it:

$$T_{ijk}\epsilon_{kji} + T_{ikj}\epsilon_{jki} = T_{ijk}\epsilon_{kji} + T_{ijk}\epsilon_{jki} =$$
$$= T_{ijk}(\epsilon_{kji} + \epsilon_{jki}) = 0$$

due to the antisymmetry of $\underset{\equiv}{\epsilon}$ in all pairs of indices. Hence there are no linear invariants for $\underset{\equiv}{d}$.

Invariants of the next higher order do exist. A third rank tensor $\underset{\equiv}{T}$ symmetric in the last two indices has five independent quadratic invariants:

$$I_1 = T_{ijj}T_{ikk} \qquad I_2 = T_{iik}T_{jjk} \qquad I_3 = T_{ijk}T_{ijk}$$
$$I_4 = T_{ijk}T_{jik} \qquad I_5 = T_{iik}T_{kjj}$$

Let's check the invariance of I_1 for $\underset{\equiv}{d}$ for point group $\bar{4}3m$ and the transformation (7.18) that takes the standard str($\underset{\smile}{d}$) to the form (7.19). In the conventional axes (before the transformation):

$$I_1 = d_{ijj}d_{ikk} = 0$$

because all terms in the left half of $\underset{\smile}{d}$ are zero. In the rotated axes:

$$I_1 = d'_{ijj}d'_{ikk} = (d'_{311} + d'_{322})(d'_{311} + d'_{322}) = \left(\frac{1}{2} - \frac{1}{2}\right)^2 = 0$$

The constancy of the remaining invariants $I_2 = 0$, $I_3 = \frac{3}{2}$, $I_4 = \frac{3}{2}$, $I_5 = 0$, can be checked in the same way.

It has recently been conjectured [9] that the total number of independent invariants is seven, i.e. another two cubic or higher degree independent invariants may exist.

The transformation (7.18) respects them as well, or else it would not have been feasible to cancel d_{14}, d_{25} and d_{36} as we did above.

To sum up, the variety of RS shapes of Figs. 7.4, 7.5, 7.6, 7.6, 7.7, 7.8, 7.9, 7.10, and 7.11 are the result of linear combinations of only two basic types of geometrical objects (Figs. 7.15 and 7.16), with coefficients given by $\mathrm{str}(\underset{\sim}{d})$. It is the particular combination of basic geometrical objects and coefficients that determine the shape and thus the symmetry of the property.

References

1. Born, M., Wolf, E.: Principles of Optics: Electromagnetic Theory of Propagation, Interference and Diffraction of Light. Elsevier, Amsterdam (2013)
2. Backus, G.: A geometrical picture of anisotropic elastic tensors. Rev. Geophys. **8**(3), 633–671 (1970)
3. Hamermesh, M.: Group Theory and its Application to Physical Problems. Dover Publications Inc., New York (2003)
4. Jerphagnon, J.: Invariants of the third-rank cartesian tensor: optical nonlinear susceptibilities. Phys. Rev. B **2**(4), 1091 (1970)
5. Dinckal, C.: Orthonormal decomposition of third rank tensors and applications. In: Lecture Notes in Engineering and Computer Science: Proceedings of the World Congress on Engineering 2013, 3–5 July, 2013, London, UK, 139, vol. 144 (2013)
6. Norris, A.N.: Quadratic invariants of elastic moduli. Q. J. Mech. Appl. Math. **60**(3), 367–389 (2007)
7. Ahmad, F.: Invariants of a cartesian tensor of rank 3. Arch. Mech. **63**(4), 383–392 (2011)
8. Desmorat, R., Auffray, N., Olive, M.: Generic separating sets for 3D elasticity tensors (2018). arXiv:1812.11380
9. Qi, L.: Eigenvalues and invariants of tensors. J. Math. Anal. Appl. **325**(2), 1363–1377 (2007)

Chapter 8
Fourth Rank Properties

The extension to fourth rank tensors of the ideas presented in the previous chapter is straightforward. Properties like magnetoresistivity ρ^{mag}, the Kerr effect coefficient $\underline{\underline{K}}$, the piezo-optic coefficient $\underline{\underline{\pi}}^{opt}$, the electrostriction coefficient $\underline{\underline{M}}$, and the elastic compliance $\underline{\underline{s}}$ and stiffness $\underline{\underline{c}}$ are fourth rank properties whose RSs are defined in the same way.

The Kerr coefficient, the electrostriction coefficient and magnetoresistivity appear in constitutive equations as phenomenological coefficients:

$$\Delta \underline{\underline{B}} = \underline{\underline{K}} : \underline{E}\,\underline{E} \qquad \underline{\underline{\epsilon}} = \underline{\underline{M}} : \underline{E}\,\underline{E} \qquad \underline{\underline{\rho}} \equiv \underline{\underline{\rho}}^{mag} : \underline{H}\,\underline{H}$$

multiplying dyadics $\underline{E}\,\underline{E}$ or $\underline{H}\,\underline{H}$ which already are purely longitudinal, i.e. "half" the longitudinal projector $: \underline{n}\,\underline{n}\,\underline{n}\,\underline{n}$. Their RSs display the projection along the same direction as the field $\underline{E}$ or $\underline{H}$. In electrostriction for example, the RS gives the dependence on orientation of the longitudinal strain in the same direction as the applied field $\underline{E}$.

The elastic stiffness $\underline{\underline{c}}$ and compliance $\underline{\underline{s}}$ that appear in Hooke's CE:

$$\underline{\underline{\tau}} = \underline{\underline{c}} : \underline{\underline{\epsilon}} \qquad \underline{\underline{\epsilon}} = \underline{\underline{s}} : \underline{\underline{\tau}} \tag{8.1}$$

are reciprocal: $\underline{\underline{c}} = \underline{\underline{s}}^{-1}$ (or $\underline{\underline{c}} = \underline{\underline{s}}^{-1}$) and count among the most important fourth rank properties in terms of widespread use and volume of applications.

Their RSs have direct interpretations in terms of elastic moduli (4.32). For these reasons, they occupy center stage in this chapter, although their RSs are also valid for the rest of fourth rank tensor properties.

© Springer Nature Switzerland AG 2020
M. Laso and N. Jimeno, *Representation Surfaces for Physical Properties of Materials*, Engineering Materials, https://doi.org/10.1007/978-3-030-40870-1_8

The detailed identification of the symmetry elements of the RS carried out in Chap. 7 for third rank properties will not be repeated here, as it is largely identical. Similarly, the basic geometrical objects from which fourth rank RSs are built will be presented and accompanied by a brief discussion only.

8.1 Linear Elasticity

Major and minor index symmetries of $\underset{=}{c}$ and $\underset{=}{s}$ have already been surveyed in Sect. 3.5. Unlike the $\underset{\equiv}{c}$ and $\underset{\equiv}{s}$ of linear elasticity, the rest of the properties mentioned above only have minor symmetry. Minor symmetry in ρ^{mag}, $\underset{=}{K}$, $\underset{=}{\pi^{opt}}$ and $\underset{=}{M}$ is due to these properties being phenomenological constants that couple second rank symmetric tensors.

If $\underset{=}{A}$ and $\underset{=}{C}$ are generic symmetric ($A_{ij} = A_{ji}$, $C_{ij} = C_{ji}$) second rank tensors, and $\underset{\equiv}{B}$ a generic fourth rank property that couples them in a linear relation:

$$\underset{=}{A} = \underset{\equiv}{B} : \underset{=}{C} \qquad A_{ij} = B_{ijkl}C_{lk}$$

the symmetry of $\underset{=}{A}$ and $\underset{=}{C}$ allows swapping their indices in the previous expression without altering the result:

$$A_{ij} = B_{ijkl}C_{lk} \quad \Rightarrow \quad A_{ji} = B_{ijkl}C_{kl}$$

This can be rewritten as:

$$A_{ji} = B_{ijkl}C_{kl} \qquad \underset{=}{A} = {}^{T}(\underset{\equiv}{B})^{T} : \underset{=}{C} \quad \text{so that} \quad \underset{\equiv}{B} = {}^{T}(\underset{\equiv}{B})^{T}$$

which is the statement of minor symmetry.

On the other hand, the major symmetry of $\underset{=}{s}$ and $\underset{=}{c}$ has a thermodynamic origin [1–3] that is absent in the other cases.

As far as the RS for fourth rank properties is concerned, the index symmetry of the fourth rank longitudinal projector $\underline{n}\,\underline{n}\,\underline{n}\,\underline{n}$ treats all fourth rank properties equally, regardless of minor or major symmetry, if any. The RSs presented in this chapter for elastic properties like $\underset{\equiv}{s}$ are valid for other fourth rank properties as well.

However, the major symmetry of $\underset{\equiv}{s}$ and $\underset{\equiv}{c}$ introduces a restriction in the shape of their RSs. For a generic fourth rank property $\underset{\equiv}{T}$ the RS is:

$$r = \underset{\equiv}{T} \vdots \underline{n}\,\underline{n}\,\underline{n}\,\underline{n} \quad r = T_{ijkl}n_{l}n_{k}n_{j}n_{i}$$

Depending on the numerical values of the components T_{ijkl}, the radius vector r may be positive or negative, or change sign. In the latter case, the RS will contain the origin.

However, the linear elastic properties $\underset{\equiv}{s}$ and $\underset{\equiv}{c}$ are also related to the volumetric density of elastic energy U_{elast} (J/m³). If stress and strain at a point in a material are $\underset{=}{\tau}$ and $\underset{=}{\epsilon}$, U_{elast} is given by:

$$U_{elast} = \frac{1}{2} \underset{=}{\tau} : \underset{=}{\epsilon} \tag{8.2}$$

which for a linear elastic material can be rewritten in terms of stress or strain by inserting Hooke's law in (8.2):

$$U_{elast} = \frac{1}{2} \underset{=}{\tau} : \underset{=}{\epsilon} = \frac{1}{2} \underset{\equiv}{s} : \underset{=}{\tau}\underset{=}{\tau} = \frac{1}{2} \underset{\equiv}{c} : \underset{=}{\epsilon}\underset{=}{\epsilon} \tag{8.3}$$

This can also be written using Voigt's notation:

$$U_{elast} = \frac{1}{2} \vec{\tau}^T \vec{\epsilon} = \frac{1}{2} \vec{\tau}^T \underset{\smile}{s} \vec{\tau} = \frac{1}{2} \vec{\epsilon}^T \underset{\smile}{c} \vec{\epsilon}$$

Since the elastic energy stored cannot be negative (assuming $U_{elast} = 0$ in the unstrained state):

$$\frac{1}{2} \vec{\tau}^T \underset{\smile}{s} \vec{\tau} > 0 \quad \forall \vec{\tau} \neq 0 \qquad \frac{1}{2} \vec{\epsilon}^T \underset{\smile}{c} \vec{\epsilon} > 0 \quad \forall \vec{\epsilon} \neq 0 \tag{8.4}$$

the matrices $\underset{\smile}{s}$ and $\underset{\smile}{c}$ that appear in these quadratic forms must be positive definite.

The RS for fourth rank tensors $r = \underset{\equiv}{T} : \underline{n}\,\underline{n}\,\underline{n}\,\underline{n}$ is of the form of (8.3), independently of the index symmetry of $\underset{\equiv}{T}$ because:

$$r = \underset{\equiv}{T} : \underline{n}\,\underline{n}\,\underline{n}\,\underline{n} = \underline{n}\,\underline{n} : \underset{\equiv}{T} : \underline{n}\,\underline{n}$$

Hence, the positive definiteness of the two quadratic forms of (8.4) implies:

$$\underset{\equiv}{s} : \underline{n}\,\underline{n}\,\underline{n}\,\underline{n} > 0 \qquad \underset{\equiv}{c} : \underline{n}\,\underline{n}\,\underline{n}\,\underline{n} > 0 \tag{8.5}$$

so that the radius vector of the RS for $\underset{\equiv}{s}$ and $\underset{\equiv}{c}$ will always be positive. The RS will not contain the origin and, being the RS of an even-rank property, will be centrosymmetric. An inspection of Figs. 8.1, 8.2, 8.3, 8.4, 8.5, 8.6, 8.7, 8.8 and 8.9 confirms this statement.

8.2   Representation Surfaces for Crystallographic and Limit Classes

Typical shapes for the RS of fourth rank properties are shown in Figs. 8.1, 8.2, 8.3, 8.4, 8.5, 8.6, 8.7, 8.8 and 8.9 using the elastic compliance $\underset{=}{s}$ as a particular example.

Each of the figures contains several views of a representative RSs for all point groups or for entire crystallographic systems that share the same $r = r(n_1, n_2, n_3)$. In a few cases, the RS depends on a single numerical parameter, so that the RS is, apart from this scale factor, the same for all classes.

For the rest of the point groups, the RSs shown in the figures are for s_{ij} of specific materials. Changes in these numerical values will make the RSs look quite different, although their shapes, and hence the symmetry of the property represented, will all belong to the same point group.

Equations for all different RS symmetries, one for each figure, are given below. Unlike in the previous chapter, $r = r(n_1, n_2, n_3)$ is written using exponents instead of expressions like $n_1 n_1 n_2 n_2$. Terms $n_1^2 + n_2^2$ have been written as $(1 - n_3^2)$ whenever convenient to emphasize axial symmetry.

- RSs for triclinic materials (Fig. 8.1) only possess an inversion center; they belong to the triclinic holoedric class $\bar{1}$. The standard does not define conventional axes because no other geometric symmetry element exists (e.g. mirror planes, axes) that could play the role of conventional axes.

$$
\begin{aligned}
r = {} & s_{11}n_1^4 + (2s_{12} + s_{66})n_1^2 n_2^2 + (2s_{13} + s_{55})n_1^2 n_3^2 + 2(s_{14} + s_{56})n_1^2 n_2 n_3 \\
& + 2s_{15}n_1^2 n_3 n_3 + 2s_{16}n_1^3 n_2 + s_{22}n_2^4 + (2s_{23} + s_{44})n_2^2 n_3^2 \\
& + 2s_{24}n_2^3 n_3 + 2(s_{25} + s_{46})n_2^2 n_1 n_3 + 2s_{26}n_2^3 n_1 + s_{33}n_3^2 \\
& + 2s_{34}n_3^3 n_2 + 2s_{35}n_3^3 n_1 + 2(s_{36} + s_{45})n_3^2 n_1 n_2
\end{aligned} \tag{8.6}
$$

- RSs for monoclinic materials (Fig. 8.2) belong to the monoclinic holoedric class $2/m$. The twofold axis, perpendicular to the mirror plane, is axis ② by convention.

$$
\begin{aligned}
r = {} & s_{11}n_1^4 + (2s_{12} + s_{66})n_1^2 n_2^2 + (2s_{13} + s_{55})n_1 n^2 n_3^2 + 2s_{15}n_1^2 n_3^2 \\
& + s_{22}n_2^4 + (2s_{23} + s_{44})n_2 2n_3 n^2 + 2(s_{25} + s_{46})n_2^2 n_1 n_3 \\
& + s_{33}n_3^4 + 2s_{35}n_3^3 n_1
\end{aligned} \tag{8.7}
$$

- RSs for orthorhombic materials (Fig. 8.3) belong to the orthorhombic holoedric class mmm. The three twofold axes, perpendicular to the three mirror planes, are axes ①②③ by convention.

$$
\begin{aligned}
r = {} & s_{11}n_1^3 + s_{22}n_2^4 + s_{33}n_3^4 \\
& + (2s_{12} + s_{66})n_1^2 n_2^2 + (2s_{13} + s_{55})n_1^2 n_3^2 + (2s_{23} + s_{44})n_2^2 n_3^2
\end{aligned} \tag{8.8}
$$

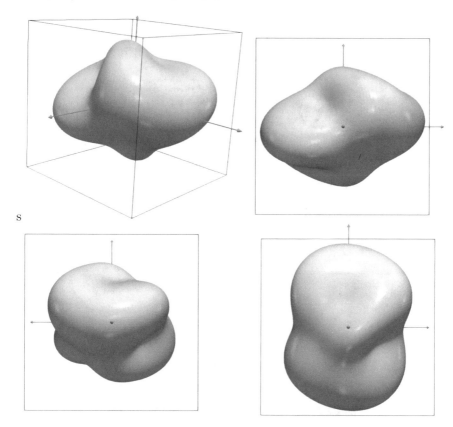

S

Fig. 8.1 Representative RS for fourth rank properties for triclinic materials (all classes), Eq. (8.6); RS for elastic compliance of monocrystalline copper sulfate pentahydrate $CuSO_4 \cdot 5H_2O$. Left to right, top to bottom, perspective view, views along conventional axes $-①$, $-②$, $-③$

- RSs for tetragonal materials of classes 4, $\bar{4}$, $4/m$ (Fig. 8.4) belong to the tetragonal holoedric class $4/mmm$. The fourfold axis is ③ by convention; but now the four symmetry planes containing this axis, and the four twofold axes perpendicular to this axis are not aligned with axes ①②③.
  The extra 7th independent component s_{16} in $\mathrm{str}(s)$ of these classes with respect to classes $4mm$, $\bar{4}2m$, 422, $4/mmm$ is responsible for the rotation about ③. Crystals of classes 4, $\bar{4}$, $4/m$ do not possess mirror planes containing the fourfold axis, nor binary axes perpendicular to it that could be assigned the role of conventional axes. Hence the arbitrariness of the rotation about ③.
  This is the same situation already encountered in trigonal classes 3 and $\bar{3}$ for third rank properties.

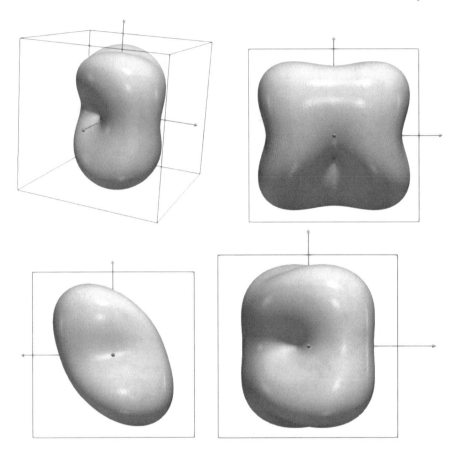

Fig. 8.2 Representative RS for fourth rank properties for monoclinic materials (all classes), Eq. (8.7); RS for elastic compliance of monocrystalline gypsum $CaSO_4 \cdot 2H_2O$. Left to right, top to bottom, perspective view, views along conventional axes $-①, -②, -③$

$$r = s_{11}(n_1^4 + n_2^4) + s_{33}n_3^4 + (2s_{12} + s_{66})n_1^2 n_2^2$$
$$+ (2s_{13} + s_{44})n_3^2(1 - n_3^2) + 2s_{16}n_1 n_2(n_1^2 - n_2^2) \tag{8.9}$$

- RSs for tetragonal materials of classes $4mm, \bar{4}2m, 422, 4/mmm$ (Fig. 8.5) also belong to the tetragonal holoedric class $4/mmm$. The fourfold axis is $③$ by convention; the four symmetry planes containing this axis, and the four twofold axes perpendicular to this axis are aligned with axes $①②③$.

$$r = s_{11}(n_1^4 + n_2^4) + s_{33}n_3^4 + (2s_{12} + s_{66})n_1^2 n_2^2$$
$$+ (2s_{13} + s_{44})n_3^2(1 - n_3^2) \tag{8.10}$$

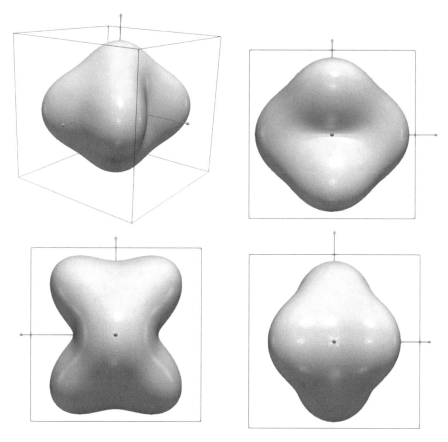

Fig. 8.3 Representative RS for fourth rank properties for orthorhombic materials (all classes), Eq. (8.8); RS for elastic compliance of monocrystalline aragonite $CaCO_3$. Left to right, top to bottom, perspective view, views along conventional axes $-①$, $-②$, $-③$

- RSs for trigonal materials of classes 3 and $\bar{3}$ (Fig. 8.6) belong to the trigonal holoedric class $\bar{3}m$. The extra 7th independent component s_{25} in str(s) of these classes with respect to classes 32, $3m$ and $\bar{3}m$ is responsible for the arbitrary rotation about ③, as in the case of tetragonal classes 4, $\bar{4}$, $4/m$.

$$r = s_{11}(1 - n_3^2)^2 + s_{33}n_3^4 + (2s_{13} + s_{44})n_3^2(1 - n_3^2)$$
$$+ 2s_{14}n_2n_3(3n_1^2 - n_2^2) + 2s_{25}n_1n_3(3n_2^2 - n_1^2) \qquad (8.11)$$

- RSs for trigonal materials of classes 32, $3m$ and $\bar{3}m$ (Fig. 8.7) also belong to the trigonal holoedric class $\bar{3}m$. The threefold axis is ③ by convention; the positions of the three symmetry planes containing this axis, and of the three twofold axes perpendicular to this axis with respect to the axes ①②③ are in agreement with the crystallographic convention.

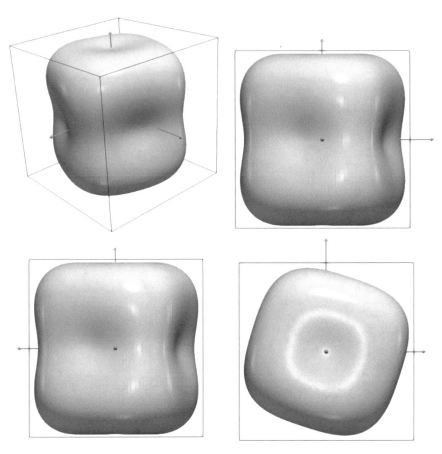

Fig. 8.4 Representative RS for fourth rank properties for tetragonal classes $4, \bar{4}, 4/m$, Eq. (8.9); RS for elastic compliance of monocrystalline silver chlorate $AgClO_3$. Left to right, top to bottom, perspective view, views along conventional axes $-①, -②, -③$

$$r = s_{11}(1 - n_3^2)^2 + s_{33}n_3^4 + (2s_{13} + s_{44})n_3^2(1 - n_3^2)$$
$$+ 2s_{14}n_2n_3(3n_1^2 - n_2^2) \tag{8.12}$$

- RSs for all hexagonal classes and for $\infty, \infty m, \infty/m, \infty 2, \infty/mm$ limit classes (Fig. 8.8) are axisymmetric, limit class ∞/mm. Polynomial (8.13) is of degree four in the trigonometric functions of the polar angles, hence insufficient to reproduce the sixfold symmetry axis typical of the hexagonal system. For fourth rank properties, hexagonal materials are axisymmetric and belong to the cylindrical holoedric class. In fourth rank there is no equivalent to $3 \perp m \Leftrightarrow \bar{6}$, which made hexagonal symmetry possible for third rank properties.

$$r = s_{11}(1 - n_3^2)^2 + s_{33}n_3^4 + (2s_{13} + s_{44})n_3^2(1 - n_3^2) \tag{8.13}$$

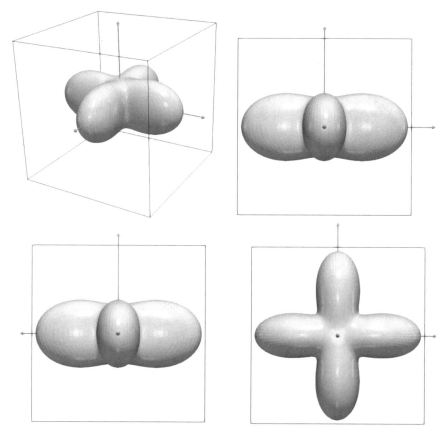

Fig. 8.5 Representative RS for fourth rank properties for tetragonal classes $4mm$, $\bar{4}2m$, 422, $4/mmm$, Eq. (8.10); RS for elastic compliance of monocrystalline tin. Left to right, top to bottom, perspective view, views along conventional axes $-①$, $-②$, $-③$

- RSs for all cubic classes (Fig. 8.9) belong to the cubic holoedric class $m3m$. The highly symmetric way Eq. (8.14) is written makes it easy to identify all 48 elements of point group $m3m$.
  Cubic classes are not isotropic for fourth rank properties. The degree of deviation from isotropy depends on the numerical value of the term $2(s_{11} - s_{12} - \frac{1}{2}s_{44})$ in (8.14).

$$r = s_{11} - 2\left(s_{11} - s_{12} - \frac{1}{2}s_{44}\right)(n_1^2 n_2^2 + n_2^2 n_3^2 + n_3^2 n_1^2) \qquad (8.14)$$

- limit classes $\infty\infty m$, $\infty\infty$ (Fig. 8.10) are isotropic for fourth rank properties; the RS for all properties is the centered sphere.

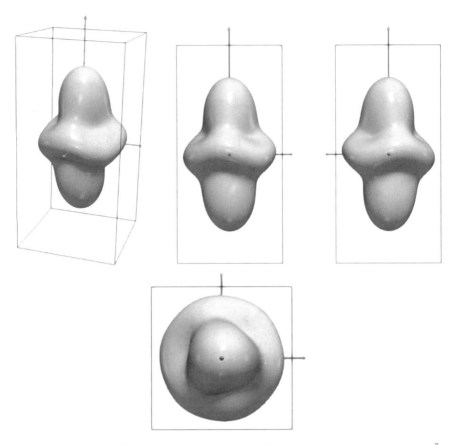

Fig. 8.6 Representative RS for fourth rank properties for trigonal materials (classes 3 and $\bar{3}$), Eq. (8.11); RS for elastic compliance of monocrystalline jarosite $KFe_3(SO_4)_2(OH)_6$. Left to right, top to bottom, perspective view, views along conventional axes $-①$, $-②$, $-③$

As can be checked in the previous list, the compulsory inversion center associated with even rank properties increases the symmetry of the RS of non-holoedric classes up to the holoedric point group.[1]

A simple illustrative example is the conical class ∞, whose only geometric element of symmetry is a directed infinite-fold axis. Adding an inversion center makes the axis undirected and also induces infinite mirror planes containing the axis, plus an additional one perpendicular to the axis. These are the geometric elements of symmetry of the cylindrical holoedric class.

---

[1]Check explicitly that adding an inversion center to the single fourfold axis of tetragonal class 4 increases the symmetry to that of point group $4/mmm$. Hint: take a generic point of coordinates (x_1, x_2, x_3), transform its coordinates by all actions of the 4 axis and of the inversion center $\bar{1}$. To which point group does the resulting set of points belong?

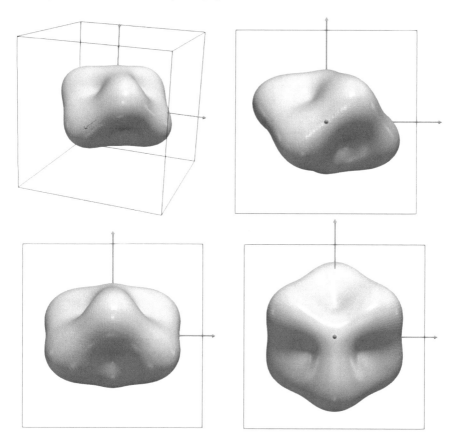

Fig. 8.7 Representative RS for fourth rank properties for trigonal materials (classes 32, 3m and $\bar{3}m$), Eq. (8.12); RS for elastic compliance of monocrystalline α-quartz. Left to right, top to bottom, perspective view, views along conventional axes $-①$, $-②$, $-③$

The situation is similar for classes like 3 and $\bar{3}$, and 4, $\bar{4}$, 4/m: the mirror planes induced by the inversion center that contain the threefold or fourfold axis have arbitrary rotation with respect to the only axis.

When the point group does have this kind of axes or mirror planes (classes 32, 3m, $\bar{3}m$, 4mm, $\bar{4}2m$, 422, 4/mmm), their positions conform to the crystallographic standard and are respected by the inversion center.

The variability of the RS shape depends on the numerical values of the components s_{ij}, or to be more precise, on how many truly independent parameters appear in the RS equation.

The grouping of s_{ij} coefficients (due both to relations between components in str(s) and to the index symmetry of the projector) is apparent for the tetragonal, trigonal, hexagonal, cubic and all limit classes. Although the number of independent material parameters for the classes belonging to these systems is not changed by

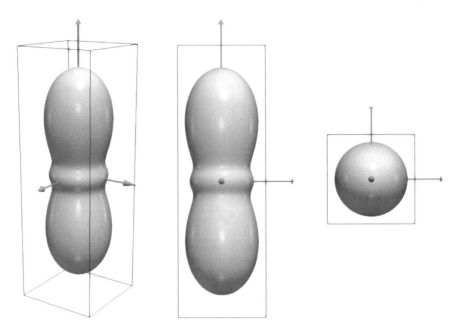

Fig. 8.8 Representative RS for fourth rank properties of all hexagonal classes, and limit classes ∞, ∞m, ∞/m, $\infty 2$, ∞/mm, Eq. (8.13); RS for elastic compliance of monocrystalline zinc. Left to right, perspective view, views along conventional axes $-①$ and $-③$

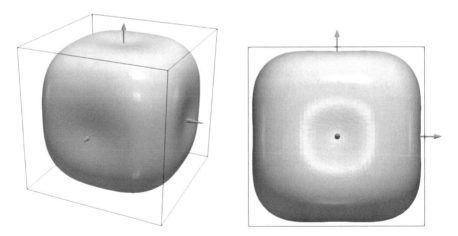

Fig. 8.9 Representative RS for fourth rank properties of cubic materials (all classes), Eq. (8.14); RS for elastic compliance of NaCl. Left to right, perspective view, view along any of the conventional axes

Fig. 8.10 Representative RS
for fourth rank properties of
limit classes $\infty\infty m$, $\infty\infty$.
For fourth rank properties,
these classes are isotropic

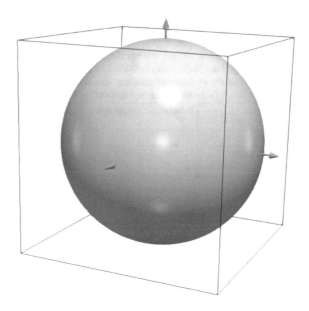

grouping (i.e. 21 for triclinic materials, 13 for monoclinic, etc.), the shape of the RS
may depend on a lower number of combinations of parameters only.

Taking cubic materials as an example, their RS (8.14) is a sphere of radius s_{11}
modified by the term in parenthesis with a weight $2(s_{11} - s_{12} - \frac{1}{2}s_{44})$. The grouping
of components reduces the number of degrees of freedom of the RS to two.

Furthermore, since the absolute size of the RS is not relevant for its shape, s_{11} can
be taken as the unit, so it is the dimensionless ratio:

$$\frac{-2(s_{11} - s_{12} - \frac{1}{2}s_{44})}{s_{11}}$$

that actually determines the RS shape. For cubic materials, all possible RSs form a
one-parameter family of surfaces.

Similar considerations apply to the other classes/systems, as was the case for
third order properties. Table 8.1 lists the number of effective degrees of freedom that
determine the shape of the RS, together with the reasons for the loss of degrees of
freedom.

There is no distinction here between tetragonal and trigonal classes with six
or seven independent components. In both systems the seventh component in s_{ij}
amounts to a rotation about the ③ axis (Figs. 8.4 and 8.6), and does not have an
influence on RS shape.

The seventh component in s_{ij} in trigonal classes 3, $\bar{3}$ (d_{25}) and tetragonal classes
$4, \bar{4}, 4/m$ (d_{16}) can be made to disappear through a rotation about ③, so that
the RS is aligned with the axes as in the six-component classes $32, \bar{3}m, 3m$ and
$4mm, \bar{4}2m, 422, 4/mmm$.

Table 8.1 Number of effective degrees of freedom that determine the shape of the RS. Only some (third column) of the original number of parameters (second column) from $\mathrm{str}(\underset{\sim}{s})$ may appear in the functional form of the RS; the number may be further reduced because absolute size does not influence shape (fourth column), because components appear grouped (fifth column, see Sect. 8.3), and by elimination of arbitrary rotations (sixth column). The net number of effective degrees of freedom is listed in the last column. Minus signs indicate loss of degrees of freedom

System/class	From str($\underset{\sim}{s}$)	In RS appear	Scale factor	Grouping of s_{ij}	Arbitrary rotation	Eff. d.o.f.
Triclinic (all)	21	21	-1	-6	-3	11
Monoclinic (all)	13	13	-1	-4	-1	7
Orthorhombic (all)	9	9	-1	-3	0	5
Tetragonal						
$4, \bar{4}, 4/m$	7	7	-1	-2	-1	3
$4mm, \bar{4}2m, 422, 4/mmm$	6	6	-1	-2	0	3
Trigonal						
$3, \bar{3}$	7	6	-1	-1	-1	3
$32, \bar{3}m, 3m$	6	5	-1	-1	0	3
Hexagonal (all)	5	4	-1	-1	0	2
$\infty, \infty m, \infty/m, \infty 2, \infty/mm$	5	4	-1	-1	0	2
Cubic (all)	3	3	-1	-1	0	1
$\infty\infty m, \infty\infty$	2	1	-1	0	0	0

The angle of this distinguished rotation is not directly related to the geometry/positions of the atoms, molecules or macroscopic components of the material,[2] but is determined by the values of the components s_{ij}.

Therefore, the rotation that reduces the seven-constant $\mathrm{str}(\underset{\sim}{s})$ to six constants will be different for each property, even for the same material. It must necessarily be so, because there is nothing in the geometry of the material (mirror planes, binary axes) that allows a priori positioning of the conventional ②③ axes; that is the reason these axes do not appear in the stereograms of these classes.[3]

8.3 The Basic Objects for the RS of Elastic Compliance and Stiffness

We saw in Sect. 7.4 that only one basic type of geometric object was required to construct the RS of a first rank or of a symmetric second rank tensor, and two for a third rank tensor with the symmetry of $\underset{\equiv}{d}$ or $\underset{\equiv}{r}$. We will address now the same question for fourth rank properties.

---

[2] It is indirectly related because these entities are responsible for the numerical values of the property components.

[3] Why have we not discussed hexagonal classes $6, \bar{6}, 6/m$, for which the convention does not define ②③ axes either?

The full index symmetry of the projector $\vdots \underline{n}\,\underline{n}\,\underline{n}\,\underline{n}$ introduces simplifications in $\mathrm{str}(\underset{\approx}{c})$ or $\mathrm{str}(\underset{\approx}{s})$. The expression[4]:

$$r = \underset{\approx}{s} \vdots \underline{n}\,\underline{n}\,\underline{n} = s_{ijkl}n_l n_k n_j n_i$$

implies that all components of $\underset{\approx}{s}$ obtained by permutation of the indices i, j, k, l of s_{ijkl} will contribute equally to the factor containing $n_l n_k n_j n_i$.

For example, $s_{66} = 4s_{1212} = 4s_{1221} = 4s_{2112} = 4s_{2121}$ and $s_{12} = s_{2211} = s_{1122}$ will all appear as coefficients of the term containing $n_1 n_1 n_2 n_2$, so that s_{66} and d_{21} will not contribute separate basic objects to the RS.

Table 8.2 shows the relationships among elements of $\underset{\approx}{s}$ that get grouped when evaluating the RS, either due to $\mathrm{str}(\underset{\approx}{s})$ or to the index symmetry of $\underline{n}\,\underline{n}\,\underline{n}\,\underline{n}$, for the general, least symmetric case (triclinic class 1).

Each line of Table 8.2 contains the indices of the coefficients that multiply the common $n_i n_j n_k n_l$ factor listed in the middle. The full index symmetry of $\underline{n}\,\underline{n}\,\underline{n}\,\underline{n}$ reduces the 21 components of $\underset{\approx}{d}$ to only 15, naturally arranged in four groups of 3, 6, 3 and 3 congruent isomers, which only differ in their orientation in space. The four types of basic objects are shown in this table at different scales to display sufficient detail, but their relative sizes are quite different (Fig. 8.15):

- the three isomers with four equal indices are shown in Fig. 8.12. The three panels of this figure show the terms $n_i n_i n_i$, $i = 1, 2, 3$ (no sum). They are surfaces of revolution about the ①, ②, ③ axes, of cross section $\cos^4 \alpha$, with α the angle between the radius vector and the axis. Their shape belongs to Curie point group ∞/mm, the holoedric cylindrical class.
- the six isomers with three equal indices (Fig. 8.12) belong to the monoclinic holoedric point group $2/m$.
- the three isomers with two pairs of equal indices. They are shown in Fig. 8.13 and belong to the tetragonal holoedric point group $4/mmm$.
- the three isomers with one pair of equal indices and two unequal indices. Figure 8.14 belong to the orthorhombic holoedric point group mmm.

The RS for fourth rank properties for any material can thus be written as a linear combination of the 15 geometrical objects of Figs. 8.11, 8.12, 8.13 and 8.14. Figure 8.15 gives a visual idea of the importance (weight) of the contributions of individual components.

We are now in a position to answer the same question of Sect. 7.4 regarding the minimum number and kind of geometrical objects needed to describe the most general (triclinic) RS: the last column of Table 8.1 gives the number of degrees of freedom that determine the shape of the RS for all point groups. The minimum number is 11 for the most general case.

---

[4]In the following we will use $\underset{\approx}{s}$ only, knowing that all results are valid for $\underset{\approx}{c}$ as well, except for the factors of 2 or 4 required by Voigt's notation.

Table 8.2 Elements of a generic fourth rank tensor $\underset{=}{T}$ and $\underset{\sim}{T}$ (left) that are gathered in a single coefficient of $n_i n_j n_k n_l$ (middle) in the RS. The type of basic geometric object that corresponds to each group is shown on the right

All indices equal	$\begin{cases} 1111(11) & n_1n_1n_1n_1 \\ 2222(22) & n_2n_2n_2n_2 \\ 3333(33) & n_3n_3n_3n_3 \end{cases}$	
Three indices equal	$\begin{cases} 1112, 2111, 1121, 1211(16) & n_1n_1n_1n_2 \\ 2223, 2232, 2322, 3222(24) & n_2n_2n_3n_3 \\ 3331, 3313, 3133, 1333(35) & n_3n_3n_3n_1 \\ 2221, 2212, 2122, 1222(26) & n_2n_2n_2n_1 \\ 3332, 3323, 3233, 2333(34) & n_3n_3n_3n_2 \\ 1113, 1131, 1311, 3111(15) & n_1n_1n_1n_3 \end{cases}$	
Two pairs of indices equal	$\begin{cases} 1122, 2211(12), 1212, \\ 2112, 2121, 1221(66) & n_1n_1n_2n_2 \\ 2233, 3322(23), 2323, \\ 3223, 3232, 2332(44) & n_2n_2n_3n_3 \\ 3311, 1133(31), 3131, \\ 1313, 1313, 3131(55) & n_3n_3n_1n_1 \end{cases}$	
Two indices equal, the other two unequal	$\begin{cases} 1123, 1132, 2311, 3211(14), \\ 1213, 1231, 2113, 2131, \\ 1312, 1321, 3112, 3121(56) & n_1n_1n_2n_3 \\ 2231, 2213, 3122, 1322(25), \\ 2321, 2312, 3221, 3212, \\ 2113, 2132, 1223, 1232(46) & n_2n_2n_3n_1 \\ 3312, 3321, 1233, 2133(36), \\ 3132, 3123, 1332, 1321, \\ 3221, 3213, 2331, 2313(45) & n_3n_3n_1n_2 \end{cases}$	

For monoclinic classes the conventional ② axis is fixed, so the remaining arbitrary rotation is about this axis, which eliminates one degree of freedom; thus, the minimum number is 7 for monoclinic classes (second line of Table 8.1). In the remaining cases, either all three conventional axes are defined, or there is one arbitrary rotation about ③.

The question remains about which components of $\underset{\sim}{s}$ can be eliminated to remove the arbitrary orientation. For the piezoelectric modulus we could eliminate the elements on the diagonal of the right 3×3 sub-block. This is not possible for $\underset{\sim}{s}$ or $\underset{\sim}{c}$ because of their positive definiteness. In more physical terms, the diagonal elements of $\underset{\sim}{s}$ are reciprocal elastic moduli, canceling s_{44}, s_{55} or s_{66}, would imply infinite longitudinal or shear moduli.

What is feasible now is to remove one or more off-diagonal elements of the bottom-right 3×3 sub-block of $\underset{\sim}{s}$. Either the d_{46} for monoclinic classes:

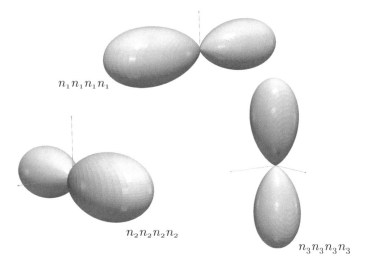

Fig. 8.11 RS for the three isomers of the form $n_i n_i n_i n_i$, all indices equal (no sum); symbol •
in (8.16) Axes ① ③ ②

$$
\begin{bmatrix}
\bullet & \bullet & \bullet & \cdot & \bullet & \cdot \\
 & \bullet & \bullet & \cdot & \bullet & \cdot \\
 & & \bullet & \cdot & \bullet & \cdot \\
 & & & \bullet & \cdot & \bullet \\
 & & & & \bullet & \cdot \\
 & & & & & \bullet
\end{bmatrix}
\Rightarrow
\begin{bmatrix}
\bullet & \bullet & \bullet & \cdot & \cdot & \bullet \\
 & \bullet & \bullet & \cdot & \cdot & \bullet \\
 & & \bullet & \cdot & \cdot & \bullet \\
 & & & \bullet & \cdot & \cdot \\
 & & & & \bullet & \cdot \\
 & & & & & \bullet
\end{bmatrix}
\tag{8.15}
$$

or the d_{45}, d_{46} and d_{56} for the most general case:

$$
\begin{bmatrix}
\bullet & \bullet & \bullet & \bullet & \bullet & \bullet \\
 & \bullet & \bullet & \bullet & \bullet & \bullet \\
 & & \bullet & \bullet & \bullet & \bullet \\
 & & & \bullet & \bullet & \bullet \\
 & & & & \bullet & \bullet \\
 & & & & & \bullet
\end{bmatrix}
\Rightarrow
\begin{bmatrix}
\bullet & \bullet & \bullet & \bullet & \bullet & \bullet \\
 & \bullet & \bullet & \bullet & \bullet & \bullet \\
 & & \bullet & \bullet & \bullet & \bullet \\
 & & & \bullet & \cdot & \cdot \\
 & & & & \bullet & \cdot \\
 & & & & & \bullet
\end{bmatrix}
\tag{8.16}
$$

In both cases the required orthogonal transformation is given by the matrix $\underset{\sim}{L}$
whose rows are the eigenvectors of the bottom-right 3×3 sub-block of $\underset{\sim}{s}$:

$$
\underset{\sim}{L} =
\begin{bmatrix}
v_1^{*①} & v_2^{*①} & v_3^{*①} \\
v_1^{*②} & v_2^{*②} & v_3^{*②} \\
v_1^{*③} & v_2^{*③} & v_3^{*③}
\end{bmatrix}
\text{ with } \vec{v}^{*①}, \vec{v}^{*②}, \vec{v}^{*③}
\tag{8.17}
$$

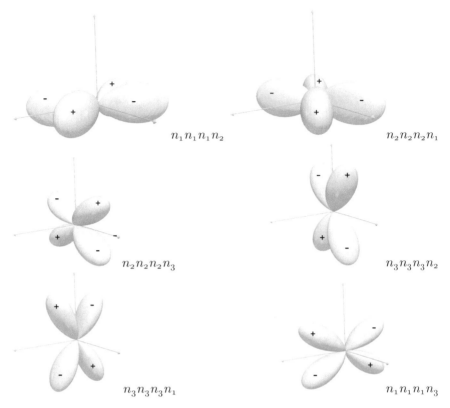

$$n_1 n_1 n_1 n_2 \qquad n_2 n_2 n_2 n_1$$

$$n_2 n_2 n_2 n_3 \qquad n_3 n_3 n_3 n_2$$

$$n_3 n_3 n_3 n_1 \qquad n_1 n_1 n_1 n_3$$

Fig. 8.12 RS for the six isomers of the form $n_i n_i n_i n_j$, three indices equal (no sum); symbol $\square$ in (8.16) Axes ① ② ③

the eigenvectors of the bottom-right 3×3 sub-block of $\underset{\sim}{s}$ $\begin{bmatrix} s_{44} & s_{45} & s_{46} \\ s_{45} & s_{55} & s_{56} \\ s_{46} & s_{56} & s_{66} \end{bmatrix}$

Eliminating the orientational degree(s) of freedom removes three of the six terms of the form $n_i n_i n_j n_k$ with one pair of equal indices (symbol ▼ in Fig. 8.16).

The implication is that even in the most reduced form of str($\underset{\sim}{s}$), all four types of basic geometric objects from Table 8.2 are necessary to build the general RS (triclinic). Recall that for third rank properties, reducing str($\underset{\sim}{s}$) also accomplished the elimination of one type of basic object, so that only two of them were necessary to build the triclinic $\underset{\sim}{d}$ (Sect. 7.4).

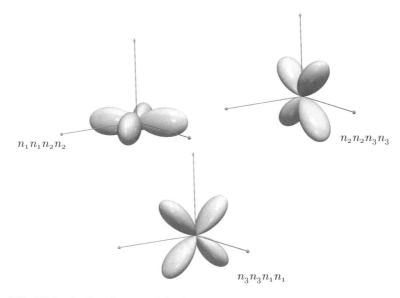

$n_1 n_1 n_2 n_2$

$n_2 n_2 n_3 n_3$

$n_3 n_3 n_1 n_1$

Fig. 8.13 RS for the three isomers of the form $n_i n_i n_j n_j$, two pairs of indices equal (no sum), symbol △ in (8.16) Axes ① ③ ②

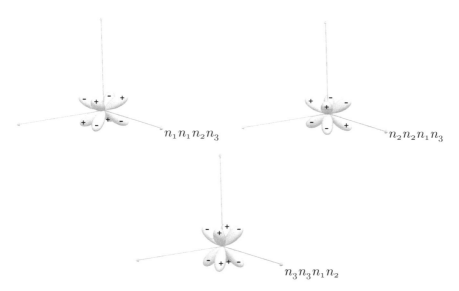

$n_1 n_1 n_2 n_3$

$n_2 n_2 n_1 n_3$

$n_3 n_3 n_1 n_2$

Fig. 8.14 RS for the three isomers of the form $n_i n_i n_j n_k$, with one pair of equal indices and two unequal indices (no sum), symbol s_{iijk} ▼ in (8.16) Axes ① ③ ②

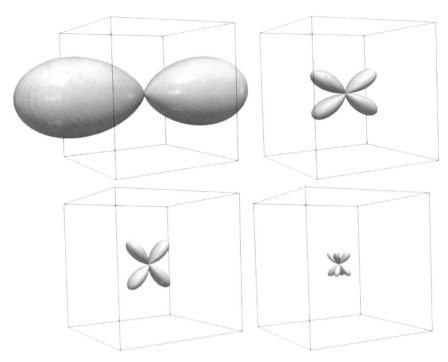

Fig. 8.15 Size comparison of the four types of geometric contributions (isomers corresponding to $n_1n_1n_1n_1$, $n_1n_1n_1n_2$, $n_1n_1n_3n_3$, and $n_3n_3n_1n_2$ left to right, top to bottom) to the RS of the longitudinal elastic modulus $\underline{\underline{s}}$ represented at the same scale. The edge of the enclosing boxes has unit length in all panels

Fig. 8.16 The four types of $n_in_jn_kn_l$ terms in $\mathrm{str}(\underline{s})$ from Table 8.2. $\bullet$ $n_in_in_in_i$ all indices equal; $\square$ $n_in_in_in_j$ three indices equal; $\triangle$ $n_in_in_jn_j$ two pairs of equal indices; $\blacktriangledown$ $n_in_in_jn_k$ one pair of equal indices

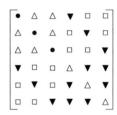

8.4 RSs for Elastic Moduli

As explained in Sect. 4.7, engineering elastic moduli are closely related to projections of $\underline{\underline{s}}$. Young moduli are reciprocal to longitudinal projections of $\underline{\underline{s}}$:

$$E_{\underline{n}} = \frac{1}{\underline{\underline{s}} : \underline{n}\,\underline{n}\,\underline{n}\,\underline{n}}$$

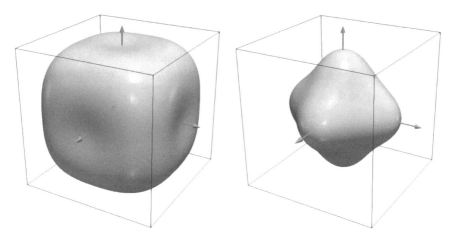

Fig. 8.17 RS for the elastic compliance (left) and for Young modulus (right) of monocrystalline NaCl. The radius vectors of the surfaces are reciprocal: in directions in which the compliance is high, the modulus is low, and viceversa

and the reciprocal of more general projections, like:

$$\underset{\equiv}{s} \vdots \underline{n}^A \, \underline{n}^B \, \underline{n}^C \, \underline{n}^C$$

for orthonormal $\underline{n}^A \, \underline{n}^B \, \underline{n}^C$, may also include, besides a Young E or a shear G modulus, a Poisson coupling, and appears off-diagonally in $\underset{\smile}{s}$. If we take $\underline{n}^A \, \underline{n}^B \, \underline{n}^C$ to be the basis vectors ①②③ (in that order) of a reference frame, then:

$$\frac{1}{2\underset{\equiv}{s} \vdots \underline{n}^A \, \underline{n}^B \, \underline{n}^C \, \underline{n}^C} = \frac{1}{2\underset{\equiv}{s} \vdots \underline{n}_1 \, \underline{n}_2 \, \underline{n}_3 \, \underline{n}_3} = \frac{1}{s_{36}} = \frac{-\nu_{63}}{G_6}$$

which represents the coupling[5] between the shear stress $\tau_6 = \tau_{12}$ and the longitudinal deformation $\epsilon_3 = \epsilon_{33}$. We will come back to this kind of transversal projections in the next chapter.

The RS for $\underset{\equiv}{s}, r = \underset{\equiv}{s} \vdots \underline{n} \, \underline{n} \, \underline{n} \, \underline{n}$ can be thus interpreted as the reciprocal of the RS for the longitudinal elastic modulus $E_{\underline{n}}$: in directions in which the radius vector of the RS for $\underset{\equiv}{s}$ is large, the modulus will be small, and viceversa. The RS gives information on the orientational dependence of the longitudinal elastic modulus.

The two reciprocal surfaces of the longitudinal elastic compliance and the longitudinal or Young modulus are shown in Fig. 8.17 for monocrystalline NaCl. The

---

[5]This shear stress-longitudinal deformation coupling is zero for the isotropic material, but non-zero in the general case.

compliance is maximum and the modulus E minimum along crystallographic directions of the type $\langle 1\,1\,1\rangle$. The opposite is true along directions of the type $\langle 1\,0\,0\rangle$.

A final question to check understanding: if the longitudinal modulus is reciprocal to the projection $\underline{\underline{s}} : n\,n\,n\,n$, and $\underline{\underline{c}}$ and $\underline{\underline{s}}$ are reciprocal, why have we not computed Young modulus in the direction $\underline{n}$ as:

$$E_{\underline{n}} = \underline{\underline{c}} : \underline{n}\,\underline{n}\,\underline{n}\,\underline{n} \; ? \qquad \text{(wrong!)}$$

(Hint: p. 101 may be of help.)

8.5 Moduli in Two Dimensions

Far from being an abstract case, many materials, such as fabrics, textile materials, films, sheets, mats, paper, biological and artificial membranes, etc. can be considered bidimensional. Even laminate composite materials, although essentially three dimensional, are reduced in Helmholtz's Classical Laminate Plate Theory [4] to two dimensions.

In this section we consider orientational dependence of properties for two-dimensional materials, focusing on engineering moduli.

In two dimensions, the elastic compliance has six independent components only. It is traditional to use a shortened version of Voigt notation in which the indices take only the values 1 and 2, and the double index 12 is condensed into 6, as if it were in three dimensions, even though any condensed index containing 3 (3, 4 and 5 do not exist) does not exist.

In two dimensions, stress, strain and elastic compliance are represented in condensed notation as:

$$\vec{\tau} = \begin{bmatrix} \tau_1 \\ \tau_2 \\ \tau_6 \end{bmatrix} \qquad \vec{\epsilon} = \begin{bmatrix} \epsilon_1 \\ \epsilon_2 \\ \epsilon_6 \end{bmatrix} \qquad \underset{\sim}{s} = \begin{bmatrix} s_{11} & s_{12} & s_{16} \\ s_{12} & s_{22} & s_{26} \\ s_{16} & s_{26} & s_{66} \end{bmatrix} \qquad (8.18)$$

where the relationships to the in-plane engineering moduli are the same as in 3D: $E_1 = \frac{1}{s_{11}}$, $E_2 = \frac{1}{s_{22}}$, $G_6 = \frac{1}{s_{66}}$, etc.

If the RS is restricted to the plane of the material, i.e. if the vector $\underline{n}$ only samples orientations contained in the plane of the material, the RS is reduced to a flat curve, the representation line (RL), in the plane of the material. Often, the material is only approximately 2D, e.g. a coarse-grained description of a more detailed, 3D structure. In this case, the RL is the intersection of the full, 3D RS of the material with the plane onto which the material structure has been coarse-grained. The RL is the polar representation of the dependence of $\underline{\underline{s}} : \underline{n}\,\underline{n}\,\underline{n}\,\underline{n}$, or its reciprocal, the longitudinal modulus E.

Fig. 8.18 Schematic of a two dimensional balanced plain weave, and unit vectors deviated φ with respect to the conventional axes ①②

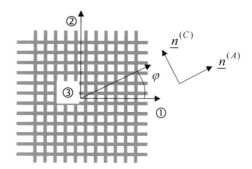

As a simple example we will consider a balanced[6] plain weave fabric of known properties. If considered as a strictly 2D material, it belongs to the holoedric $4mm$ plane group; the structures str( ) of its properties will be given by those of the tetragonal holoedric class $4/mmm$, reduced to two dimensions, as in (8.18).

Figure 8.18 schematically shows conventional axes ①②③, the last one perpendicular to the plane of the material and irrelevant in what follows.

Restriction to the plane reduces the orientational degrees of freedom to a single angle, the azimuth φ in this example. The two unit vectors represented in the axes of Fig. 8.18 are:

$$[\underline{n}^A] = \begin{bmatrix} \cos \varphi \\ \sin \varphi \\ 0 \end{bmatrix} \qquad [\underline{n}^C] = \begin{bmatrix} -\sin \varphi \\ \cos \varphi \\ 0 \end{bmatrix}$$

so that the longitudinal modulus of the material is given by its projection in one direction: $\underline{\underline{s}} \vdots \underline{n}^A \underline{n}^A \underline{n}^A \underline{n}^A$. Evaluating this expression with str($\underline{\underline{s}}$) for class $4/mmm$ and $\underline{n}^A$ above yields:

$$\underline{\underline{s}} \vdots \underline{n}^A \underline{n}^A \underline{n}^A \underline{n}^A = s_{1111}(\sin^4 \varphi + \cos^4 \varphi) + 2s_{1122}\sin^2 \varphi \cos^2 \varphi + 4s_{1212}\sin^2 \varphi \cos^2 \varphi$$

or in condensed notation:

$$\underline{\underline{s}} \vdots \underline{n}^A \underline{n}^A \underline{n}^A \underline{n}^A = s_{11}(\sin^4 \varphi + \cos^4 \varphi) + 2s_{12}\sin^2 \varphi \cos^2 \varphi + s_{66}\sin^2 \varphi \cos^2 \varphi =$$

$$s_{11} + (s_{66} - 2s_{11} + 2s_{12})\sin^2 \varphi \cos^2 \varphi \tag{8.19}$$

i.e. a constant (a circle of radius s_{11}) modulated by a $\sin^2 2\varphi$ term. The dependence of Young modulus on angle:

$$E = \frac{1}{s_{\underline{n}^A \underline{n}^A \underline{n}^A \underline{n}^A}} = \frac{1}{s_{11} + \frac{1}{4}(s_{66} - 2s_{11} + 2s_{12})\sin^2 2\varphi} \tag{8.20}$$

---

[6]Threads of the same weight (size) and the same number of ends as picks per unit length [5].

As the direction $(\underline{n}^A)$ changes, the modulus passes through maxima and minima:

$$\frac{d}{d\varphi}\left(s_{11} + (s_{66} - 2s_{11} + 2s_{12})\sin^2 2\varphi\right) = 0 \quad \Rightarrow$$

$$\sin\varphi\cos^3\varphi - \cos\varphi\sin^3\varphi = \sin\varphi\cos\varphi(\cos^2\varphi - \sin^2\varphi) = 0$$

which is fulfilled for $\varphi = 0°$, $\varphi = 90°$ (minimum of $\underset{\equiv}{s} \vdots \underline{n}^A\underline{n}^A\underline{n}^A\underline{n}^A$ and maxima for

E, if $s_{66} - 2s_{11} + 2s_{12} > 0$); or $\varphi = 45°$, $\varphi = 135°$ (maxima for $\underset{\equiv}{s} \vdots \underline{n}^A\underline{n}^A\underline{n}^A\underline{n}^A$ and

minima for E under the same condition).

The shear modulus is obtained as $\frac{1}{4}$ the reciprocal of the projection $\underset{\equiv}{s} \vdots \underline{n}^C\underline{n}^A\underline{n}^C\underline{n}^A$:

$$s_{\underline{n}^C\underline{n}^A\underline{n}^C\underline{n}^A} = \underset{\equiv}{s} \vdots \underline{n}^A\underline{n}^C\underline{n}^A\underline{n}^C = \frac{1}{4}\left(s_{66} - (s_{66} - 2s_{11} + 2s_{12})\sin^2 2\varphi\right) \qquad (8.21)$$

$$G = \frac{1}{s_{66} - (s_{66} - 2s_{11} + 2s_{12})\sin^2 2\varphi} \qquad (8.22)$$

The angular dependence of E and G on φ is qualitatively shown in Fig. 8.19, both of them for the same material.

A first order semiquantitative estimation of the RL for this fabric can be obtained by making these admittedly coarse simplifications:

- warp and weft identical in all respects,
- warp and weft are mechanically independent when the material is subjected to traction along the threads. According to this assumption, if, for example, traction is applied in the direction of the warp, the separation of the weft threads remains the same as before the load,
- under a shear load, there is only weak interaction between warp and weft threads (e.g. mediated by contact and friction). This is tantamount to assuming the shear modulus being much smaller than $G \ll E$.

Under these assumptions, the independence of warp and weft implies vanishing Poisson's ratio:

$$\nu_{12} = \nu_{21} \simeq 0 \quad \Rightarrow \quad s_{12} = \frac{-\nu_{21}}{E_2} \simeq 0$$

while the assumption $E \gg G$ leads to $s_{11} \ll s_{66}$. Taking as a rough estimate $s_{66} = 10s_{11}$, and inserting in (8.20) and (8.22):

$$E(\varphi) \simeq \frac{1}{s_{11}}\frac{1}{1 + 2\sin^2 2\varphi} \qquad (8.23)$$

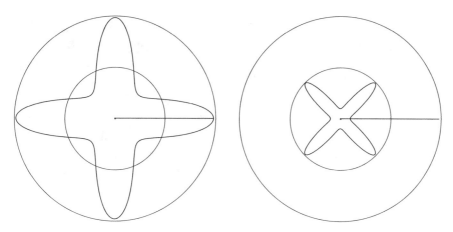

Fig. 8.19 Angular dependence of the dimensionless Young modulus E (left) and shear modulus G (right). The horizontal line is the origin of azimuth φ; both moduli have been scaled by s_{11}. Outer circle radius is 1 in both panels

The angular dependence (8.23) is represented in the left panel of Fig. 8.19. As expected, in the directions of the threads (warp and weft $\varphi = 90$ y $\varphi = 180°$) the modulus is largest and equal to $E = \dfrac{1}{s_{11}}$.

In an analogous fashion the shear modulus dependence on orientation is:

$$G(\varphi) \simeq \frac{1}{s_{11}} \frac{1}{10 - 8\sin^2 2\varphi} = \frac{1}{s_{66}} \frac{1}{1 - 0.8\sin^2 2\varphi} \tag{8.24}$$

The scale of both curves in Fig. 8.19 is the same. They have been scaled by s_{11}, i.e. the plots represent Es_{11} and Gs_{11}. Notice also that $E(\varphi) \leq \dfrac{1}{s_{11}}$ $\forall \varphi$ while $G(\varphi) \geq \dfrac{1}{s_{66}}$ $\forall \varphi$.

The RL on the left panel is obtained as the intersection of a tetragonal $4/mmm$ RS, like that of Fig. 8.5, with the plane ①②.

For this simplified fabric model the angular dependence is very strong; real fabrics have less pronounced maxima and minima. But even in real fabrics E and G are strongly anisotropic, and if the fabric is balanced, maxima and minima of E and G are shifted by $45°$.

While the meaning of "direction" is clear for a longitudinal projection or a longitudinal modulus E, in the definition of G as $\frac{1}{4}$ the reciprocal of $\underline{\underline{s}} \vdots \underline{n}^C \underline{n}^A \underline{n}^C \underline{n}^A$ we encounter two projecting directions, one of them $\underline{n}^C$ transversal to the direction of the radius vector $\underline{n}^A$. What is then to be understood by the "direction" of the shear modulus G?

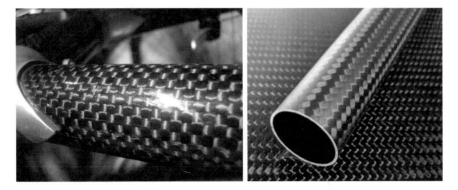

Fig. 8.20   Tubes made of a plain weave of carbon fiber oriented at 0° (left) and 45° (right)

In the case of an experimental measurement of the shear modulus G of a 2D material, the vectors $\underline{n}^A$ and $\underline{n}^C$ would be the directions in which the force is exerted, and the normal to the side of the surface element on which the force is exerted (see the definition in Sect. 4.5). In this sense there is no single direction for G.

The question of "a direction" also arises when representing the angular dependence of G. The angle φ directly gives the orientation of $\underline{n}^A$, whereas $\underline{n}^C$ is advanced by 90° (Fig. 8.19) with respect to $\underline{n}^A$. In a plot of G as a function of φ the vector $\underline{n}^A$ the value of G is read along the direction of $\underline{n}^A$. Although the same value of G would be obtained if it were read off along $\underline{n}^C$ because of the minor symmetries of $\underline{\underline{s}}$.

As Fig. 8.19 shows, the material has highest E moduli in the directions of the two sets of threads. Minima correspond to a 45° deviation of the load.

The shear modulus G however is highest at 45°, i.e. when $\underline{n}^A$ and $\underline{n}^C$ point along the diagonal of the squares defined by warp and weft.

Knowledge of this orientational dependence allows optimal use of a fabric for a particular use. A typical application of plain weave carbon composite is in structures like bicycle frames. The tube acts as a beam under flexural load, which leads to one half of the tube cross section being in traction, the other half in compression. Shear plays an insignificant role.

The axes ①② of Fig. 8.18 match the 2D coordinate system of Fig. 8.21. Written in this 2D coordinate system the stress has one normal component[7]:

$$\llbracket \underline{\tau} \rrbracket = \begin{bmatrix} \tau_{11} & 0 \\ 0 & 0 \end{bmatrix}$$

Optimal material use is achieved when it is oriented so that the maximum elongational modulus is attained in the direction of the tube axis, which corresponds to 0°, i.e. fibers parallel to the tube axis. This is indeed the arrangement found in the majority of straight members of composite bicycle frames (Fig. 8.20, left).

---

[7] τ_{11} is not homogeneous. Can you plot τ_{11} as a function of distance from the mid-plane?

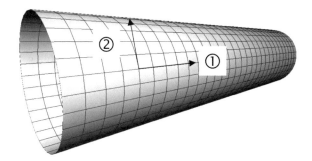

Fig. 8.21 Axes on the surface of a hollow cylinder. Axis ① (axial) points along the cylinder generatrix, axis ② (tangential) is tangent to the tube and perpendicular to ①

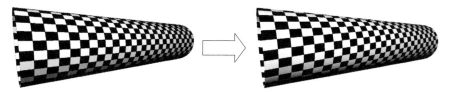

Fig. 8.22 A local homogeneous angular deformation causes global torsion in the tube

However, if a thin walled[8] hollow composite tube is to be used as a power transmission shaft, a torque will be applied to it and it will operate under torsion.

A torque (global) produces a shear stress (local) in the tube wall. Similarly, the (global) torsion of the tube corresponds to a (local) angular deformation. Written in the same 2D coordinate system of Fig. 8.21, the (homogeneous) stress and strain are:

$$[\![\underline{\underline{\tau}}]\!] = \begin{bmatrix} 0 & \tau_{12} \\ \tau_{12} & 0 \end{bmatrix} \qquad [\![\underline{\underline{\epsilon}}]\!] = \begin{bmatrix} 0 & \epsilon_{12} \\ \epsilon_{12} & 0 \end{bmatrix}$$

In this case, optimal material use is attained by maximizing the shear modulus G. According to the RL of Fig. 8.19, G goes through maxima when the directions of $\underline{n}^A$ and $\underline{n}^C$ are at $\varphi = 45°$ to the axes, i.e. to the threads of warp and weft.[9]

Maximizing G is the reason why the fibers in fabric composite members subjected to torsion are oriented at $\varphi = 45°$ with respect to the axial direction, although at this angle the longitudinal modulus E is minimum.

There is no fabric orientation for which both longitudinal and shear modulus are at a maximum simultaneously. A compromise between both has to be found if the tube is subjected to a combined flexion and torsion (or traction and torsion) load (Fig. 8.22).

---

[8]So that strain and stress can be considered constant through the thickness.

[9]For tetragonal point group $4/mmm$ it is also correct to align the axes ①② along the diagonals of the squares formed by warp and weft threads.

8.6   Isotropy

The fourth rank analog of (3.7) for isotropic properties with minor or major index symmetry is:

$$\underline{\underline{T}} = A\underline{\delta}\,\underline{\delta} + B\left(\underline{\underline{I}} + \underline{\underline{I}}^T\right) \qquad T_{ijkl} = A\delta_{ij}\delta_{kl} + B(\delta_{il}\delta_{kj} + \delta_{ik}\delta_{jl}) \qquad (8.25)$$

Symmetry under index swapping is the reason for having to use the combination $\underline{\underline{I}} + \underline{\underline{I}}^T$. The addition of two (or more) tensors, each of which alone does not have the desired symmetry, is called *symmetrization*.

The RS for the basic components of (8.25) are:

$$\underline{\delta}\,\underline{\delta} : \underline{u}\,\underline{u}\,\underline{u}\,\underline{u} = \delta_{ij}\delta_{kl}u_l u_k u_j u_i = u_i u_i u_j u_j = 1 \qquad (8.26)$$

$$\underline{\underline{I}} : \underline{u}\,\underline{u}\,\underline{u}\,\underline{u} = \delta_{il}\delta_{kj}u_l u_k u_j u_i = u_l u_j u_j u_l = 1 \qquad (8.27)$$

$$\underline{\underline{I}}^T : \underline{u}\,\underline{u}\,\underline{u}\,\underline{u} = \delta_{ik}\delta_{jl}u_l u_k u_j u_i = u_l u_j u_l u_j = 1 \qquad (8.28)$$

i.e. a centred sphere of unit radius in each case. For the particular case of the elastic stiffness $\underline{\underline{c}}$ we have:

$$\underline{\underline{c}} = \lambda\underline{\delta}\,\underline{\delta} + \mu\left(\underline{\underline{I}} + \underline{\underline{I}}^T\right) \qquad (8.29)$$

where λ and μ are the Lamé constants of the material. Thus, the RS of $\underline{\underline{c}}$ for an isotropic material is a centered sphere of radius $\lambda + 2\mu$.

Using Voigt notation, the expression (8.25) fills in the matrix $\underset{\smile}{c}$:

$$\begin{bmatrix} \lambda + 2\mu & \lambda & \lambda & 0 & 0 & 0 \\ \lambda & \lambda + 2\mu & \lambda & 0 & 0 & 0 \\ \lambda & \lambda & \lambda + 2\mu & 0 & 0 & 0 \\ 0 & 0 & 0 & \mu & 0 & 0 \\ 0 & 0 & 0 & 0 & \mu & 0 \\ 0 & 0 & 0 & 0 & 0 & \mu \end{bmatrix} \qquad (8.30)$$

From (8.30), the independent elements of $\underset{\smile}{c}$ are related to the Lamé constants by:

$$c_{12} = \lambda \qquad c_{11} = \lambda + 2\mu$$

As expected, (8.30) conforms with the $str(\underset{\sim}{c})$ from Table 3.14. Since the matrices $\underset{\sim}{c}$ and $\underset{\sim}{s}$ are inverse to one another:

$$
\begin{bmatrix}
\lambda+2\mu & \lambda & \lambda & 0 & 0 & 0 \\
\lambda & \lambda+2\mu & \lambda & 0 & 0 & 0 \\
\lambda & \lambda & \lambda+2\mu & 0 & 0 & 0 \\
0 & 0 & 0 & \mu & 0 & 0 \\
0 & 0 & 0 & 0 & \mu & 0 \\
0 & 0 & 0 & 0 & 0 & \mu
\end{bmatrix}
=
\begin{bmatrix}
\frac{1}{E} & \frac{-\nu}{E} & \frac{-\nu}{E} & 0 & 0 & 0 \\
\frac{-\nu}{E} & \frac{1}{E} & \frac{-\nu}{E} & 0 & 0 & 0 \\
\frac{-\nu}{E} & \frac{-\nu}{E} & \frac{1}{E} & 0 & 0 & 0 \\
0 & 0 & 0 & \frac{2(1+\nu)}{E} & 0 & 0 \\
0 & 0 & 0 & 0 & \frac{2(1+\nu)}{E} & 0 \\
0 & 0 & 0 & 0 & 0 & \frac{2(1+\nu)}{E}
\end{bmatrix}^{-1}
$$

$$(8.31)$$

Inverting $\underset{\sim}{s}$ and identifying terms:

$$
c_{11} = \frac{s_{11}+s_{12}}{(s_{11}-s_{12})(s_{11}+2s_{12})} \qquad
c_{12} = \frac{-s_{12}}{(s_{11}-s_{12})(s_{11}+2s_{12})} \tag{8.32}
$$

from which the well-known relationships between Lamé's constants and engineering moduli are obtained:

$$
E = \frac{\mu(3\lambda+2\mu)}{\lambda+\mu} \qquad \nu = \frac{\lambda}{2(\lambda+\mu)}
$$

$$
\lambda = \frac{E\nu}{(1+\nu)(1-2\nu)} \qquad \mu = \frac{E}{2(1+\nu)} = G
$$

Since isotropic second properties have a diagonal matrix representation (6.5), it is reasonable to ask now why the matrices (8.31) are not diagonal for isotropic properties.

In the elastic case, from the physical point of view, Poisson's coefficient is responsible for the non-diagonality of $\underset{\sim}{s}$ or $\underset{\sim}{c}$: a normal stress such as τ_{11} produces a longitudinal strain ϵ_{11} in the same direction but also two equal transversal strains $\epsilon_{12} = \epsilon_{13} = -\nu\epsilon_{11}$ as well.

Similarly, a plane state of stress like:

$$
[\![\underline{\underline{\tau}}]\!] =
\begin{bmatrix}
\tau_{11} & 0 & 0 \\
0 & \tau_{22} & 0 \\
0 & 0 & 0
\end{bmatrix}
\quad \Rightarrow \quad
\vec{\tau} =
\begin{bmatrix}
\tau_1 \\
\tau_2 \\
0 \\
0 \\
0 \\
0
\end{bmatrix}
$$

will produce a strain:

$$\vec{\epsilon} = \begin{bmatrix} \dfrac{1}{E}\tau_1 - \dfrac{\nu}{E}\tau_2 \\ \dfrac{1}{E}\tau_2 - \dfrac{\nu}{E}\tau_1 \\ -\dfrac{\nu}{E}(\tau_1 + \tau_2) \\ 0 \\ 0 \\ 0 \end{bmatrix}$$

which is neither proportional to $\vec{\tau}$ nor plane.

Although the analogy between the second and fourth rank cases exists, there are two differences: the similarity is based on the use of Voigt's notation that makes the fourth rank $\underset{\equiv}{s}$ look like a matrix $\underset{\sim}{s}$. Secondly, even using Voigt notation, $\underset{\sim}{s}$ is 6×6, i.e. it is defined in a 6-dimensional space, not in 3D Euclidean space.

Thus, a stress like:

$$\vec{\tau} = \begin{bmatrix} \tau_1 & \tau_2 & 0 & 0 & 0 & 0 \end{bmatrix}^T$$

is defined in a space in which the base vectors are the three (unit) normal stresses and the three (unit) tangential stresses. Similarly, the three (unit) longitudinal strains and the three (unit) angular strains are the base vectors for strain. From the non-diagonal form of $\underset{\sim}{s}$ we can conclude that three of the base vectors are not eigenvectors of $\underset{\sim}{s}$. If they were, a τ_1 would exclusively produce a ϵ_1, which is not the case.

It is possible to diagonalise $\underset{\sim}{s}$ or $\underset{\sim}{c}$ in this 6-dimensional space:

- the simple eigenvalue $\dfrac{1-2\nu}{E}$ with associated eigenvector $\begin{bmatrix} 1 \\ 1 \\ 1 \\ 0 \\ 0 \\ 0 \end{bmatrix}$

- the double eigenvalue $\dfrac{1+\nu}{E}$ with eigenvectors $\begin{bmatrix} 1 \\ -1 \\ 0 \\ 0 \\ 0 \\ 0 \end{bmatrix}$ and $\begin{bmatrix} 1 \\ 0 \\ -1 \\ 0 \\ 0 \\ 0 \end{bmatrix}$

- the triple eigenvalue $\dfrac{1}{G}$ with eigenvectors $\begin{bmatrix} 0 \\ 0 \\ 0 \\ 1 \\ 0 \\ 0 \end{bmatrix}$, $\begin{bmatrix} 0 \\ 0 \\ 0 \\ 0 \\ 1 \\ 0 \end{bmatrix}$ and $\begin{bmatrix} 0 \\ 0 \\ 0 \\ 0 \\ 0 \\ 1 \end{bmatrix}$

so that the matrix $\underset{\sim}{s}^*$ referred to the eigenvector base:

$$\tau_1^* = \begin{bmatrix} 1 \\ 1 \\ 1 \\ 0 \\ 0 \\ 0 \end{bmatrix} \quad \tau_2^* = \begin{bmatrix} 1 \\ -1 \\ 0 \\ 0 \\ 0 \\ 0 \end{bmatrix} \quad \tau_3^* = \begin{bmatrix} 1 \\ 0 \\ -1 \\ 0 \\ 0 \\ 0 \end{bmatrix} \quad \tau_4^* = \begin{bmatrix} 0 \\ 0 \\ 0 \\ 1 \\ 0 \\ 0 \end{bmatrix} \quad \tau_5^* = \begin{bmatrix} 0 \\ 0 \\ 0 \\ 0 \\ 1 \\ 0 \end{bmatrix} \quad \tau_6^* = \begin{bmatrix} 0 \\ 0 \\ 0 \\ 0 \\ 0 \\ 1 \end{bmatrix}$$

is diagonal:

$$\underset{\sim}{s}^* = \begin{bmatrix} \frac{1-2\nu}{E} & 0 & 0 & 0 & 0 & 0 \\ 0 & \frac{1+\nu}{E} & 0 & 0 & 0 & 0 \\ 0 & 0 & \frac{1+\nu}{E} & 0 & 0 & 0 \\ 0 & 0 & 0 & \frac{1}{G} & 0 & 0 \\ 0 & 0 & 0 & 0 & \frac{1}{G} & 0 \\ 0 & 0 & 0 & 0 & 0 & \frac{1}{G} \end{bmatrix} = \begin{bmatrix} \frac{1-2\nu}{E} & 0 & 0 & 0 & 0 & 0 \\ 0 & \frac{1+\nu}{E} & 0 & 0 & 0 & 0 \\ 0 & 0 & \frac{1+\nu}{E} & 0 & 0 & 0 \\ 0 & 0 & 0 & \frac{2(1+\nu)}{E} & 0 & 0 \\ 0 & 0 & 0 & 0 & \frac{2(1+\nu)}{E} & 0 \\ 0 & 0 & 0 & 0 & 0 & \frac{2(1+\nu)}{E} \end{bmatrix}$$

Notice that the last five of the six eigenvectors produce traceless deformations ($\epsilon_{ii} = 0$), i.e. imply no change in volume; while the first one produces isotropic expansion:

- this first eigenvector τ_1^* consists of three equal normal stress, i.e. it is a hydrostatic state of stress. The effect $\epsilon_1^* = \underset{\sim}{s}^* \tau_1^*$ is an isotropic deformation:

$$\underset{\sim}{s}^* \tau_1^* = \underset{\sim}{s}^* \begin{bmatrix} 1 \\ 1 \\ 1 \\ 0 \\ 0 \\ 0 \end{bmatrix} \quad \Rightarrow \quad \epsilon_1^* = \frac{1-2\nu}{E} \begin{bmatrix} 1 \\ 1 \\ 1 \\ 0 \\ 0 \\ 0 \end{bmatrix} = \frac{1-2\nu}{E} \tau_1^*$$

that is, the effect is a scalar copy of the cause with a scale factor $\frac{1-2\nu}{E}$.
- τ_2^* and τ_3^* are combinations of a positive (tensile) and a negative (compressive) stresses, so that their Poisson effects cancel mutually:

$$
\underset{\smile}{s}{}^{*}\underset{\smile}{\tau}{}_{2}^{*} = \underset{\smile}{s}
\begin{bmatrix} 1 \\ -1 \\ 0 \\ 0 \\ 0 \\ 0 \end{bmatrix}
\quad \Rightarrow \quad
\epsilon_2^{*} = \frac{1+\nu}{E}
\begin{bmatrix} 1 \\ -1 \\ 0 \\ 0 \\ 0 \\ 0 \end{bmatrix}
= \frac{1+\nu}{E} \tau_2^{*}
$$

and analogously for τ_3^{*}. Again, cause and effect are proportional to each other, the proportionality constant being $\frac{1+\nu}{E}$.

- the three remaining eigenvectors are shear stresses:

$$
\underset{\smile}{s}{}^{*}\underset{\smile}{\tau}{}_{4}^{*} = \underset{\smile}{s}
\begin{bmatrix} 0 \\ 0 \\ 0 \\ 1 \\ 0 \\ 0 \end{bmatrix}
\quad \Rightarrow \quad
\epsilon_4^{*} = \frac{1}{G}
\begin{bmatrix} 0 \\ 0 \\ 0 \\ 1 \\ 0 \\ 0 \end{bmatrix}
= \frac{1}{G} \tau_4^{*}
$$

and similarly for the remaining two: in an isotropic material a shear stress in a given plane produces an angular deformation in the same plane. There is no angular Poisson effect for the isotropic material. The proportionality between cause and effect is $\dfrac{1}{G} = \dfrac{1+2\nu}{E}$ in this case.

In summary, for the six individual different stresses described above, cause (stress) and effect (strain) are colinear. In the six-dimensional base formed by $\tau_1^{*}, \ldots, \tau_6^{*}, \underset{\smile}{s}$ is diagonal. However, the diagonal elements are not all identical, as is the case for second rank properties for an isotropic material. This is not surprising, since fourth rank isotropy involves two independent numerical constants, instead of just one as in second rank. The analogy with second rank isotropy cannot be pushed too far.

8.7   Transversal Isotropy

Transversal isotropy for fourth rank properties with minor and major index symmetry can be constructed using the same scheme as for second rank properties (Sect. 6.3). It is somewhat more involved, since the number of basic building blocks turns out to be larger.

The building blocks will be all independent fourth rank tensors that have minor and major index symmetries, and that are constructed from either directionally neutral objects, like $\underset{=}{\underset{=}{\delta}}\,\underset{=}{\delta}$, $\underset{\equiv}{I}$ and $\underset{\equiv}{I}^{T}$, or from $\underline{u}$, which contains the directional information of the normal to the isotropy plane:

1. $\underset{=}{\delta}\,\underset{=}{\delta} \equiv \underline{\delta}_i\underline{\delta}_j\underline{\delta}_k\underline{\delta}_l \delta_{ij}\delta_{kl}$
2. $\underset{\equiv}{I} + \underset{\equiv}{I}^{T} = \underline{\delta}_i\underline{\delta}_j\underline{\delta}_k\underline{\delta}_l(\delta_{il}\delta_{jk} + \delta_{ik}\delta_{jl})$
3. in addition to the isotropic ones, the next simplest option based on $\underline{u}$ is:

$$\underline{u}\,\underline{u}\,\underline{u}\,\underline{u} = \delta_i\delta_j\delta_k\delta_l u_i u_j u_k u_l$$

which has the correct symmetry.

4. using $\underline{u}\,\underline{u}$ and $\underline{\underline{\delta}}$ it is possible to build $\underline{u}\,\underline{u}\,\underline{\underline{\delta}}$, which has minor but lacks major symmetry: $u_i u_j \delta_{kl} \neq u_k u_l \delta_{ij}$. This drawback is easily solved by taking its symmetrized form as third building block:

$$\underline{u}\,\underline{u}\,\underline{\underline{\delta}} + \underline{\underline{\delta}}\,\underline{u}\,\underline{u} = \delta_i\delta_j\delta_k\delta_l (u_i u_j \delta_{kl} + \delta_{ij} u_k u_l)$$

Under the index swapping demanded by major symmetry, both terms exchange their identities:

$$\underline{u}\,\underline{u}\,\underline{\underline{\delta}} \Leftrightarrow \underline{\underline{\delta}}\,\underline{u}\,\underline{u}$$

so that the sum $(\underline{u}\,\underline{u}\,\underline{\underline{\delta}} + \underline{\underline{\delta}}\,\underline{u}\,\underline{u})$ remains unchanged and thus symmetric by major index swapping.

5. from $\underline{u}\,\underline{u}$ and $\underline{\underline{\delta}}$ it is also possible to form $\underline{u}\,\underline{\underline{\delta}}\,\underline{u}$. It does not fulfill major nor minor index symmetry. But the following symmetrized form does:

$$\underline{u}\,\underline{\underline{\delta}}\,\underline{u} +^T(\underline{u}\,\underline{\underline{\delta}}\,\underline{u}) + (\underline{u}\,\underline{\underline{\delta}}\,\underline{u})^T +^T(\underline{u}\,\underline{\underline{\delta}}\,\underline{u})^T =$$
$$\delta_i\delta_j\delta_k\delta_l (u_i \delta_{jk} u_l + u_j \delta_{ik} u_l + u_i \delta_{jl} u_k + u_j \delta_{il} u_k)$$

Under major and minor symmetry index swapping the four terms transform among themselves, so that the sum again remains unchanged, as required:

$$\underline{u}\,\underline{\underline{\delta}}\,\underline{u} \quad \Leftrightarrow \quad (\underline{u}\,\underline{\underline{\delta}}\,\underline{u})^T$$

$$\Updownarrow \qquad\qquad\qquad \Updownarrow$$

$$^T(\underline{u}\,\underline{\underline{\delta}}\,\underline{u}) \quad \Leftrightarrow \quad {}^T(\underline{u}\,\underline{\underline{\delta}}\,\underline{u})^T$$

The five basic building blocks for transversal isotropy are thus:

$$\underline{\underline{\delta}}\,\underline{\underline{\delta}}$$
$$\underline{\underline{\underline{I}}} + \underline{\underline{\underline{I}}}^T$$
$$\underline{u}\,\underline{u}\,\underline{u}\,\underline{u}$$
$$\underline{u}\,\underline{u}\,\underline{\underline{\delta}} + \underline{\underline{\delta}}\,\underline{u}\,\underline{u}$$
$$\underline{u}\,\underline{\underline{\delta}}\,\underline{u} +^T(\underline{u}\,\underline{\underline{\delta}}\,\underline{u}) + (\underline{u}\,\underline{\underline{\delta}}\,\underline{u})^T +^T(\underline{u}\,\underline{\underline{\delta}}\,\underline{u})^T$$

Their respective RSs are obtained, as usual, by projecting with $\vdots\,\underline{n}\,\underline{n}\,\underline{n}\,\underline{n}$.

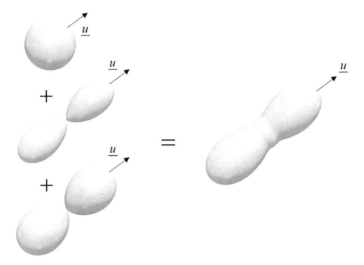

Fig. 8.23 The general form of the RS of a fourth rank transversally isotropic tensor (right) is the sum of the RSs of the five basic objects (8.33). Two of them are spheres, one is a surface of revolution obtained by rotating $\cos^4 \alpha$ about $\underline{u}$ (middle left), the last two are surfaces of revolution obtained by rotating $\cos^2 \alpha$ about $\underline{u}$ (bottom left)

- $\underline{\delta}\,\underline{\delta} : \underline{n}\,\underline{n}\,\underline{n}\,\underline{n} = \delta_{ij} n_i n_j \delta_{kl} u_k u_l = (\underline{n} \cdot \underline{n})^2 = 1$ is a centred sphere of unit radius.
- $\left(\underline{\underline{I}} + \underline{\underline{I}}^T \right) : \underline{n}\,\underline{n}\,\underline{n}\,\underline{n} = 2$ is a centred sphere of radius 2,
- $\underline{u}\,\underline{u}\,\underline{u}\,\underline{u} : \underline{n}\,\underline{n}\,\underline{n}\,\underline{n} = \cos^4 \alpha$ (Fig. 8.23, left) where α is the angle between the longitudinal direction $\underline{u}$ and the radius vector $\underline{n}$, as in Fig. 5.13. It is a surface of revolution obtained by rotating the curve $\cos^4 \alpha$ about the vector $\underline{u}$.
- $(\underline{u}\,\underline{u}\,\underline{\delta} + \underline{\delta}\,\underline{u}\,\underline{u}) : \underline{n}\,\underline{n}\,\underline{n}\,\underline{n} = 2 \cos^2 \alpha$ which we have already met (Fig. 6.1),
- $\left[\underline{u}\,\underline{\delta}\,\underline{u} +^T (\underline{u}\,\underline{\delta}\,\underline{u}) + (\underline{u}\,\underline{\delta}\,\underline{u})^T +^T (\underline{u}\,\underline{\delta}\,\underline{u})^T \right] : \underline{n}\,\underline{n}\,\underline{n}\,\underline{n} = 4 \cos^2 \alpha$, similar to the previous one.

The most general form for a transversally isotropic fourth rank property is a linear combination of these five building blocks:

$$\underline{\underline{T}} = D\underline{\delta}\,\underline{\delta} + E \left(\underline{\underline{I}} + \underline{\underline{I}}^T \right) + A\underline{u}\,\underline{u}\,\underline{u}\,\underline{u} + B(\underline{u}\,\underline{u}\,\underline{\delta} + \underline{\delta}\,\underline{u}\,\underline{u})$$
$$+ C \left[\underline{u}\,\underline{\delta}\,\underline{u} +^T (\underline{u}\,\underline{\delta}\,\underline{u}) + (\underline{u}\,\underline{\delta}\,\underline{u})^T +^T (\underline{u}\,\underline{\delta}\,\underline{u})^T \right] \quad (8.33)$$

in which five material-specific parameters A, B, C, D, E appear. This number is consistent with the number of independent components in str( ).

The vector radius for (8.33) will be the sum of these five contributions. The RS (Fig. 8.23, right) belongs to the limit class ∞/mm point group, which in agreement with Neumann's principle, includes or is equal to the point groups of the fourth rank transversally isotropic hexagonal classes and limit classes ∞, ∞m, ∞/m, $\infty 2$, ∞/mm.

In spite of there being five independent components in str( ), the RS is built from only three basic geometric objects. If the radius of the sphere is taken as length unit, the shape (not the size) of the RS depends on two parameters only (see Table 8.1).

In the particular case of the elastic stiffness $\underset{\equiv}{c}$ the constants D and E could be assimilated to Lamé's constants, so that:

$$\underset{\equiv}{c} = \lambda \underline{\underline{\delta}} \, \underline{\underline{\delta}} + \mu \left(\underline{\underline{I}} + \underline{\underline{I}}^T \right) + A \underline{u} \, \underline{u} \, \underline{u} \, \underline{u} + B(\underline{u} \, \underline{u} \, \underline{\underline{\delta}} + \underline{\underline{\delta}} \, \underline{u} \, \underline{u})$$

$$+ C \left[\underline{u} \, \underline{\underline{\delta}} \, \underline{u} +^T (\underline{u} \, \underline{\underline{\delta}} \, \underline{u}) + (\underline{u} \, \underline{\underline{\delta}} \, \underline{u})^T +^T (\underline{u} \, \underline{\underline{\delta}} \, \underline{u})^T \right] \qquad (8.34)$$

Notice also that up to this point no reference to any kind of axes has been made. The construction of the transversally isotropic fourth rank tensor is exclusively based on $\underline{u}$.

The structure $\text{str}(\underset{\smile}{c})$ follows from (8.34) by "filling in" the elements of $\underset{\smile}{c}$ with the contributions from each of the five terms:

1. $\lambda \underline{\underline{\delta}} \, \underline{\underline{\delta}}$ contributes the following components to $\underset{\smile}{c}$:

$$\lambda \begin{bmatrix} 1 & 1 & 1 & 0 & 0 & 0 \\ & 1 & 1 & 0 & 0 & 0 \\ & & 1 & 0 & 0 & 0 \\ & & & 0 & 0 & 0 \\ & & & & 0 & 0 \\ & & & & & 0 \end{bmatrix}$$

2. $\mu \left(\underline{\underline{I}} + \underline{\underline{I}}^T \right)$ contribution:

$$\mu \begin{bmatrix} 2 & 0 & 0 & 0 & 0 & 0 \\ & 2 & 0 & 0 & 0 & 0 \\ & & 2 & 0 & 0 & 0 \\ & & & 1 & 0 & 0 \\ & & & & 1 & 0 \\ & & & & & 1 \end{bmatrix}$$

3. $A \underline{u} \, \underline{u} \, \underline{u} \, \underline{u}$ contribution:

$$
A
\begin{bmatrix}
u_1^4 & u_1^2 u_2^2 & u_1^2 u_3^2 & u_1^2 u_2 u_3 & u_1^3 u_3 & u_1^3 u_2 \\
 & u_2^4 & u_2^2 u_3^2 & u_2^3 u_3 & u_2^2 u_1 u_3 & u_2^3 u_1 \\
 & & u_3^4 & u_3^3 u_2 & u_3^3 u_1 & u_3^2 u_1 u_2 \\
 & & & u_2^2 u_3^2 & u_2 u_3^2 u_1 & u_2^2 u_3 u_1 \\
 & & & & u_1^2 u_3^2 & u_1^2 u_2 u_3 \\
 & & & & & u_1^2 u_2^2
\end{bmatrix}
$$

4. $B \left(\underline{u}\,\underline{u}\,\underline{\underline{\delta}} + \underline{\underline{\delta}}\,\underline{u}\,\underline{u} \right)$ contribution:

$$
B
\begin{bmatrix}
2u_1^2 & u_1^2 + u_2^2 & u_1^2 + u_3^2 & u_2 u_3 & u_1 u_3 & u_1 u_2 \\
 & 2u_2^2 & u_2^2 + u_3^2 & u_2 u_3 & u_1 u_3 & u_1 u_2 \\
 & & 2u_3^2 & u_2 u_3 & u_1 u_3 & u_1 u_2 \\
 & & & 0 & 0 & 0 \\
 & & & & 0 & 0 \\
 & & & & & 0
\end{bmatrix}
$$

5. $C \left[\underline{u}\,\underline{\underline{\delta}}\,\underline{u} + {}^T(\underline{u}\,\underline{\underline{\delta}}\,\underline{u}) + (\underline{u}\,\underline{\underline{\delta}}\,\underline{u})^T + {}^T(\underline{u}\,\underline{\underline{\delta}}\,\underline{u})^T \right]$ contribution:

$$
C
\begin{bmatrix}
4u_1^2 & 0 & 0 & 0 & 2u_1 u_3 & 2u_1 u_2 \\
 & 4u_2^2 & 0 & 2u_2 u_3 & 0 & 2u_1 u_2 \\
 & & 4u_3^2 & 2u_2 u_3 & 2u_1 u_3 & 0 \\
 & & & u_2^2 + u_3^2 & u_2 u_1 & u_3 u_1 \\
 & & & & u_1^2 + u_3^2 & u_2 u_3 \\
 & & & & & u_1^2 + u_2^2
\end{bmatrix}
$$

The resulting $\underset{\sim}{c}$ is:

$$
\underset{\sim}{c} =
\begin{bmatrix}
M1 & M2 \\
M2 & M3
\end{bmatrix}
\tag{8.35}
$$

where the blocks $M1$ (symmetric), $M2$ and $M3$ (symmetric) are:

$$
M1 =
\begin{bmatrix}
\lambda + 2\mu + (Au_1^2 + 2B + 4C)u_1^2 & \lambda + Au_1^2 u_2^2 + B(u_1^2 + u_2^2) & \lambda + Au_1^2 u_3^2 + B(u_1^2 + u_3^2) \\
 & \lambda + 2\mu + (Au_2^2 + 2B + 4C)u_2^2 & \lambda + Au_2^2 u_3^2 + B(u_2^2 + u_3^2) \\
 & & \lambda + 2\mu + (Au_3^2 + 2B + 4C)u_3^2
\end{bmatrix}
$$

$$
M2 =
\begin{bmatrix}
(Au_1^2 + B)u_2 u_3 & (Au_1^2 + 2C + B)u_1 u_3 & (Au_1^2 + 2C + B)u_1 u_2 \\
(Au_2^2 + 2C + B)u_2 u_3 & (Au_2^2 + B)u_1 u_3 & (Au_2^2 + 2C + B)u_1 u_2 \\
(Au_3^2 + 2C + B)u_2 u_3 & (Au_3^2 + 2C + B)u_1 u_3 & (Au_3^2 + B)u_1 u_2
\end{bmatrix}
$$

$$
M3 =
\begin{bmatrix}
\mu + Au_2^2 u_3^2 + C(u_2^2 + u_3^2) & (Au_3^2 + C)u_1 u_2 & (Au_2^2 + C)u_3 u_1 \\
 & \mu + Au_1^2 u_3^2 + C(u_1^2 + u_3^2) & (Au_1^2 + C)u_2 u_3 \\
 & & \mu + Au_1^2 u_2^2 + C(u_1^2 + u_2^2)
\end{bmatrix}
$$

Similarly to transversal isotropy in second rank (6.9), the str($\underset{\sim}{c}$) seems to be

triclinic:
$$\begin{bmatrix} \cdot & \cdot & \cdot & \cdot & \cdot & \cdot \\ & \cdot & \cdot & \cdot & \cdot & \cdot \\ & & \cdot & \cdot & \cdot & \cdot \\ & & & \cdot & \cdot & \cdot \\ & & & & \cdot & \cdot \\ & & & & & \cdot \end{bmatrix}.$$
Again, the reason is that the longitudinal ("l") vector $\underline{u}$ has

arbitrary orientation. If, according to convention, we set $\underline{u} = ③$, then $[\![\underline{u}]\!] = \begin{bmatrix} 0 \\ 0 \\ 1 \end{bmatrix}$.

Substituting these values of u_i in (8.35):

$$\underset{\sim}{c} = \begin{bmatrix} \lambda + 2\mu & \lambda & \lambda + B & 0 & 0 & 0 \\ & \lambda + 2\mu & \lambda + B & 0 & 0 & 0 \\ & & \lambda + 2\mu + A & 0 & 0 & 0 \\ & & & \mu + C & 0 & 0 \\ & & & & \mu + C & 0 \\ & & & & & \mu \end{bmatrix}$$

which has the str($\underset{\sim}{c}$):
$$\begin{bmatrix} \ddots & \vdots & \vdots & \cdot & \cdot & \cdot \\ & \cdot & \vdots & \cdot & \cdot & \cdot \\ & & \ddots & \cdot & \cdot & \cdot \\ & & & \ddots & \cdot & \cdot \\ & & & & \nearrow & \cdot \\ & & & & & \times \end{bmatrix}$$
expected for transversally isotropic properties.

The relationship indicated by the "$\times$", $c_{66} = \dfrac{1}{2}(c_{11} - c_{12})$ is fulfilled as well. Finally, and not surprisingly, if the three basic objects that contain directional information are deleted (by making $A = B = C = 0$) the isotropic str($\underset{\sim}{c}$) is recovered:

$$\underset{\sim}{c} = \begin{bmatrix} \lambda + 2\mu & \lambda & \lambda & 0 & 0 & 0 \\ & \lambda + 2\mu & \lambda & 0 & 0 & 0 \\ & & \lambda + 2\mu & 0 & 0 & 0 \\ & & & \mu & 0 & 0 \\ & & & & \mu & 0 \\ & & & & & \mu \end{bmatrix}$$

8.8 Polycrystalline Averages

Most materials, especially in bulk applications, are polycrystalline: a huge assembly of non-eumorphic crystals or grains. Very frequently, both the point group and the properties of the grains and of the polycrystalline material are different. Polycrystalline averaging consists in obtaining the properties of the (macroscopic) material from the properties of the (microscopic) grains and from information on how these

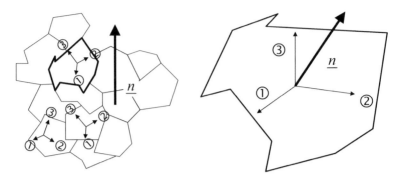

Fig. 8.24 Grains in a bulk polycrystalline material. The macroscopic properties of the bulk material (left) are obtained by averaging the contributions of each grain to the laboratory frame. In the case of a uniaxially oriented, or fiber texture (right), the vector $\underline{n}$ is the external, fixed orientation direction, which looks different in the reference frame of each grain

grains are oriented. With these two basic ingredients, bulk properties are calculated as averages over all grain orientations.

Grain orientation and texture can be modified by processing conditions. Knowledge of the particular texture of a material is a prerequisite for polycrystalline averaging.[10] In this section we will limit ourselves to applying RSs to the evaluation of basic polycrystalline orientational averages.

Taking as an illustrative example a unidirectionally oriented material (a fiber texture), a processing operation like filament drawing gives grains a tendency to orient along an external direction, (vector $\underline{n}$ in Fig. 8.24).

From the point of view of the bulk material (the laboratory, Fig. 8.24, left), each grain has a particular orientation e.g. defined by the Euler angles that transform the grain frame of reference to the laboratory frame. Each grain makes a contribution to the macroscopic value of the property that depends on its orientation with respect to the laboratory.

From the point of view of an individual grain (Fig. 8.24, right) it is the laboratory that is rotated. The contribution of each grain to the external direction $\underline{n}$ is given by the projection of the property along $\underline{n}$, as seen from the grain. The bulk property is calculated as the average of all grain contributions.

Since the number of grains is so large, it makes sense to define their orientation by a continuous *orientational distribution function $F(\ )$*. This function quantitatively describes how probable a particular orientation of the grain is.

If we want $F(\ )$ to describe the orientation of a single vector,[11] the distribution $F(\ )$ depends on two angles only. $F(\varphi, \theta)$ is then defined on the surface of the sphere S^2: $\varphi \in [0, 2\pi)$, $\theta \in [0, \pi]$. Every point on the surface corresponds to the orientation of

---

[10]References [6, 7] are excellent references.

[11]E.g. if the conventional ③ axes of the metal grains in a wire have a tendency to be oriented along the drawing direction.

a vector that goes from the center of sphere to the point. $F(\varphi, \theta)$ gives the probability of that particular orientation of the ③ axis.

If we need to describe the orientation of two axes or of a reference frame,[12] the distribution $F(\ )$ depends on three angles: φ, θ, ψ is then defined on $S^2 \times L^1$: $\varphi \in [0, 2\pi)$, $\theta \in [0, \pi]$, $\psi \in [0, 2\pi)$. For every point on the sphere (defined by fixing φ, θ), we have the additional ψ degree of freedom.

We will shorten expressions using the notation dA in integrals. So $\int \ldots dA$ is equivalent to $\int_0^\pi \ldots \sin\theta d\theta \int_0^{2\pi} \ldots d\varphi$ if only two angles are necessary to specify a direction, or to $\int_0^\pi \ldots \sin\theta d\theta \int_0^{2\pi} \ldots d\varphi \int_0^{2\pi} \ldots d\psi$ if three angles are necessary to specify a reference frame, so that:

$$\int_{S^2} dA = \int_0^\pi \sin\theta d\theta \int_0^{2\pi} d\varphi = 4\pi$$

$$\int_{S^2 \times L^1} dA = \int_0^\pi \sin\theta d\theta \int_0^{2\pi} d\varphi \int_0^{2\pi} d\psi = 8\pi^2$$

The orientational distribution function is a probability distribution and, as such, normalized:

$$\int_{S^2} F(\varphi, \theta) dA = \int_0^\pi \sin\theta d\theta \int_0^{2\pi} d\varphi F(\varphi, \theta) = 1$$

$$\int_{S^2 \times L^1} F(\varphi, \theta, \psi) dA = \int_0^\pi \sin\theta d\theta \int_0^{2\pi} d\varphi \int_0^{2\pi} d\psi = 1$$

Let's first gain some visual impression on the simplest orientational distribution functions. The uniform, constant orientational distribution function on the sphere is:

$$F(\varphi, \theta) = \frac{1}{\int_0^{2\pi} d\varphi \int_0^\pi \sin\theta d\theta} = \frac{1}{4\pi} \tag{8.36}$$

Unit vectors drawn from this distribution point in all possible 3D directions with equal probability.

Figure 8.25 gives an intuitive idea of orientational uniformity. The left panel contains 4000 segments of equal length whose orientations have been sampled from (8.36), located at random within a cubic sample volume.

In the right panel, for each of the segments, one of its ends has been placed at the origin while maintaining its orientation, so that the 4000 segments radiate from a common center. The other end of each segment has been drawn as a point, the segment itself has not been drawn.

Since they are of unit length, all segment ends lie on the surface of a sphere. The uniformity of the orientational distribution (8.36) is easily recognized as the uniform density of points on the surface of the unit sphere.

---

[12]E.g. the orientation of the PVDF molecules in the poling process of Fig. 4.9.

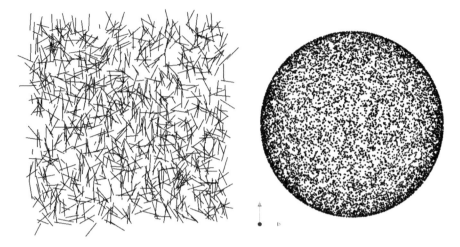

Fig. 8.25 Uniform orientational distribution of segments. 4000 segments whose orientations have been sampled from the distribution (8.36) are shown in the left panel. In the right panel the segment (not drawn!) joining the center of the sphere and a given point on the surface of the sphere represents the orientation of one of the segments of the left panel

A non-uniform distribution which favors certain orientations is shown in Fig. 8.26. Segment orientations that are more parallel to a given, fixed external direction are more probable. If the angle between a given segment and the fixed orientation (up-down on the page) is θ, the probability is proportional to $\cos^2 \theta$: $F(\varphi, \theta) \propto \cos^2 \theta$. The distribution is independent of φ and has axial symmetry.

Segments (Fig. 8.26, left panel) whose orientation is $\theta \simeq 0$ or $\theta \simeq 180$ are more probable than those perpendicular to the external direction ($\theta \simeq 90°$).

Figure 8.26 gives an intuitive idea of the concentration of orientations towards the poles of the sphere, i.e. orientations parallel to the external one (up-down on the paper).

Simple functional forms for distributions that tend to favor orientations perpendicular to the external direction ($\theta \simeq 90°$) are shown in Fig. 8.29.[13] Segment ends tend to cluster in the equator of the sphere. Notice that for the same value of the exponent in $F(\varphi, \theta, \psi)$, i.e. $\sin^n \theta$ versus $\cos^n \theta$, orientations seem to be more sharply concentrated for $\cos^n \theta$ than for $\sin^n \theta$ (compare the right panels of Figs. 8.28 and 8.29). Can you explain why (Fig. 8.27)?

Once the orientation of the monocrystals is known through $F(\ )$, the property of the polycrystalline material is obtained by averaging the contributions of all monocrystals or grains, weighted by the orientational probability $F(\ )$.

In the following we will assume the general case of the individual grains being triclinic and we will first focus on the electric resistivity $\underline{\underline{\rho}}$ as a representative second rank property. The conventional axes of the material and all components of $\underline{\underline{\rho}}$ for the monocrystalline material in these axes are known.

---

[13] Why are only even exponents used in these examples?

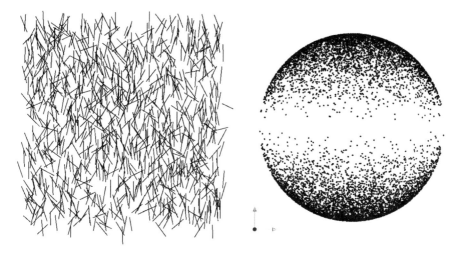

Fig. 8.26 Orientational distribution function $F(\varphi, \theta, \psi) \propto \cos^2 \theta$ (8.40). Segment orientations parallel to the external direction (up-down on the page) are more probable

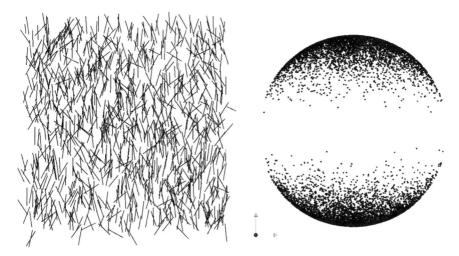

Fig. 8.27 Orientational distribution function $F(\varphi, \theta, \psi) \propto \cos^4 \theta$

Since the individual grains are electrically neither in series nor in parallel, exactly adding up the contributions of the grains is a complex task that requires numerical methods. A quite acceptable result can however be obtained by the empirical Voigt–Reuss–Hill (VRH) approximation [8]:

$$\text{VRH} = \frac{\text{series} + \text{parallel}}{2}$$

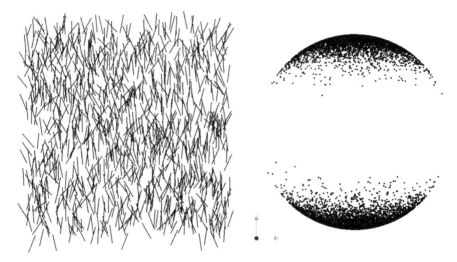

Fig. 8.28 Orientational distribution function $F(\varphi, \theta, \psi) \propto \cos^8 \theta$. As the exponent increases the orientation along the external direction is more pronounced

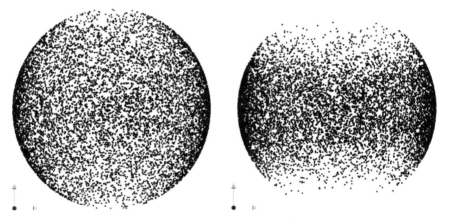

Fig. 8.29 Orientational distribution function $F(\varphi, \theta, \psi) \propto \sin^2 \theta$ (left) and $F(\varphi, \theta, \psi) \propto \sin^8 \theta$ (right). Segment orientations perpendicular to the external direction are more probable

that is, the average between the value of the property computed under the assumption that all grains are in series and under the assumption that they are in parallel. VRH is the arithmetic average of the averages.

For generalized resistivities, the series polycrystalline average will be an upper bound, the parallel polycrystalline average a lower bound. The opposite will be true for generalized conductivities.

Assuming then that all grains are in series, the contributions from the grains can be added linearly. In the frame of the laboratory, the reference frame (conventional axes) of a given grain has a random orientation sampled from the orientational distribution function $F()$. Seen *from the point of view of the grain*, a given, fixed external direction is given by a unit vector $\underline{n}$. The contribution of a single grain to the macroscopic resistivity will be the projection of the grain resistivity along the external direction:

$$\rho_{\underline{n}} = \underline{\underline{\rho}} : \underline{n}\,\underline{n}$$

so that the (macroscopic) resistivity of the polycrystalline material will be the average:

$$\langle \rho_{\underline{n}} \rangle = \langle \underline{\underline{\rho}} : \underline{n}\,\underline{n} \rangle = \underline{\underline{\rho}} : \langle \underline{n}\,\underline{n} \rangle$$

where the meaning of the angle brackets is:

$$\langle X \rangle = \int_{S^2} X F() dA$$

i.e. the weighted average of any observable X of any tensorial rank, with $F()$ as weighting function. Thus:

$$\langle \underline{n}\,\underline{n} \rangle = \int_{S^2} \underline{n}\,\underline{n} F() dA \tag{8.37}$$

For a symmetric second rank property such as $\underline{\underline{\rho}}$, the integral (8.37) can be carried out by expanding $\rho_{\underline{n}} = \rho_{ij} n_i n_j$:

$$\rho_{\underline{n}} = \underline{\underline{\rho}} : \underline{n}\,\underline{n} = \rho_{11} n_1^2 + 2\rho_{12} n_1 n_2 + 2\rho_{13} n_1 n_3 + \rho_{22} n_2^2 + 2\rho_{23} n_2 n_3 + \rho_{33} n_3^2$$

so that:

$$\langle \rho'_{\underline{n}} \rangle = \langle \underline{\underline{\rho}} : \underline{n}\,\underline{n} \rangle = \underline{\underline{\rho}} : \langle \underline{n}\,\underline{n} \rangle =$$
$$\rho_{11} \langle n_1^2 \rangle + 2\rho_{12} \langle n_1 n_2 \rangle + 2\rho_{13} \langle n_1 n_3 \rangle + \rho_{22} \langle n_2^2 \rangle + 2\rho_{23} \langle n_2 n_3 \rangle + \rho_{33} \langle n_3^2 \rangle \tag{8.38}$$

The prime $'$ (as in ρ' in Eq. (8.38)) will be used to distinguish properties of the monocrystal (unprimed) and those of the polycrystal (primed). The averages in (8.38) are:

$$\langle n_1^2 \rangle = \frac{1}{4\pi} \int_0^{2\pi} d\varphi \int_0^{\pi} \sin\theta d\theta \sin^2\theta \cos^2\varphi = \frac{1}{3}$$

$$\langle n_2^2 \rangle = \frac{1}{4\pi} \int_0^{2\pi} d\varphi \int_0^{\pi} \sin\theta d\theta \sin^2\theta \sin^2\varphi = \frac{1}{3}$$

$$\langle n_3^2 \rangle = \frac{1}{4\pi} \int_0^{2\pi} d\varphi \int_0^{\pi} \sin\theta \cos^2\theta d\theta = \frac{1}{3}$$

while the averages of the cross terms vanish. For example:

$$\langle n_1 n_2 \rangle = \frac{1}{4\pi} \int_0^{2\pi} d\varphi \int_0^{\pi} \sin\theta d\theta \sin^2\theta \sin\varphi \cos\varphi = 0$$

The polycrystalline average is then:

$$\langle \rho' \rangle = \rho_{11}\langle n_1^2 \rangle + \rho_{22}\langle n_2^2 \rangle + \rho_{33}\langle n_3^2 \rangle = \frac{1}{3}(\rho_{11} + \rho_{22} + \rho_{33}) \tag{8.39}$$

so that the resistivity of the polycrystalline material is the arithmetic average of the diagonal elements of $\underline{\underline{\rho}}$.

If the grains are not uniformly oriented but have a preferential orientation (fiber texture), the procedure is identical. The only difference is the orientational distribution that describes the texture.

Let's assume that, e.g. due to processing, the conventional ③ axis of the grains have a tendency to be aligned along a fixed external direction, the tendency being described by a simple $\cos^2\theta$ orientational distribution function of the ③ axes of the grains:

$$F(\varphi, \theta, \psi) = \frac{3}{8\pi^2}\cos^2\theta \tag{8.40}$$

where the numerical prefactor $\frac{3}{8\pi^2}$ is the normalization constant of the probability distribution:

$$\int_{S^2} F(\varphi, \theta, \psi) dA = \int_0^{2\pi} d\varphi \int_0^{\pi} \sin\theta d\theta \int_0^{2\pi} d\psi F(\varphi, \theta, \psi) = 1$$

The polycrystalline material is not isotropic anymore. According to the axes convention, the external direction will become axis ③' and the polycrystalline material will belong to point group ∞/mm, with two independent components in $\text{str}(\underline{\underline{\rho}})$: $\langle \rho'_{11} \rangle = \langle \rho'_{22} \rangle \neq \langle \rho'_{33} \rangle$. Each of them is to be computed by the same procedure as before.

There is however a small difference: in the unoriented material only one vector $\underline{n}$ was necessary to describe the orientation of the external direction in the frame of the monocrystal. Any of the rows or columns of Euler's $\underline{\underline{L}}$ or the $\underline{n}^A$ of (4.4) would do.

For the oriented material it is necessary to describe *two* orientations: that of the external, fixed direction, and the direction transversal to it. This can be accomplished by taking any two rows or any two columns of $\underline{\underline{L}}$. Taking any two of $\underline{n}^A$, $\underline{n}^B$ or $\underline{n}^C$ of (4.4) would be incorrect, because only $\underline{n}^A$ samples all possible orientations; neither $\underline{n}^B$ nor $\underline{n}^C$ sample all orientations.

Taking the first and third rows of Euler's $\underline{\underline{L}}$ as $\underline{n}_1$ and $\underline{n}_3$ $(n_{ij} = (\underline{\underline{L}})_{ij})$:

$$\langle \rho'_{33} \rangle = \rho_{11}\langle n_{31}^2 \rangle + 2\rho_{12}\langle n_{31}n_{32} \rangle + 2\rho_{13}\langle n_{31}n_{33} \rangle + \rho_{22}\langle n_{32}^2 \rangle + 2\rho_{23}\langle n_{32}n_{33} \rangle + \rho_{33}\langle n_{33}^2 \rangle$$

and

$$\langle \rho'_{11} \rangle = \rho_{11} \langle n^2_{11} \rangle + 2\rho_{12} \langle n_{11}n_{12} \rangle + 2\rho_{13} \langle n_{11}n_{13} \rangle + \rho_{22} \langle n^2_{12} \rangle + 2\rho_{23} \langle n_{12}n_{13} \rangle + \rho_{33} \langle n^2_{13} \rangle$$

we obtain for $\langle \rho'_{33} \rangle$:

$$\langle n^2_{31} \rangle = \frac{3}{8\pi^2} \int_0^{2\pi} d\varphi \int_0^{\pi} d\theta \int_0^{2\pi} d\psi \sin\theta \cos^2\theta (\sin\theta \sin\varphi)^2 = \frac{1}{5} = \langle n^2_{32} \rangle \quad (8.41)$$

$$\langle n^2_{33} \rangle = \frac{3}{8\pi^2} \int_0^{2\pi} d\varphi \int_0^{\pi} d\theta \int_0^{2\pi} d\psi \sin\theta \cos^2\theta (\cos\theta)^2 = \frac{3}{5}$$

$$\langle n_{3i}n_{3j} \rangle = 0 \quad \text{if } i \neq j$$

and for $\langle \rho'_{11} \rangle = \langle \rho'_{22} \rangle$:

$$\langle n^2_{11} \rangle = \frac{3}{8\pi^2} \int_0^{2\pi} d\varphi \int_0^{\pi} d\theta \int_0^{2\pi} d\psi \sin\theta \cos^2\theta (\cos\psi \cos\varphi - \cos\theta \sin\varphi \sin\psi)^2 = \frac{2}{5} = \langle n^2_{12} \rangle$$

$$\langle n^2_{13} \rangle = \frac{3}{8\pi^2} \int_0^{2\pi} d\varphi \int_0^{\pi} d\theta \int_0^{2\pi} d\psi \sin\theta \cos^2\theta (\sin\psi \sin\theta)^2 = \frac{1}{5}$$

$$\langle n_{1i}n_{1j} \rangle = 0 \quad \text{if } i \neq j$$

so that:

$$\langle \rho' \rangle = \begin{bmatrix} \frac{1}{5}(2\rho_{11} + 2\rho_{22} + \rho_{33}) & 0 & 0 \\ 0 & \frac{1}{5}(2\rho_{11} + 2\rho_{22} + \rho_{33}) & 0 \\ 0 & 0 & \frac{1}{5}(\rho_{11} + \rho_{22} + 3\rho_{33}) \end{bmatrix}$$
$$(8.42)$$

As expected, the orientation of the grain axis ③ towards the external direction increases the weight of the contribution of the monocrystal ρ_{33} to the polycrystal's ρ'_{33} from $\frac{1}{3}$ to $\frac{3}{5}$, and reduces its contribution to the transversal ρ'_{11} from $\frac{1}{3}$ to $\frac{1}{5}$. This effect is reversed for the contributions of ρ_{11} and ρ_{22}.

For second rank properties, this direct method in which the integral average (8.37) is split in a moderate number of integrals is amenable to hand calculation. Let us now see an alternative method which will prove essential when dealing with polycrystalline averages of fourth rank properties.

As a first step, let us visualize the meaning of the integral average (8.37) $\int \underline{n}\,\underline{n}\, dA$: this integral is the sum of all the $\underline{n}\,\underline{n}$ contributions of all surface elements dA of the sphere.

At the location of each dA, $\underline{n}$ points radially to dA on the sphere. At dA the dyadic $\underline{n}\,\underline{n}$ is defined, with $\underline{n}$ taking on its local value at dA. We already know what $\underline{n}\,\underline{n}$ looks like (Fig. 6.1). In Fig. 8.30 a surface element and the RS of its $\underline{n}\,\underline{n}$ have been drawn.

Since the lobes of the RS of $\underline{n}\,\underline{n}$ are positive, when the RS of all dA elements are added, the result has spherical symmetry $\infty\infty m$, i.e. the result of the integral has

Fig. 8.30 The representation
surface of $\underline{n}\,\underline{n}$ that belongs to
the surface element dA; $\underline{n}$ is
the local unit normal. RS are
uniformly distributed over
the surface of the unit sphere

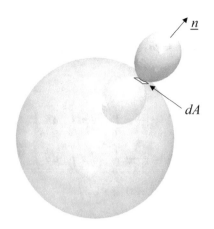

to be an isotropic second rank tensor.[14] Figure 8.31 tells us all we need to find the
solution.

Calling the result of the integral $\underline{\underline{X}}$:

$$\underline{\underline{X}} = \int \underline{n}\,\underline{n}\,dA = K\underline{\underline{\delta}} \tag{8.43}$$

we just have to determine the value of K. The most direct way is to project both
members of (8.43) onto $\underline{\underline{\delta}}$:

LHS: $\underline{\underline{\delta}} : \int \underline{n}\,\underline{n}\,dA = \int \underline{\underline{\delta}} : \underline{n}\,\underline{n}\,dA = \int n_i n_i \, dA = \int dA = 4\pi$

RHS: $K\underline{\underline{\delta}} : \underline{\underline{\delta}} = 3K$

so that $K = \dfrac{4\pi}{3}$ and the integral is:

$$\underline{\underline{X}} = \int \underline{n}\,\underline{n}\,dA = \frac{4\pi}{3}\underline{\underline{\delta}}$$

The polycrystalline average is then:

$$\langle \underline{\underline{\rho}} : \underline{n}\,\underline{n}\rangle = \underline{\underline{\rho}} : \langle \underline{n}\,\underline{n}\rangle = \underline{\underline{\rho}} : \frac{1}{4\pi}\int \underline{n}\,\underline{n}\,dA = \frac{1}{3}\underline{\underline{\rho}} : \underline{\underline{\delta}} = \frac{1}{3}\mathrm{tr}(\underline{\underline{\rho}})$$

which agrees with (8.39).

The same technique can be applied to compute the polycrystalline average of an
oriented material with a fiber texture such as (8.40). The average is now:

---

[14]What would be the value of the integral if the integrand were $\underline{n}$? If $\underline{n}\,\underline{n}\,\underline{n}$? If $\underline{n} \cdot \underline{n}$?

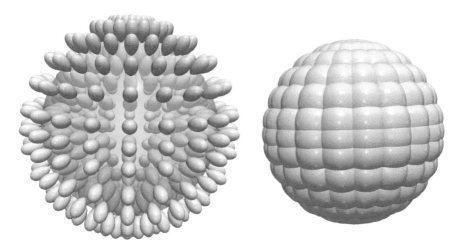

Fig. 8.31 The representation surfaces of $\underline{n}\,\underline{n}$ for 300 surface elements dA placed at the position given by $\underline{n}$ (left); the same 300 RSs placed at the common origin. The integral of $\underline{n}\,\underline{n}$ with uniform orientational distribution function $F(\varphi, \theta) = \frac{1}{4\pi}$ over the surface of the sphere is the result of adding these RSs up. In the limit, the integral has $\infty\infty m$ symmetry

$$\langle \underline{\rho} : \underline{n}\,\underline{n} \rangle = \underline{\rho} : \langle \underline{n}\,\underline{n} \rangle = \underline{\rho} : \frac{3}{4\pi} \int \underline{n}\,\underline{n} \cos^2 \theta \, dA$$

where θ is the angle between the direction of orientation $\underline{u}$ (which will be the conventional axis ③' of the oriented material) and the unit vector $\underline{n}$ that samples the surface of the sphere. Unlike for the unoriented material, for which all directions of $\underline{n}$ are equally likely (Fig. 8.31), $\underline{n}$ now has a greater tendency to align with $\underline{u}$.

From the point of view of the RS, the surface elements corresponding to these directions are more densely populated by RSs than those perpendicular to $\underline{u}$ (Fig. 8.32).

Again, the RSs of Fig. 8.31 tells us that the integral has to be transversal isotropic. Since $\cos \theta = \underline{n} \cdot \underline{u}$:

$$\langle \underline{\rho} : \underline{n}\,\underline{n} \rangle = \underline{\rho} : \langle \underline{n}\,\underline{n} \rangle = \frac{3}{4\pi}\underline{\rho} : \int \underline{n}\,\underline{n}\,(\underline{n} \cdot \underline{u})(\underline{n} \cdot \underline{u})dA = \frac{3}{4\pi}\underline{\rho} : \int \underline{n}\,\underline{n}\,\underline{n}\,\underline{n} : \underline{u}\,\underline{u}\,dA$$

and the integral has to be transversally isotropic:

$$\int \underline{n}\,\underline{n}\,\underline{n}\,\underline{n} : \underline{u}\,\underline{u}\,dA = B\underline{\delta} + C\underline{u}\,\underline{u} \tag{8.44}$$

where B and C are constants to be determined. Projecting (8.44) onto $\underline{\delta}$ and $\underline{u}\,\underline{u}$ yields the system:

Fig. 8.32 The representation surfaces of $\underline{n}\,\underline{n}$ for 300 surface elements dA with fibre texture orientational distribution function $F(\varphi, \theta, \psi) = \frac{3}{8\pi^2}\cos^2\theta$. Preferential orientation is top to bottom on the page. In the limit the integral of $\underline{n}\,\underline{n}$ over the surface of the sphere has ∞/mm symmetry. According to convention, the external direction of orientation $\underline{n}$ becomes the ③' axis

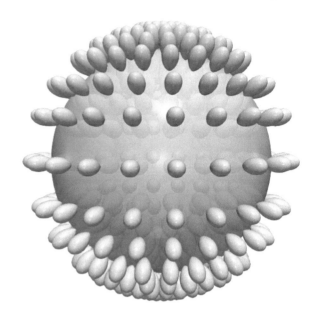

$$
\begin{cases}
\underline{\delta} : \int \underline{n}\,\underline{n}\,\underline{n}\,\underline{n} : \underline{u}\,\underline{u}\,dA = \int \underline{\underline{\delta}} : \underline{n}\,\underline{n}\,\underline{n}\,\underline{n} : \underline{u}\,\underline{u}\,dA = \int n_i n_i n_j n_j n_k u_k dA = \\
\hspace{6cm} \int \cos^2\theta\, dA = \dfrac{4\pi}{3} \\[2mm]
\underline{\delta} : (B\underline{\delta} + C\underline{u}\,\underline{u}) = 3B + C \\[2mm]
\underline{u}\,\underline{u} : \int \underline{n}\,\underline{n}\,\underline{n}\,\underline{n} : \underline{u}\,\underline{u}\,dA = \int \underline{u}\,\underline{u} : \underline{n}\,\underline{n}\,\underline{n}\,\underline{n} : \underline{u}\,\underline{u}\,dA = \int n_i u_i n_j u_j n_k u_k n_l u_l dA = \\
\hspace{6cm} \int \cos^4\theta\, dA = \dfrac{4\pi}{5} \\[2mm]
\underline{u}\,\underline{u} : (B\underline{\delta} + C\underline{u}\,\underline{u}) = B + C
\end{cases}
$$

$$
\begin{cases}
3B + C = \dfrac{4\pi}{3} \\[3mm]
B + C = \dfrac{4\pi}{5}
\end{cases}
\tag{8.45}
$$

so that $B = \dfrac{4}{15}\pi$ and $C = \dfrac{8}{15}\pi$. The polycrystalline average is then:

$$
\langle \underline{\rho} : \underline{n}\,\underline{n} \rangle = \underline{\rho} : \langle \underline{n}\,\underline{n} \rangle = \frac{3}{4\pi}\frac{4}{15}\pi\underline{\rho} : (B\underline{\delta} + C\underline{u}\,\underline{u}) = \frac{1}{5}\mathrm{tr}(\rho) + \frac{2}{5}\underline{\rho} : \underline{u}\,\underline{u}
$$

which is valid for any orientation $\underline{u}$. If, according to crystallographic convention, this orientation direction (the longitudinal or "l" direction of the transversally isotropic material) is declared to be the axis ③': $\underline{u} = \underline{\delta}_3$. Substituting above:

$$(\rho'_c)_{33} = \langle \rho'_{33} \rangle = \frac{1}{5}(\rho_{11} + \rho_{22} + 3\rho_{33})$$

which agrees with (8.42).

The transversal components $(\rho'_c)_{11} = (\rho'_c)_{22}$ can be calculated in a similar fashion. The polycrystalline average of the transversal part is obtained by projecting transversally:

$$\langle \underline{\underline{\rho}} : (\underline{\underline{\delta}} - \underline{n}\,\underline{n}) \rangle = \underline{\underline{\rho}} : \langle \underline{\underline{\delta}} \rangle - \underline{\underline{\rho}} : \langle \underline{n}\,\underline{n} \rangle = \underline{\underline{\rho}} : \underline{\underline{\delta}} \frac{3}{4\pi} \int \cos^2 \theta \, dA - \underline{\underline{\rho}} : \frac{3}{4\pi} \int \underline{n}\,\underline{n} \cos^2 \theta \, dA$$

where the second integral has been computed above. The result is:

$$\langle \underline{\underline{\rho}} : (\underline{\underline{\delta}} - \underline{n}\,\underline{n}) \rangle = \frac{4}{5}\mathrm{tr}(\rho) - \frac{2}{5}\underline{\underline{\rho}} : \underline{u}\,\underline{u}$$

which obviously satisfies $\langle \underline{\underline{\rho}} : \underline{n}\,\underline{n} \rangle + \langle \underline{\underline{\rho}} : (\underline{\underline{\delta}} - \underline{n}\,\underline{n}) \rangle = \langle \underline{\underline{\rho}} : \underline{\underline{\delta}} \rangle$. If, again, we identify $\underline{u} = \underline{\delta}'_3$:

$$\langle \underline{\underline{\rho}} : (\underline{\underline{\delta}} - \underline{n}\,\underline{n}) \rangle = \frac{1}{5}(4\rho_{11} + 4\rho_{22} + 2\rho_{33})$$

which is the correct result for the transversal part, and is *double* the value of $(\rho'_c)_{11}$ or $(\rho'_c)_{22}$ in (8.42). Why?

Notice that the coefficients in the linear system (8.45) come from the projections $\underline{\underline{\delta}} : (B\underline{\underline{\delta}} + C\underline{u}\,\underline{u})$ and $\underline{u}\,\underline{u} : (B\underline{\underline{\delta}} + C\underline{u}\,\underline{u})$ and are independent of the specific orientational distribution function. It is the RHS of the system (the terms $\frac{4\pi}{3}$, $\frac{4\pi}{5}$) that depends on the particular material texture. For different $F(\ )$, either analytical or numerical, only the integrals that yield the RHS of the system need to be carried out.

The method just described is particularly efficient in the calculation of elastic moduli for polycrystalline materials. Again, the VRH empirical rule is used to obtain an estimate as the arithmetic average of the series and parallel averages. When dealing with elastic moduli the series or linear average is usually called the Voigt average, the parallel or harmonic average is called the Reuss average.

Series (Voigt) average implies linear addition of elastic stiffnesses, parallel (Reuss) average implies harmonic addition of elastic compliances. Since the engineering moduli are reciprocal to compliances, Reuss averages of moduli are immediately obtained as $E = \dfrac{1}{\langle s'_{11} \rangle}$ etc. Voigt averages require the intermediate step of inverting $\underset{\sim}{c}'$.

Figure 8.33 describes the calculation of both averages for the unoriented polycrystal.

The direct calculation of the polycrystalline average $\langle c'_{11} \rangle$ implies expanding $c_{\underline{n}} = c_{ijkl}n_i n_k n_i n_j$ and integrating:

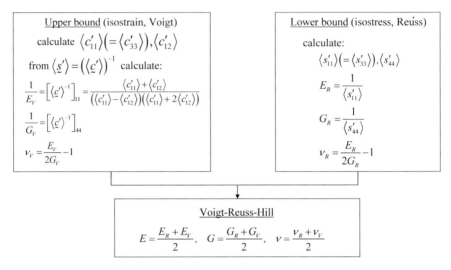

Fig. 8.33 Calculation of polycrystalline average engineering moduli; unoriented material. Subindices "V" and "R" refer to Voigt (linear) and Reuss (harmonic) averages

$$
\langle c'_{11}\rangle = \langle c'_{33}\rangle = \langle c'_{3333}\rangle = \langle n_{3p}n_{3q}n_{3r}n_{3s}c_{pqrs}\rangle =
$$

$$
c_{1111}\langle n_{31}^4\rangle + c_{2222}\langle n_{32}^4\rangle + c_{3333}\langle n_3^4\rangle +
$$

$$
2c_{1122}\langle n_{31}^2 n_{32}^2\rangle + 2c_{1133}\langle n_{31}^2 n_{33}^2\rangle + 2c_{2233}\langle n_{32}^2 n_{33}^2\rangle +
$$

$$
4c_{2223}\langle n_{32}^3 n_{33}\rangle + 4c_{2213}\langle n_{32}^2 n_{31}n_{33}\rangle + 4c_{2212}\langle n_{32}^3 n_{31}\rangle +
$$

$$
4c_{3323}\langle n_{33}^3 n_{32}\rangle + 4c_{3313}\langle n_{33}^3 n_{31}\rangle + 4c_{3312}\langle n_{33}^2 n_{31}n_{32}\rangle + \qquad (8.46)
$$

$$
4c_{1123}\langle n_{31}^2 n_{32}n_{33}\rangle + 4c_{1113}\langle n_{31}^3 n_{33}\rangle + 4c_{1112}\langle n_{31}^3 n_{32}\rangle +
$$

$$
4c_{2323}\langle n_{32}^2 n_{33}^2\rangle + 4c_{1313}\langle n_{31}^2 n_{33}^2\rangle + 4c_{1212}\langle n_{31}^2 n_{32}^2\rangle +
$$

$$
8c_{2313}\langle n_{31}n_{32}n_{33}^2\rangle + 8c_{2312}\langle n_{31}n_{32}^2 n_{33}\rangle + 8c_{1312}\langle n_{31}^2 n_{32}n_{33}\rangle
$$

where the n_{3j} are the three elements of the third row of $\underset{\sim}{L}$.[15]

These integrals are straightforward because they always factor in separate integrals for the three Euler angles, but are cumbersome because of their large number. Carrying them out in the usual way we obtain:

---

[15] Why the third row? Could we have taken any other row or column of $\underset{\sim}{L}$?

$$\langle n_{31}^4 \rangle = \langle n_{32}^4 \rangle = \langle n_{33}^4 \rangle = \frac{1}{5}$$

$$\langle n_{31}^2 n_{32}^2 \rangle = \langle n_{32}^2 n_{33}^2 \rangle = \langle n_{33}^2 n_{31}^2 \rangle = \frac{1}{15}$$

$$\langle n_{31}^2 n_{32} n_{33} \rangle = \langle n_{31} n_{32}^2 n_{33} \rangle = \langle n_{31} n_{32} n_{33}^2 \rangle = 0$$

$$\langle n_{31}^3 n_{32} \rangle = \langle n_{31}^3 n_{33} \rangle = \langle n_{32}^3 n_{33} \rangle = \langle n_{32}^3 n_{31} \rangle = \langle n_{33}^3 n_{32} \rangle = \langle n_{33}^3 n_{31} \rangle = 0$$

so that:

$$\langle c_{11}' \rangle = \langle c_{33}' \rangle = \frac{3}{15}(c_{1111} + c_{2222} + c_{3333}) + \frac{2}{15}(c_{1122} + c_{1133} + c_{2233}) +$$
$$\frac{4}{15}(c_{2323} + c_{1313} + c_{1212}) =$$
$$= \frac{3}{15}(c_{11} + c_{22} + c_{33}) + \frac{2}{15}(c_{12} + c_{13} + c_{23}) + \frac{4}{15}(c_{44} + c_{55} + c_{66})$$

$$(8.47)$$

Similar, but even lengthier, calculation yields:

$$\langle c_{12}' \rangle = \frac{1}{15}(c_{11} + c_{22} + c_{33}) + \frac{4}{15}(c_{12} + c_{13} + c_{23}) - \frac{2}{15}(c_{44} + c_{55} + c_{66})$$

$$(8.48)$$

$$\langle c_{44}' \rangle = \frac{1}{15}(c_{11} + c_{22} + c_{33}) - \frac{1}{15}(c_{12} + c_{13} + c_{23}) + \frac{3}{15}(c_{44} + c_{55} + c_{66})$$

$$(8.49)$$

It is immediate to check that these averaged stiffnesses satisfy the relationship between the element c_{44} (represented as $\times$ in str($\underset{\sim}{c}$)) and $c_{11} - c_{12}$:

$$\langle c_{44}' \rangle = \frac{1}{2}\left(\langle c_{11}' \rangle - \langle c_{12}' \rangle\right)$$

$$(8.50)$$

as they must for the isotropic polycrystal. According to the scheme of Fig. 8.33, the Voigt averages of the moduli are then:

$$\frac{1}{E_V} = \left[(\underset{\sim}{c}')^{-1}\right]_{11} = \frac{\langle c_{11}' \rangle + \langle c_{12}' \rangle}{(\langle c_{11}' \rangle - \langle c_{12}' \rangle)(\langle c_{11}' \rangle + 2\langle c_{12}' \rangle)}$$

$$\frac{1}{G_V} = \left[(\underset{\sim}{c}')^{-1}\right]_{44} = \frac{1}{\langle c_{44}' \rangle}$$

$$\nu_V = \frac{E_V}{2G_V} - 1$$

The calculation of the Reuss averages does not require a separate calculation, because only the name of the variable to be averaged changes: s instead of $\underset{\sim}{c}$, the integrals are the same. Taking into account the factors of 2 and 4 in $\underset{\sim}{s}$ in Voigt notation:

$$\frac{1}{E_R} = \langle s'_{33} \rangle = \frac{3}{15}(s_{11} + s_{22} + s_{33}) + \frac{2}{15}(s_{12} + s_{13} + s_{23}) + \frac{1}{15}(s_{44} + s_{55} + s_{66})$$

$$\frac{1}{G_R} = \langle s'_{44} \rangle = \frac{4}{15}(s_{11} + s_{22} + s_{33}) - \frac{4}{15}(s_{12} + s_{13} + s_{23}) + \frac{3}{15}(s_{44} + s_{55} + s_{66})$$

$$\langle s'_{12} \rangle = \frac{1}{15}(s_{11} + s_{22} + s_{33}) + \frac{4}{15}(s_{12} + s_{13} + s_{23}) - \frac{1}{30}(s_{44} + s_{55} + s_{66})$$

$$\nu_R = \frac{E_R}{2G_R} - 1 \tag{8.51}$$

The Voigt–Reuss–Hill empirical estimate for the moduli of the polycrystalline, unoriented material is obtained as[16]:

$$E = \frac{E_R + E_V}{2} \qquad G = \frac{G_R + G_V}{2} \qquad \nu = \frac{\nu_R + \nu_V}{2}$$

We can now switch to a textured material: the five independent engineering moduli of an oriented polycrystal with fiber texture ∞/mm described by (8.40) are obtained in a similar fashion. The corresponding VRH averaging scheme is shown in Fig. 8.34.

$$\langle c'_{11} \rangle = \langle c'_{22} \rangle = \langle c'_{1111} \rangle =$$
$$= \frac{1}{35}(9c_{11} + 9c_{22} + 3c_{33}) + \frac{1}{35}(6c_{12} + 4c_{13} + 4c_{23}) + \frac{1}{35}(8c_{44} + 8c_{55} + 12c_{66}) =$$

$$\langle c'_{33} \rangle = \langle c'_{3333} \rangle =$$
$$= \frac{1}{35}(3c_{11} + 3c_{22} + 15c_{33}) + \frac{1}{35}(2c_{12} + 6c_{13} + 6c_{23}) + \frac{1}{35}(12c_{44} + 12c_{55} + 4c_{66}) \tag{8.52}$$

$$\langle c'_{44} \rangle = \langle c'_{2323} \rangle =$$
$$= \frac{1}{35}(2c_{11} + 2c_{22} + 2c_{33}) - \frac{1}{35}(c_{12} + 3c_{13} + 3c_{23}) + \frac{1}{35}(8c_{44} + 8c_{55} + 5c_{66})$$

$$\langle c'_{12} \rangle = \langle c'_{1122} \rangle =$$
$$= \frac{1}{35}(3c_{11} + 3c_{22} + c_{33}) + \frac{1}{35}(16c_{12} + 6c_{13} + 6c_{23}) - \frac{1}{35}(2c_{44} + 2c_{55} + 10c_{66})$$

---

[16]Does the VRH-averaged material satisfy (8.50)?

Upper bound (isostrain, Voigt)

calculate $\langle c'_{11}\rangle, \langle c'_{33}\rangle, \langle c'_{12}\rangle, \langle c'_{13}\rangle, \langle c'_{44}\rangle$

from $\quad \langle \underline{s}'\rangle = \left(\langle \underline{c}'\rangle\right)^{-1}\quad$ obtain:

$$\Delta \equiv \langle c'_{33}\rangle\left(\langle c'_{11}\rangle^2 - \langle c'_{12}\rangle^2\right) + 2\langle c'_{13}\rangle^2\left(\langle c'_{12}\rangle - \langle c'_{11}\rangle\right)$$

$$\frac{1}{(E_t)_V} = \frac{1}{(E_1)_V} = \left[\langle \underline{c}'\rangle^{-1}\right]_{11} = \frac{\langle c'_{11}\rangle\langle c'_{33}\rangle - \langle c'_{13}\rangle^2}{\Delta}$$

$$\frac{1}{(E_l)_V} = \frac{1}{(E_3)_V} = \left[\langle \underline{c}'\rangle^{-1}\right]_{33} = \frac{\langle c'_{11}\rangle^2 - \langle c'_{33}\rangle^2}{\Delta}$$

$$\frac{1}{(G_{tl})_V} = \frac{1}{(G_{44})_V} = \left[\langle \underline{c}'\rangle^{-1}\right]_{44} = \frac{1}{\langle c'_{44}\rangle}$$

$$\frac{-(\nu_{21})_V}{(E_2)_V} = \frac{-(\nu_{tt})_V}{(E_t)_V} = \left[\langle \underline{c}'\rangle^{-1}\right]_{12} = \frac{\langle c'_{13}\rangle^2 - \langle c'_{12}\rangle\langle c'_{33}\rangle}{\Delta}$$

$$\frac{-(\nu_{31})_V}{(E_3)_V} = \frac{-(\nu_{tl})_V}{(E_t)_V} = \left[\langle \underline{c}'\rangle^{-1}\right]_{13} = \frac{\langle c'_{13}\rangle\left(\langle c'_{12}\rangle - \langle c'_{11}\rangle\right)}{\Delta}$$

Lower bound (isostress, Reuss)

calculate $\langle s'_{11}\rangle, \langle s'_{33}\rangle, \langle s'_{12}\rangle, \langle s'_{13}\rangle, \langle s'_{44}\rangle$

$$\frac{1}{(E_t)_R} = \frac{1}{(E_1)_R} = \langle s'_{11}\rangle$$

$$\frac{1}{(E_l)_R} = \frac{1}{(E_3)_R} = \langle s'_{33}\rangle$$

$$\frac{1}{(G_{tl})_R} = \frac{1}{(G_{44})_R} = \langle s'_{44}\rangle$$

$$\frac{-(\nu_{21})_R}{(E_2)_R} = \frac{-(\nu_{tt})_R}{(E_t)_R} = \langle s'_{12}\rangle$$

$$\frac{-(\nu_{31})_R}{(E_3)_R} = \frac{-(\nu_{tl})_R}{(E_t)_R} = \langle s'_{13}\rangle$$

Voigt-Reuss-Hill

$$E_t = \frac{(E_t)_R + (E_t)_V}{2} \quad E_l = \frac{(E_l)_R + (E_l)_V}{2} \quad G_{tl} = \frac{(G_{tl})_R + (G_{tl})_V}{2} \quad \nu_{tl} = \frac{(\nu_{tl})_R + (\nu_{tl})_V}{2} \quad \nu_{tt} = \frac{(\nu_{tt})_R + (\nu_{tt})_V}{2}$$

Fig. 8.34 Voigt–Reuss–Hill estimation of elastic moduli for an oriented polycrystal with fiber texture given by $F(\varphi, \theta, \psi)$ from (8.40)

$$\langle c'_{13}\rangle = \langle c'_{1133}\rangle =$$
$$= \frac{1}{35}\left(2c_{11} + 2c_{22} + 3c_{33}\right) + \frac{1}{35}\left(6c_{12} + 11c_{13} + 11c_{23}\right) - \frac{1}{35}\left(6c_{44} + 6c_{55} + 2c_{66}\right)$$

The effort required to obtain these results is very considerable. Each of the 21 integrals appearing in (8.46) splits in several more. For example, the integral of a term like $\langle n_{11}^2 n_{12}^2\rangle$:

$$\langle n_{11}^2 n_{12}^2\rangle = \left\langle \left(\cos\psi\cos\varphi - \cos\theta\sin\varphi\sin\psi\right)^2\left(\cos\psi\sin\varphi + \cos\theta\cos\varphi\sin\psi\right)^2\right\rangle \tag{8.53}$$

splits in another nine integrals. The evaluation of the numerical coefficients in the previous formulae for $\langle c'_{ij}\rangle$ thus involves calculating several hundred integrals. As a

Fig. 8.35 300 representation surfaces of $\underline{n}\,\underline{n}\,\underline{n}\,\underline{n}$ for fibre texture orientational distribution function $F(\varphi, \theta, \psi) = \frac{3}{8\pi^2}\cos^2\theta$ over the surface of the sphere. The resulting integral has ∞/mm symmetry. The individual RS are surfaces of revolution about the radial $\underline{n}$ with cross section $\cos^4\alpha$ whereas in Fig. 8.32 it was $\cos^2\alpha$. Preferential orientation is top to bottom on the page

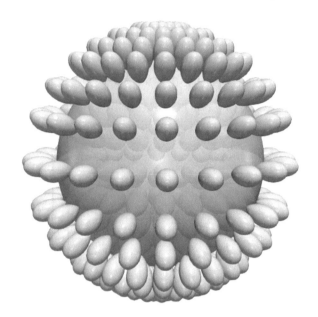

result, it is easy for errors to creep in reported results. This is clearly not the way to go.

The projection method for fourth rank properties is more efficient and much less error-prone. The polycrystalline average of a longitudinal modulus involves calculating an average of the type $\left\langle \underline{\underline{c}} : \underline{n}\,\underline{n}\,\underline{n}\,\underline{n} \right\rangle = \underline{\underline{c}} : \left\langle \underline{n}\,\underline{n}\,\underline{n}\,\underline{n} \right\rangle$. The RS for $\underline{n}\,\underline{n}\,\underline{n}\,\underline{n}$ is qualitatively similar to that of $\underline{n}\,\underline{n}$. Again, and most importantly, the lobes of the RS are both positive. If there is no tendency for the grains to be oriented, Fig. 8.35 illustrates the integration procedure. As in second rank, the result is isotropic.

Since we already know how to write an isotropic fourth rank tensor (8.7), calling $\underline{\underline{X}} \equiv \int \underline{n}\,\underline{n}\,\underline{n}\,\underline{n}\,dA$, $\underline{\underline{X}}$ must be of the form:

$$\underline{\underline{X}} = A\underline{\delta}\,\underline{\delta} + B(\underline{\underline{I}} + \underline{\underline{I}}^T) \tag{8.54}$$

where A y B to be determined by projection on the unit (isotropic) objects:

$$\underline{\delta}\,\underline{\delta} : \int \underline{n}\,\underline{n}\,\underline{n}\,\underline{n}\,dA = \underline{\delta}\,\underline{\delta} : \left(A\underline{\delta}\,\underline{\delta} + B(\underline{\underline{I}} + \underline{\underline{I}}^T) \right) \tag{8.55}$$

$$\left(\underline{\underline{I}} + \underline{\underline{I}}^T \right) : \int \underline{n}\,\underline{n}\,\underline{n}\,\underline{n}\,dA = \left(\underline{\underline{I}} + \underline{\underline{I}}^T \right) : \left(A\underline{\delta}\,\underline{\delta} + B(\underline{\underline{I}} + \underline{\underline{I}}^T) \right) \tag{8.56}$$

When multiplying the parentheses, the following products appear:

$$\underline{\underline{\delta}}\,\underline{\underline{\delta}}{:}\underline{\underline{\delta}}\,\underline{\underline{\delta}} = \delta_{ij}\delta_{kl}\delta_{lk}\delta_{ji} = \delta_{ii}\delta_{kk} = 9 \qquad \underline{\underline{I}}{:}\underline{\underline{I}} = \delta_{il}\delta_{jk}\delta_{kj}\delta_{li} = \delta_{ii}\delta_{jj} = 9$$

$$\underline{\underline{I}}^T{:}\underline{\underline{I}}^T = \delta_{ik}\delta_{jl}\delta_{lj}\delta_{ki} = \delta_{ii}\delta_{jj} = 9 \qquad \underline{\underline{\delta}}\,\underline{\underline{\delta}}{:}\underline{\underline{I}} = \delta_{ij}\delta_{kl}\delta_{kj}\delta_{li} = \delta_{ii} = 3$$

$$\underline{\underline{\delta}}\,\underline{\underline{\delta}}{:}\underline{\underline{I}}^T = \delta_{ij}\delta_{kl}\delta_{lj}\delta_{ki} = \delta_{ii} = 3 \qquad \underline{\underline{I}}{:}\underline{\underline{I}}^T = \delta_{il}\delta_{jk}\delta_{lj}\delta_{ki} = \delta_{ii} = 3$$

The LHS:

$$\underline{\underline{\delta}}{:}\int \underline{n}\,\underline{n}\,\underline{n}\,\underline{n}\, dA = \int \underline{\underline{\delta}}{:}\underline{n}\,\underline{n}\,\underline{n}\,\underline{n}\, dA = \int \delta_{ij}\delta_{kl}n_i n_k n_j n_i\, dA$$

$$= \int n_k n_k n_i n_i\, dA = \int dA = 4\pi$$

$$\left(\underline{\underline{I}} + \underline{\underline{I}}^T\right){:}\int \underline{n}\,\underline{n}\,\underline{n}\,\underline{n}\, dA = \int \left(\underline{\underline{I}} + \underline{\underline{I}}^T\right){:}\underline{n}\,\underline{n}\,\underline{n}\,\underline{n}\, dA =$$

$$\int \delta_{il}\delta_{jk}n_i n_k n_j n_i\, dA + \int \delta_{ik}\delta_{jl}n_i n_k n_j n_i\, dA =$$

$$\int n_i n_k n_k n_i\, dA + \int n_j n_i n_j n_i\, dA = 2\int dA = 8\pi$$

Equating LHS and RHS the following system is obtained:

$$\begin{cases} 9A + 6B = 1 \\ 6A + 24B = 2 \end{cases} \Rightarrow A = B = \frac{1}{15} \tag{8.57}$$

from which:

$$\frac{1}{4\pi}\int \underline{n}\,\underline{n}\,\underline{n}\,\underline{n}\, dA = \frac{1}{15}\left(\underline{\underline{\delta}}\,\underline{\underline{\delta}} + \underline{\underline{I}} + \underline{\underline{I}}^T\right) \tag{8.58}$$

$$\langle c_{11}' \rangle = \left\langle \underline{\underline{c}}{:}\underline{n}\,\underline{n}\,\underline{n}\,\underline{n}\right\rangle = \frac{1}{15}[c_{iijj} + c_{ijji} + c_{ijij}]$$

$$= \frac{3}{15}(c_{11} + c_{22} + c_{33}) + \frac{2}{15}(c_{12} + c_{13} + c_{23}) + \frac{4}{15}(c_{44} + c_{55} + c_{66})$$

which is the desired result (8.47). Unlike in the brute force method, here it was never necessary to write the components of $\underline{n}$ explicitly, i.e. referred to a specific frame. Dealing with $\underline{n}$ and $\underline{\underline{X}}$ as primary objects avoids integrals of lengthy trigonometric expressions.

The terms $\langle c_{12}' \rangle = \langle c_{1122}' \rangle$ and $\langle c_{44}' \rangle = \langle c_{2323}' \rangle$ involve a Poisson relation and a shear modulus, which we also know how to write in a frame-free fashion:

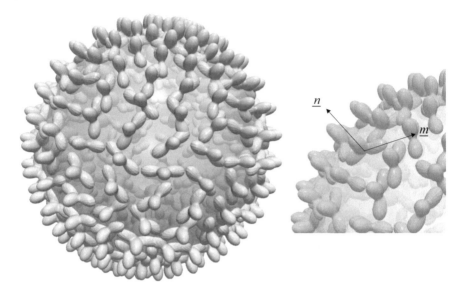

Fig. 8.36 Representation surfaces of $\underline{n}\,\underline{n}\,\underline{m}\,\underline{m}$ for 150 surface elements dA. At each element dA (right) the RS is drawn for fixed $\underline{n}$ (radial) and $\underline{m}$ (perpendicular to $\underline{n}$)

$$\langle c'_{1122}\rangle \quad \Rightarrow \quad \frac{1}{4\pi}\int \underline{n}\,\underline{n}\,\underline{m}\,\underline{m}\,dA$$

$$\langle c'_{2323}\rangle \quad \Rightarrow \quad \frac{1}{4\pi}\int \underline{n}\,\underline{m}\,\underline{n}\,\underline{m}\,dA \qquad (8.59)$$

with orthogonal $\underline{n}$ and $\underline{m}$.

Notice that for the longitudinal modulus the integrand $\underline{n}\,\underline{n}\,\underline{n}\,\underline{n}$ only involved the radial vector $\underline{n}$ at each dA. Now a second vector $\underline{m}$ appears, which is perpendicular to $\underline{n}$ and hence *not* radial. But, where does $\underline{m}$ then point to?

More importantly, if we visualize the collection of RS for $\underline{n}\,\underline{n}\,\underline{m}\,\underline{m}$ over the surface of the sphere (8.36) for some representative $\underline{n}$ and $\underline{m}$, it is not obvious at all that the result of the integral will be isotropic. On the contrary, it seems that the integral leading to $\langle c'_{1122}\rangle$ will *not* be isotropic. Since all properties of a material must be at least as symmetric as the material, does this contradict Neumann's principle?

The answer to these two questions is that the unit vector $\underline{n}$, begin radial, will sample of orientations of space as φ and θ sweep their ranges in the integration. For a fixed $\underline{n}$, the vector $\underline{m}$ must also sample (point in) all possible directions perpendicular to $\underline{n}$. Figure 8.36 is misleading because it draws a single $\underline{n}$ at a dA (correct), and a single vector $\underline{m}$ perpendicular to $\underline{n}$ (incorrect!).

The integrand has to be represented at each point dA on the surface of the sphere by a $\underline{n}$ and by all possible $\underline{m}$ perpendicular to that $\underline{n}$, i.e. by an axisymmetric object with its axis of symmetry along $\underline{n}$. Each of the axisymmetric RSs of Fig. 8.37 is the result of the integration over ψ:

Fig. 8.37 Representation surfaces of $\underline{n}\,\underline{n}\,\underline{m}\,\underline{m}$ for 150 surface elements dA with uniform orientational distribution function. At each dA, i.e. for a fixed $\underline{n}$, the vector $\underline{m}$ points in all possible directions perpendicular to $\underline{n}$

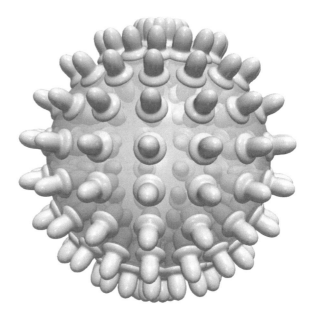

$$\int_0^{2\pi} \underline{n}\,\underline{n}\,\underline{m}\,\underline{m}\,d\psi$$

while the integrations over the remaining angles $\int_0^\pi \ldots \sin\theta d\theta \int_0^{2\pi} \ldots d\varphi$ are carried out as $\underline{n}$ samples the surface of the unit sphere.

In this way, the integration over the surface of the unit sphere is done for all possible orientations of $\underline{n}$ and $\underline{m}$. From Fig. 8.37 it is now clear that the integral will be isotropic. From Neumann's principle we knew that the result had to be isotropic, the RS of Fig. 8.37 helps us *see* it.

The need to sample all orientations with both[17] unit vectors is the ultimate reason why all three Euler angles are necessary, why two[18] unit vectors (rows or columns of Euler's $\underset{\sim}{L}$) have to be used in the averages, why lengthy terms such as (8.53) appear and why a brute force approach is not the optimal strategy.

Needless to say, the same comments apply to the integration of $\underline{n}\,\underline{m}\,\underline{n}\,\underline{m}$: we again project onto the isotropic objects, as in (8.56). For $\frac{1}{4\pi}\int \underline{n}\,\underline{n}\,\underline{m}\,\underline{m}\,dA$ we obtain the system:

$$\begin{cases} 9A + 6B = 1 \\ 6A + 24B = 0 \end{cases} \quad \Rightarrow \quad A = \frac{2}{15} \quad B = -\frac{1}{30} \tag{8.60}$$

---

[17] Or all three, if an average like $\langle c'_{1223}\rangle$ were to be computed.

[18] Again, maybe three for $\langle c'_{1223}\rangle$.

so:

$$\frac{1}{4\pi} \int \underline{n}\,\underline{n}\,\underline{m}\,\underline{m}\, dA = \frac{1}{15}\left(2\underline{\underline{\delta}}\,\underline{\underline{\delta}} - \frac{1}{2}(\underline{\underline{I}} + \underline{\underline{I}}^T)\right) \tag{8.61}$$

$$\langle c'_{12}\rangle = \left\langle \underline{\underline{c}} : \underline{n}\,\underline{n}\,\underline{m}\,\underline{m}\right\rangle = \frac{1}{15}\left[2c_{iijj} - \frac{1}{2}(c_{ijji} + c_{ijij})\right] =$$
$$= \frac{1}{15}(c_{11} + c_{22} + c_{33}) + \frac{4}{15}(c_{12} + c_{13} + c_{23}) - \frac{2}{15}(c_{44} + c_{55} + c_{66}) \tag{8.62}$$

which matches (8.48).

For $\frac{1}{4\pi}\int \underline{n}\,\underline{m}\,\underline{n}\,\underline{m}\, dA$:

$$\begin{cases} 9A + 6B = 0 \\ 6A + 24B = 1 \end{cases} \quad \Rightarrow \quad A = -\frac{1}{30} \quad B = \frac{1}{20} \tag{8.63}$$

so:

$$\frac{1}{4\pi}\int \underline{n}\,\underline{m}\,\underline{n}\,\underline{m}\, dA = \frac{1}{15}\left(-\frac{1}{2}\underline{\underline{\delta}}\,\underline{\underline{\delta}} + \frac{3}{4}(\underline{\underline{I}} + \underline{\underline{I}}^T)\right) \tag{8.64}$$

$$\langle c'_{44}\rangle = \left\langle \underline{\underline{c}} : \underline{n}\,\underline{m}\,\underline{n}\,\underline{m}\right\rangle = \frac{1}{15}\left[-\frac{1}{2}c_{iijj} + \frac{3}{4}(c_{ijji} + c_{ijij})\right]$$
$$= \frac{1}{15}(c_{11} + c_{22} + c_{33}) - \frac{1}{15}(c_{12} + c_{13} + c_{23}) + \frac{3}{15}(c_{44} + c_{55} + c_{66}) \tag{8.65}$$

which agrees with (8.49). Checking that $\langle c'_{44}\rangle = \frac{1}{2}\left(\langle c'_{11}\rangle = -\langle c'_{12}\rangle\right)$ can directly and more easily be done with (8.58), (8.61) and (8.64) than using the explicit components.

Again, notice that the coefficients of the LHS of (8.57), (8.63) and (8.60) are the same for all $\langle c'_{ij}\rangle$ averages. These coefficients are specific for the point group of the material ($\infty \infty m$) and its representation (8.54). On the other hand the RHS depend on the particular average $\langle c'_{ij}\rangle$ to be calculated.

The difference in effort between the direct approach and the projection method is even more marked for fiber textured materials. We will only calculate $\langle c'_{33}\rangle$ (8.52) for illustration purposes.

Since the material is transversally isotropic, the polycrystalline $\langle \underline{\underline{c}}'\rangle$ must be of the form:

$$\langle \underline{\underline{c}}'\rangle = A\underline{\underline{\delta}}\,\underline{\underline{\delta}} + B\left(\underline{\underline{I}} + \underline{\underline{I}}^T\right) + C\underline{u}\,\underline{u}\,\underline{u}\,\underline{u} + D(\underline{u}\,\underline{u}\,\underline{\underline{\delta}} + \underline{\underline{\delta}}\,\underline{u}\,\underline{u}) +$$
$$+ E\left(\underline{u}\,\underline{\underline{\delta}}\,\underline{u} +^T(\underline{u}\,\underline{\underline{\delta}}\,\underline{u}) + (\underline{u}\,\underline{\underline{\delta}}\,\underline{u})^T +^T(\underline{u}\,\underline{\underline{\delta}}\,\underline{u})^T\right)$$

Table 8.3 Products of the five basic objects for transversally isotropic fourth rank properties. The last line ("integrand") gives the result of contracting the basic object at the top of the table with the integrand $\underline{n}\,\underline{n}\,\underline{n}\,\underline{n}\cos^2\theta\,dA$ in (8.66)

	$\underline{\underline{\delta}}\,\underline{\underline{\delta}}$	$\underline{\underline{I}}$	$\underline{\underline{I}}^T$	$\underline{u}\,\underline{u}\,\underline{u}\,\underline{u}$	$\underline{u}\,\underline{u}\,\underline{\underline{\delta}}$	$\underline{\underline{\delta}}\,\underline{u}\,\underline{u}$	$\underline{u}\,\underline{\underline{\delta}}\,\underline{u}$	$^T(\underline{u}\,\underline{\underline{\delta}}\,\underline{u})$	$(\underline{u}\,\underline{\underline{\delta}}\,\underline{u})^T$	$^T(\underline{u}\,\underline{\underline{\delta}}\,\underline{u})^T$
$\underline{\underline{\delta}}\,\underline{\underline{\delta}}$	9	3	3	1	3	3	1	1	1	1
$\underline{\underline{I}}$		9	3	1	1	1	3	1	1	3
$\underline{\underline{I}}^T$			9	1	1	1	1	3	3	1
$\underline{u}\,\underline{u}\,\underline{u}\,\underline{u}$				1	1	1	1	1	1	1
$\underline{u}\,\underline{u}\,\underline{\underline{\delta}}$					1	3	1	1	1	1
$\underline{\underline{\delta}}\,\underline{u}\,\underline{u}$						1	1	1	1	1
$\underline{u}\,\underline{\underline{\delta}}\,\underline{u}$							3	1	1	1
$^T(\underline{u}\,\underline{\underline{\delta}}\,\underline{u})$								1	3	1
$(\underline{u}\,\underline{\underline{\delta}}\,\underline{u})^T$									1	1
$^T(\underline{u}\,\underline{\underline{\delta}}\,\underline{u})^T$										3
Integrand	1	1	1	$\cos^2\theta$	$\cos^2\theta$	$\cos^4\theta$	$\cos^2\theta$	$\cos^2\theta$	$\cos^2\theta$	$\cos^2\theta$

where $\underline{u}$ gives the external, fixed direction of orientation. The longitudinal "l" averaged projection of $\langle\underline{\underline{c}}'\rangle$ is $\left\langle\underline{\underline{c}}'\,\vdots\,\underline{u}\,\underline{u}\,\underline{u}\,\underline{u}\right\rangle$.

The five constants A, B, C, D y E are determined by projecting on the five basic objects from which transversally symmetric properties are built, as we did in (8.56). We need projection onto $\underline{\underline{\delta}}\,\underline{\underline{\delta}}$:

$$\underline{\underline{\delta}}\,\underline{\underline{\delta}}\,\vdots\int\underline{n}\,\underline{n}\,\underline{n}\,\underline{n}\cos^2\theta\,dA = \underline{\underline{\delta}}\,\underline{\underline{\delta}}\,\vdots\left[A\underline{\underline{\delta}}\,\underline{\underline{\delta}} + B\left(\underline{\underline{I}} + \underline{\underline{I}}^T\right) + C\underline{u}\,\underline{u}\,\underline{u}\,\underline{u} + \right.$$
$$\left. + D(\underline{u}\,\underline{u}\,\underline{\underline{\delta}} + \underline{\underline{\delta}}\,\underline{u}\,\underline{u}) + E\left(\underline{u}\,\underline{\underline{\delta}}\,\underline{u} + {}^T(\underline{u}\,\underline{\underline{\delta}}\,\underline{u}) + (\underline{u}\,\underline{\underline{\delta}}\,\underline{u})^T + {}^T(\underline{u}\,\underline{\underline{\delta}}\,\underline{u})^T\right)\right]$$
$$(8.66)$$

and similar projections onto the other four basic objects. In these projections a number of products appear, such as $\underline{\underline{\delta}}\,\underline{\underline{\delta}}\,\vdots\,\underline{\underline{\delta}}\,\underline{\underline{\delta}}$, $\underline{\underline{\delta}}\,\underline{\underline{\delta}}\,\vdots\,\underline{\underline{I}}$ and so on. Table 8.3 sums up the results of the fourfold contractions of all basic objects. The last line is the contraction of the basic objects with the integrand $\underline{n}\,\underline{n}\,\underline{n}\,\underline{n}\cos^2\theta\,dA$ in (8.66).

By equating both sides of (8.66) and the other four projection equations:

$$\begin{cases} 9A + 6B + C + 6D + 4E = 1 \\ 3A + 12B + C + 2D + 8E = 1 \\ A + 2B + C + 2D + 4E = \dfrac{3}{7} \\ 3A + 2B + C + 4D + 4E = \dfrac{3}{5} \\ A + 4B + C + 2D + 6E = \dfrac{3}{5} \end{cases} \Rightarrow \begin{cases} A = \dfrac{1}{35} \\ B = \dfrac{1}{35} \\ C = 0 \\ D = \dfrac{2}{35} \\ E = \dfrac{2}{35} \end{cases}$$

so that the general form of $\langle c'_{33} \rangle$ is:

$$\langle c'_{33} \rangle = \frac{1}{35} \underline{\underline{c}} : \left[\underline{\delta}\,\underline{\delta} + \left(\underline{\underline{I}} + \underline{\underline{I}}^T \right) + 2(\underline{u}\,\underline{u}\,\underline{\delta} + \underline{\delta}\,\underline{u}\,\underline{u}) + \right.$$
$$\left. + 2 \left(\underline{u}\,\underline{\delta}\,\underline{u} + ^T(\underline{u}\,\underline{\delta}\,\underline{u}) + (\underline{u}\,\underline{\delta}\,\underline{u})^T + ^T(\underline{u}\,\underline{\delta}\,\underline{u})^T \right) \right]$$

for an arbitrary $\underline{u}$ giving the orientation. If, according to convention, the direction of $\underline{u}$ is that of ③', then $\underline{u} = \underline{\delta}'_3$ and:

$$\langle c'_{33} \rangle = \frac{1}{35} \left[c_{iijj} + c_{ijji} + c_{ijij} + 4c_{ii33} + 8c_{3ii3} \right] =$$
$$= \frac{1}{35} (3c_{11} + 3c_{22} + 15c_{33}) + \frac{1}{35} (2c_{12} + 6c_{13} + 6c_{23}) + \frac{1}{35} (12c_{44} + 12c_{55} + 4c_{66})$$

in agreement with (8.52).

The reader is encouraged to use the projection method to compute a term like $\langle c'_{14} \rangle = \langle c'_{1123} \rangle$ both for unoriented and oriented polycrystals. The key point is to visualize the set of RSs that make up the integral, as in Figs. 8.35 and 8.37. To that end, first write in a frame-free fashion (as in Eqs. (8.59)) the integrand of[19]:

$$\langle c'_{1123} \rangle \quad \Rightarrow \quad \frac{1}{4\pi} \int ? \, dA$$

then draw a single RS of the integrand, as in Fig. 8.30. Once you have the RS of integrand, the sum of all the RSs is immediate.

8.9 Higher Rank Properties

Properties of rank higher than four are quite uncommon. Among other reasons, the number of components that have to be determined in an experimental measurement of the property grows (if we do not consider symmetries) exponentially. But once the values of the components are known, constructing RSs for higher rank properties is a straightforward task.

The necessity of considering properties of higher rank crops up in many contexts, including elastic properties. A material like that of Fig. 8.38 belongs to the hexagonal holoedral class $6/mmm$ (whether we consider it 2D or 3D is immaterial now). As such, it is transversally isotropic both for second and fourth rank properties, i.e. independent of the angle in which the property is measured.

On the other hand, as the material is formed by three sets of threads at 120°, one would expect its properties to show at least a certain degree of sixfold angular dependence. In a mechanical experiment, should we not measure a higher E modulus

---

[19]Hint: you will need three orthonormal vectors.

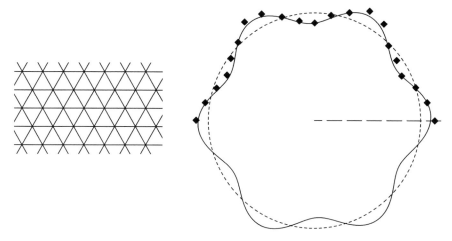

Fig. 8.38 Angular dependence (RL) of Young modulus E for the three-directional weave on the left. The scale of the horizontal dashed line 1 GPa. The dotted circle is Hooke's law prediction with E fitted to the experimental points. The experimental plot on the right is centrosymmetric, only the upper part is shown. Why only the upper half?

along the directions of the fibers? Yet the RL given by the intersection of the RS $\underset{\equiv}{s} : \underline{n}\,\underline{n}\,\underline{n}\,\underline{n}$ with the plane perpendicular to ③ is a circumference.

Figure 8.38 shows experimental measurements of E [9] as a function of the orientation angle. These measurements do show deviation from the expected independence on orientation (dotted circle in the figure). What is wrong or missing here?

The explanation of this apparent contradiction is that we are enforcing linearity through the CE on a phenomenon that is nonlinear. Elastic compliance $\underset{\equiv}{s}$ is the phenomenological coefficient that relates two second rank tensors in a Hooke's CE; since Hooke's law is *linear* in strain and stress, $\underset{\equiv}{s}$ must be fourth rank (the sum of the ranks of the gradient and flux it couples).

We now know that fourth degree polynomials in sine and cosine (Eq. (8.8)) are unable to reproduce hexagonal symmetry. The rank (four) versus symmetry (sixfold) conflict is resolved by turning into a Curie class (Fig. 8.8), so Hooke's law necessarily leads to axial symmetry for fourth rank properties and cannot possibly reproduce the experimental angular dependence of Fig. 8.38, not even qualitatively.

Although contradictory, both the calculated RL and intuition are correct. The origin of the conflict is our assumption that Hooke's linear law applies to the material. It does, to first order, but a nonlinear CE is necessary to reproduce sixfold symmetry. This can be considered as a second order correction to linear elasticity, which may take one of the forms:

$$\underline{\epsilon} = \underset{\equiv}{s} : \underline{\tau} + \underset{\equiv}{s} \vdots \underline{\tau}\,\underline{\tau} \qquad \underline{\tau} = \underset{\equiv}{c} : \underline{\epsilon} + \underset{\equiv}{c} \vdots \underline{\epsilon}\,\underline{\epsilon}$$

The sixth rank tensors $\underset{\equiv}{s}$ and $\underset{\equiv}{c}$ are often called third order elastic constants. They have been measured for a few selected materials [10]. The main reason for their introduction was to account for nonlinearity in the stress-strain relation.

But as an added bonus, the RSs of $\underset{\equiv}{s}$ and $\underset{\equiv}{c}$ are written in terms of polynomials of degree six in sine and cosine, which are naturally suited to reproduce sixfold symmetry. Experimental data like those of Fig. 8.38 can be used to obtain numerical values of the components s_{ijklmn} and c_{ijklmn}. A great experimental effort is required to determine all 56 independent components[20] of $\underset{\equiv}{s}$, so in practice it is very rare for all of them to be known. Once known, constructing the RS of third order elastic constants is simple. They do display the expected sixfold angular dependence.

The situation just described is not limited to elasticity. High rank properties have been introduced, very often in connection with surface physics and chemistry [8], in an attempt to account for deviations from linear behavior, where tensors of rank up to 10 are not unknown.

References

1. Nye, J.F.: Physical Properties of Crystals. Oxford University Press, Oxford (1985)
2. Coleman, B.D., Noll, W.: The thermodynamics of elastic materials with heat conduction and viscosity. The Foundations of Mechanics and Thermodynamics, pp. 145–156. Springer, Berlin (1974)
3. Weiner, J.H.: Statistical Mechanics of Elasticity. Wiley, New York (1983)
4. Jones, R.M.: Mechanics of Composite Materials. Taylor & Francis, Boca Raton (1999)
5. Collier, A.M.: A Handbook of Textiles. Pergamon Press, Oxford (1970)
6. Kocks, F., Tomé, C.N., Wenk, H.-R.: Texture and Anisotropy: Preferred Orientations in Polycrystals and Their Effect on Materials Properties. Cambridge University Press, Cambridge (1998)
7. Bunge, H.-J.: Texture Analysis in Materials Science: Mathematical Methods. Elsevier, Amsterdam (2013)
8. Newnham, R.E.: Properties of Materials: Anisotropy, Symmetry, Structure. Oxford University Press, Oxford (2005)
9. Gurvich, M.R., Skudra, A.M.: Effect of the geometry of the structure on the strength distribution of multilaminate tridirectional reinforced plastics. Mech. Compos. Mater. 24(5), 602–609 (1989)
10. Bechmann, R., Hearmon, R.F.S.: The third-order elastic constants. Landolt-Bornstein, Numerical Data and Functional Relationships in Science and Technology, vol. 2, p. 112. Springer, Berlin (1969)

---

[20]Can you show why $\underset{\equiv}{s}$ has 56 independent components?

Chapter 9
Transversality and Skew Symmetric Properties

The reader may have wondered why almost only RS corresponding to projections such as $: \underline{n}\,\underline{n}$, $\vdots\,\underline{n}\,\underline{n}\,\underline{n}$, $\vdots\,\underline{n}\,\underline{nn}\,\underline{n}$ have appeared up to now. As a matter of fact, we have also encountered projections along more than one direction: the calculation of the angular dependence of the shear modulus G (8.22) of a 2D fabric involved the use of two vectors (8.18). The calculation of polycrystalline averages also involved terms like $\underline{\underline{c}}\vdots\,\underline{n}\,\underline{n}\,\underline{m}\,\underline{m}$ and $\underline{\underline{c}}\vdots\,\underline{n}\,\underline{m}\,\underline{n}\,\underline{m}$ (8.59). All these expressions come from the definition of engineering moduli (Sect. 4.7) or similar magnitudes that involve more than one direction in their projection.

In spite of that, we have devoted minimal attention to projections which are not purely longitudinal. This chapter is dedicated to more general projections, and to skew-symmetric properties, in which transversality is an inherent feature.

9.1 General Projections and Representation Solids

Contractions with objects such as $:\,\underline{n}$, $:\,\underline{n}\,\underline{n}$, $\vdots\,\underline{n}\,\underline{n}\,\underline{n}$ and $\vdots\,\underline{n}\,\underline{n}\,\underline{n}\,\underline{n}$ are usually called "longitudinal"; so Young's modulus is also referred to as the "longitudinal modulus" of the material:

$$E = \frac{1}{\underline{\underline{s}}\vdots\,\underline{n}^{A}\underline{n}^{A}\underline{n}^{A}\underline{n}^{A}} \tag{9.1}$$

It relates a normal stress and a longitudinal strain, both calculated as longitudinal projections *along the same direction*. The modulus in the direction of $\underline{n}^{A}$ is the ratio between:

$$\underline{\underline{\tau}} : \underline{n}^{A}\underline{n}^{A} \quad \text{and} \quad \underline{\underline{\epsilon}} : \underline{n}^{A}\underline{n}^{A}$$

© Springer Nature Switzerland AG 2020
M. Laso and N. Jimeno, *Representation Surfaces for Physical Properties of Materials*, Engineering Materials, https://doi.org/10.1007/978-3-030-40870-1_9

Similarly, it is also usual to refer to the longitudinal (direct) piezoelectric effect as the projection along a given direction $\underline{n}^A$ of the polarization that appears in a material that is subjected to a normal stress along the same direction. I.e. it is the ratio between:

$$\underline{p} \cdot \underline{n}^A \quad \text{and} \quad \underline{\underline{\tau}} : \underline{n}^A \underline{n}^A$$

Should a modulus that relates two longitudinal projections along *two different* directions[1]:

$$\underline{\underline{\epsilon}} : \underline{n}^A \underline{n}^A \qquad \underline{\underline{\tau}} : \underline{n}^B \underline{n}^B$$

also be called longitudinal? Similar remarks can be made about the use of "transversal" (or "shear" in the mechanical context). The shear modulus G:

$$G = \frac{1}{4\underline{\underline{s}} \vdots \underline{n}^A \underline{n}^B \underline{n}^A \underline{n}^B} \tag{9.2}$$

relates a tangential stress and an angular strain, both calculated as projections along the same two directions. The G modulus in the directions $\underline{n}^A, \underline{n}^B$ is the ratio between:

$$\underline{\underline{\tau}} : \underline{n}^A \underline{n}^B \quad \text{and} \quad \underline{\underline{\epsilon}} : \underline{n}^A \underline{n}^B$$

What name should then be used for a modulus that relates the following shear stress and angular deformation[2]:

$$\underline{\underline{\tau}} : \underline{n}^A \underline{n}^B \qquad \underline{\underline{\epsilon}} : \underline{n}^A \underline{n}^C \ ?$$

It is clear that the adjectives "longitudinal" and "transversal" are perfectly adequate for the usual moduli, piezoelectric effects, etc. but are inadequate in more general situations. Calling $G = \dfrac{1}{4\underline{\underline{s}} \vdots \underline{n}^A \underline{n}^B \underline{n}^A \underline{n}^B}$ a "longitudinal-transversal-longitudinal-transversal" modulus is at the very least impractical. By far the simplest and clearest solution is to use the unit vectors $\underline{n}^A, \underline{n}^B, \underline{n}^C$ to specify what kind of projection is meant.

However, the main point about general projections is not the denomination, but the fact that there is no single representation surface for a material property that is a general, and not just a longitudinal, projection.

For a longitudinal projection like (9.1) the radius vector in the direction of the only $\underline{n}$ is the value (or proportional to the value) of the property in that direction.

---

[1] What kind of engineering modulus, or which combination of them would this modulus correspond to?

[2] Same question!

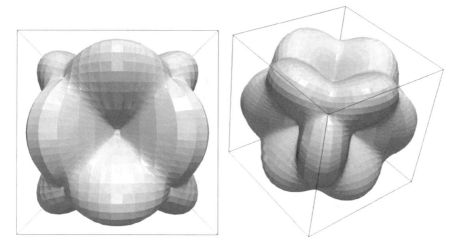

Fig. 9.1 Representation solid of $\underline{\underline{s}} \vdots \underline{n}^A \underline{n}^B \underline{n}^A \underline{n}^B$ for a cubic composite with $s_{11} = 23$, $s_{44} = 42$ and $s_{12} = 0.22$ GPa^{-1}, obtained by letting all three Euler angles φ, θ and ψ vary in their full ranges. Left: view along any of the conventional axes; right: view along on of the four three-fold axes

Defining the orientation of this single vector in 3D space only requires two angles, so that these angles parametrize the RS: it will be a two-dimensional set of points, i.e. a surface.

When a more general type of projection is used in which two directions $\underline{n}^A$, $\underline{n}^B$ (like Eq. 9.2) or three directions $\underline{n}^A$, $\underline{n}^B$, $\underline{n}^C$ appear, all three Euler angles are necessary. The resulting set of points is now parametrized by *three* angles, and it is not a surface anymore, but a 3D solid object.

Figure 9.1 shows such a representation solid for the shear modulus:

$$r(\varphi, \theta, \psi) = \underline{\underline{s}} \vdots \underline{n}^A \underline{n}^B \underline{n}^A \underline{n}^B \tag{9.3}$$

for a cubic material.

When the expressions of the components of $\underline{n}^A$ and $\underline{n}^B$ are substituted in (9.3) the radius vector depends on three angles: $r(\varphi, \theta, \psi)$ and there is an infinity of values of $\underline{\underline{s}} \vdots \underline{n}^A \underline{n}^B \underline{n}^A \underline{n}^B$ for every radial direction.

This also rises the question of the "radial" direction or the "radius vector" of a RS for a property that involves a projection in two directions, like $\underline{\underline{s}} \vdots \underline{n}^A \underline{n}^B \underline{n}^A \underline{n}^B$, or in three, like $\underline{\underline{s}} \vdots \underline{n}^A \underline{n}^B \underline{n}^A \underline{n}^C$. Is the radial direction along which the scalar value $r(\varphi, \theta, \psi)$ is to be placed, that of $\underline{n}^A$, of $\underline{n}^B$ or of $\underline{n}^C$?

To answer this question let's review in which way the three unit vectors formed by the three rows of Euler's $\underline{\hat{L}}$ sample orientations. If the Euler angles are allowed to vary in their full ranges, all three unit vectors along ①***, ②*** and ③*** (see Fig. 4.6)

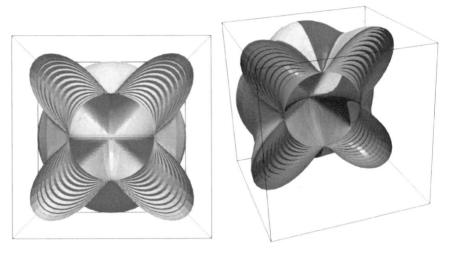

Fig. 9.2 Two views of the section of the representation solid of Fig. 9.1 through a plane perpendicular to any of the three conventional axes. The representation solid was discretized by drawing the 30 RSs corresponding to $\psi = i\frac{\pi}{30}$, $i = 1, \ldots, 30$. On the right panel different values of ψ have been shaded with grey levels proportional to the value of ψ

sample all directions of 3D space. Equation (9.3) then describes the representation solid of Fig. 9.1.

This solid can be thought of as the union of a (continuous) infinite number of individual RS obtained by fixing the value of one of Euler's angles and letting the other two vary freely.

In Fig. 9.2 the representation solid has been discretized by drawing 30 individual RSs, each of them for a fixed, given value of $\psi = i\frac{\pi}{30}$, $i = 1, \ldots, 30$ and letting φ and θ vary in their full ranges. The individual RSs may intersect one another, they do not conform to the idea of "onion" layers. On the right panel of Fig. 9.2 RSs for the different values of ψ have been shaded with different grey levels to show that RSs for individual values of ψ may intersect.

The panels of Fig. 9.3 show eight RS obtained by fixing Euler's third angle at the values $\psi = (i-1)\frac{\pi}{8}$, $i = 1, \ldots, 8$ and letting the other two vary in $\varphi \in [0, 2\pi)$, $\theta \in [0, \pi]$. In these RSs the third row of Euler's $\underset{\smile}{L}$ (i.e. the axis ③*** in Fig. 4.6) was taken as the vector $\underline{n}^A$ in $\underset{\equiv}{s} \vdots \underline{n}^A \underline{n}^B \underline{n}^A \underline{n}^B$, the second row of $\underset{\smile}{L}$ (i.e. the axis ①*** in Fig. 4.6) as $\underline{n}^B$.

The last panel corresponds to $\psi = \frac{7}{8}\pi$. For $\psi = \pi$ the same RS as in the first panel is obtained. Pairs of panels in this figure (i.e. 2–8, 3–7, 4–6) are mirror images.

There is something striking about the set of RS of Fig. 9.3: they do not have cubic symmetry, as they should according to Neumann's principle.

The full cubic symmetry of the orientational dependence of G is only achieved when all Euler angles are allowed to vary in their full ranges. While none of the

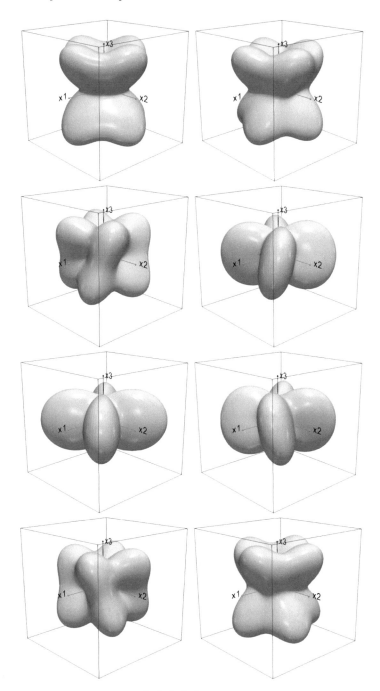

Fig. 9.3 Representation surfaces of $\underline{\underline{s}} \vdots \underline{n}^A \underline{n}^B \underline{n}^A \underline{n}^B$ for a cubic material. Each panel is the surface obtained from (9.3) by fixing the value of ψ and letting φ and θ vary in their full ranges. Top to bottom and left to right $\psi = (i-1)\frac{\pi}{8}$, $i = 1, \ldots, 8$

Fig. 9.4 Orientation sampled by base vectors taken from the rows of Euler's $\underset{\sim}{L}$ if ψ is fixed at $\psi = 0$

Fig. 9.5 Orientation sampled by base vectors taken from the rows of Euler's $\underset{\sim}{L}$ if ψ is fixed at an angle $0 \le \psi \le \frac{\pi}{2}$

Fig. 9.6 Orientation sampled by base vectors taken from the rows of Euler's $\underset{\sim}{L}$ if ψ is fixed at an angle $\frac{\pi}{2} < \psi \le \pi$

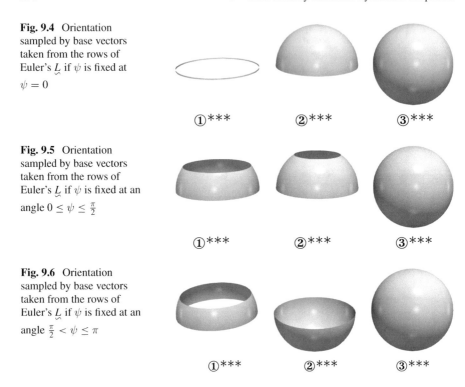

individual RS of Fig. 9.3 is cubic, the union of all of them, i.e. the representation solid of Fig. 9.1 does have the symmetry of the cubic holoedric point group $m3m$.

The lower symmetry of the individual RSs is a consequence of fixing the value of one of the angles, ψ in this case (Figs. 9.4, 9.5 and 9.6):

- If ψ is fixed at $\psi = 0$ and φ and θ are free to vary (first panel of Fig. 9.3), the axis ③*** will correctly sample all directions. But the axis ①*** will only sample directions in the equator of the unit sphere, i.e. the plane defined by the original, unrotated ①②. The second base vector, along ②***, will sample the upper hemisphere:
- If ψ is fixed at an angle $0 \le \psi \le \frac{\pi}{2}$, $\underline{n}^A \equiv$ ③*** will still sample all 3D directions, and the axes ①*** and ②*** will sample orientations that lie between the equator and two parallels of the northern hemisphere.[3]
- If ψ is fixed at an angle $\frac{\pi}{2} < \psi \le \pi$ the axis ①*** will still sample orientations that lie between the equator and a parallel on the northern hemisphere. But the axis ②*** will sample from the equator down to a parallel below the equator. In both cases, the axis ③*** samples all directions.
- Due to centrosymmetry, larger values $\psi \ge \pi$ lead to the same sampling as for $\psi - \pi$.

---

[3]Can you calculate the latitudes of these parallels?

Fig. 9.7 Open RS of $\underline{\underline{s}} : \underline{n}^A \underline{n}^B \underline{n}^A \underline{n}^B$ for a cubic material, as in the first panel of Fig. 9.3, but choosing as radius vector $\underline{n}^A = ②***$ and $\underline{n}^B = ①***$. The RS fails to close because the radial direction $\underline{n}^A = ②***$ does not sample all possible orientations in 3D space

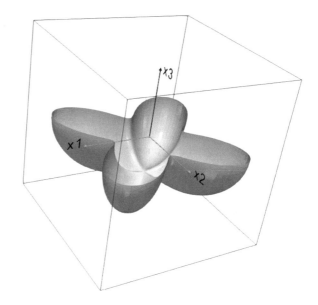

As soon as ψ is fixed, the set of orientations simultaneously sampled by the three axes $①***$, $②***$ and $③***$ loses its spherical symmetry. As a consequence, the symmetry of the resulting RS is lowered, and does not conform to the point group of the material.

The same also happens if we fix φ or θ and let the remaining two Euler angles vary freely. There is simply no way to have any two of the three axes $①***$, $②***$ and $③***$ simultaneously sample all orientations by fixing one and varying the other two Euler angles. Furthermore, if ψ is fixed and the scalar value of $r(\varphi, \theta, \psi)$ is not plotted along the $③***$ axis, the resulting RS will not be closed (Fig. 9.7).

Taking this into account, the natural choice for the vector radius direction is $\underline{n}^A = ③***$ because it is the only axis that samples all orientations in all cases, and the other two rows as transversal directions for projection, as required.[4] Figure 9.3 was drawn with $\underline{n}^A = ③***$; in addition, $\underline{n}^B$ was set to $①***$.

Even though they do not display the full symmetry of the representation solid, these RSs (which are "slices" of the representation solid) obtained by fixing one of Euler's angles, give correct information on the directional dependence of the transversal projection.

---

[4]Because the angle ψ does not appear in the third row of Euler's $\underline{\underline{L}}$. A similar discussion as above can be made for the use of the *columns* of $\underline{\underline{L}}$ as unit projection vectors. In this case, it is φ the one that does not appear in the third column.

On the other hand, the usefulness of representations solids is in practice limited to the determination of the point group of the property.

9.2 Representation Lines

Another useful and simplified way of presenting the dependence of properties which are not purely longitudinal, but general projections, is to reduce the dimensionality of the representation down to a flat curve or representation line (RL). We have already encountered this in Sect. 8.5 as we calculated the orientational dependence of the E and G moduli in the plane of a two-dimensional fabric material.

For 2D materials, the only meaningful dependence on orientation of projections like $\underline{\underline{s}} \vdots \underline{n}^A \underline{n}^A \underline{n}^A \underline{n}^A$ or $\underline{\underline{s}} \vdots \underline{n}^A \underline{n}^B \underline{n}^A \underline{n}^B$ is within the plane of the material. Moduli (or any other property) are obtained as a function of a single angle (e.g. Eq. 8.21):

$$G = \frac{1}{s_{66} - (s_{66} - 2s_{11} + 2s_{12})\sin^2 2\varphi}$$

The plane defined by $\underline{n}^A$ and $\underline{n}^C$ in Sect. 8.5 corresponds to $\theta = 0$. In this case, the other two Euler angles collapse on a single one, which we took to be φ for simplicity.

But the calculation of angular dependence is also meaningful for 3D materials in which it is required to know how a given property/projection like $\underline{\underline{\rho}} : \underline{n}^A \underline{n}^B$, $\underline{\underline{s}} \vdots \underline{n}^A \underline{n}^B \underline{n}^A \underline{n}^B$, ... varies when $\underline{n}^A$ and $\underline{n}^B$ are constrained to lie on an arbitrarily oriented plane.

Typical applications are hardness and catalytic activity of crystal faces, surface electric conductivity, elastic moduli of wood products like sliced veneer, friction coefficient, reflectivity, surface diffusion, etc.

The variation of a property on an arbitrarily oriented plane of a 3D material is simply obtained by calculating the value of $r(\varphi, \theta, \psi)$ for the property for fixed values of $\varphi = \varphi_0$ and $\theta = \theta_0$ that define the orientation of the plane, and variable ψ, that performs rotation within the plane. The RL is the resulting one-parameter line, with ψ as parameter.

Figure 9.8 shows the orientational dependence of G of the cubic material of Fig. 9.1 on the plane defined by $\varphi = 1.8$ and $\theta = 2.0$ rad as a thick line (tube). As the two vectors $\underline{n}^A$ and $\underline{n}^B$ that define G are rotated by ψ within the plane, $\underline{\underline{s}} \vdots \underline{n}^A \underline{n}^B \underline{n}^A \underline{n}^B$ goes through four maxima and minima (corresponding to four minima and maxima of G).

This curve is *not* the intersection of the plane $\varphi = 1.8$ and $\theta = 2.0$ with any of the RSs of Fig. 9.3. The RSs of Fig. 9.3 are drawn at constant ψ and variable φ and θ, whereas the curve of Fig. 9.8 has fixed φ and θ, and variable ψ.

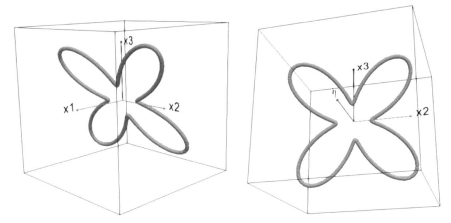

Fig. 9.8 Representation curve of $G = \left(4\underline{\underline{s}} \vdots \underline{n}^A \underline{n}^B \underline{n}^A \underline{n}^B\right)^{-1}$ for the cubic material described in Fig. 9.1, in the plane defined by $\varphi = 1.8$ and $\theta = 2.0$ rad; the view direction in the right panel is perpendicular to the plane of the curve. To improve visibility, the 2D curve has been drawn as a tube

9.3 Antisymmetric Properties

While the majority of the properties considered in this book are symmetric in one or more pairs of indices, a few others may be *antisymmetric* or *skew symmetric* in a pair of indices.

Some second rank properties like the Seebeck $\underline{\underline{\Sigma}}$, Peltier $\underline{\underline{\Pi}}$ and Thomson $\underline{\underline{\gamma}}$ coefficients for some point groups (tetragonal 4, $\bar{4}$, $4/m$, 3, $\bar{3}$, 6, $\bar{6}$, $6/m$, ∞, ∞/m classes) have a str( ) that is skew symmetric (Table 9.1). Skew symmetric properties can also be visualized by RSs, although, as we will now see, the procedure is slightly different as for symmetric properties.

A second rank tensor $\underline{\underline{A}}$ is skew symmetric when $\underline{\underline{A}}^T = -\underline{\underline{A}}$, meaning that $A_{ij} = -A_{ji}$ and therefore $A_{ii} = 0$ (no sum). Represented as a matrix, a skew symmetric second rank tensor has the form:

$$[\![\underline{\underline{A}}]\!] = \begin{bmatrix} 0 & A_{12} & A_{13} \\ -A_{12} & 0 & A_{23} \\ -A_{13} & -A_{23} & 0 \end{bmatrix} \tag{9.4}$$

in which only three components are independent, instead of the six independent components of a symmetric second rank tensor, or the nine components of a general, asymmetric second rank tensor.

It is wasteful for the nine slots of a second rank skew symmetric tensor to only hold three independent components, e.g. A_{12}, A_{13} and A_{23} in (9.4), just as it can be

Tabla 9.1 Structures of the Seebeck $\underset{\sim}{\Sigma}$, Peltier $\underset{\sim}{\Pi}$, Thomson $\underset{\sim}{\gamma}$ and Hall $\underline{\underline{R}}$ coefficients (dual condensed of $\underline{\underline{R}}$). The integer to the right of str( ) is the number of independent components of the property

·	zero component
•	non-zero component
•——•	components equal
•——○	same absolute value, opposite signs

system(s)	class(es)	str ()
triclinic	all	$\begin{bmatrix} \bullet & \bullet & \bullet \\ \bullet & \bullet & \bullet \\ \bullet & \bullet & \bullet \end{bmatrix}_9$
monoclinic	all	$\begin{bmatrix} \bullet & \cdot & \bullet \\ \cdot & \bullet & \cdot \\ \bullet & \cdot & \bullet \end{bmatrix}_5$
orthorhombic	all	$\begin{bmatrix} \bullet & \cdot & \cdot \\ \cdot & \bullet & \cdot \\ \cdot & \cdot & \bullet \end{bmatrix}_3$
tetragonal trigonal hexagonal limit classes	$4mm,\ \bar{4}2m,\ 422,\ 4/mmm$ $3m,\ 32,\ \bar{3}m$ $6mm,\ \bar{6}m2,\ 622,\ 6/mmm$ $\infty m, \infty 2, \infty/mm$	$\begin{bmatrix} \bullet & \searrow & \cdot \\ & \bullet & \cdot \\ \cdot & \cdot & \bullet \end{bmatrix}_2$
tetragonal trigonal hexagonal limit classes	$4,\ \bar{4},\ 4/m$ $3,\ \bar{3}$ $6,\ \bar{6},\ 6/m$ $\infty, \infty/m$	$\begin{bmatrix} \bullet & \times & \cdot \\ \circ & & \cdot \\ \cdot & \cdot & \bullet \end{bmatrix}_3$
cubic limit classes	all $\infty\infty m, \infty\infty$	$\begin{bmatrix} \bullet & \cdot & \cdot \\ \cdot & \searrow & \cdot \\ \cdot & \cdot & \searrow \bullet \end{bmatrix}_1$

considered inefficient to use nine slots to store the six independent components of a second rank symmetric tensor.

Motivated by Voigt's notation condensing a 3×3 symmetric tensor $\underline{\underline{S}}$ into a zero-waste 6×1 vector $\vec{S}$, it is natural to look for a way to present the three independent components of $\underline{A}$ in a more compact form.

In $d = 3$ dimensions, a second rank skew symmetric tensor has three independent components, which (in 3D!) happens to be the number of components of a rank one tensor. It is quite tempting to use a 3×1 vector to represent a skew symmetric second rank tensor $\underset{\sim}{\underline{A}}$.

As a matter of fact the situation is even better: it is possible to pack the three independent components of $\underline{\underline{A}}$ into a rank one *tensor*, say $\underline{u}$, that has all[5] usual bona fide tensor transformation properties. We will call this conversion $\underline{\underline{A}} \leftrightarrow \underline{u}$ the *dual condensation of a skew symmetric tensor*.[6]

The original skew symmetric tensor $\underline{\underline{A}}$ and the condensed, rank one $\underline{u}$ are related by:

$$u_k \equiv \frac{1}{2!} \epsilon_{ijk} A_{ij} \qquad \left(u_k = -\frac{1}{2} \epsilon_{kji} A_{ij} \Rightarrow \underline{u} = -\frac{1}{2} \underline{\underline{\epsilon}} : \underline{\underline{A}} \right) \tag{9.5}$$

where $\underline{\underline{\epsilon}}$ is the completely antisymmetric third rank tensor, whose components are given by the Levi-Civita symbol ϵ_{ijk} (3.4). According to definition (9.5), the components of $\underline{u}$ are related to those of $\underline{\underline{A}}$ by:

$$u_1 = \frac{1}{2}(\epsilon_{231} A_{23} + \epsilon_{321} A_{32}) = \frac{1}{2}(\epsilon_{231} A_{23} - \epsilon_{321} A_{23}) = A_{23}$$

$$u_2 = \frac{1}{2}(\epsilon_{132} A_{13} + \epsilon_{312} A_{31}) = \frac{1}{2}(\epsilon_{132} A_{13} - \epsilon_{312} A_{13}) = -A_{13}$$

$$u_3 = \frac{1}{2}(\epsilon_{123} A_{12} + \epsilon_{213} A_{21}) = \frac{1}{2}(\epsilon_{123} A_{12} - \epsilon_{213} A_{12}) = A_{12}$$

The three independent components of $\underline{\underline{A}}$ show up in $\underline{u}$ thus:

$$[\![\underline{\underline{A}}]\!] = \begin{bmatrix} 0 & A_{12} & A_{13} \\ -A_{12} & 0 & A_{23} \\ -A_{13} & -A_{23} & 0 \end{bmatrix} \quad \Rightarrow \quad [\![\underline{u}]\!] = \begin{bmatrix} A_{23} \\ -A_{13} \\ A_{12} \end{bmatrix} \tag{9.6}$$

The more compact $\underline{u}$ contains the same information as $\underline{\underline{A}}$ and is used in practice to represent and report numerical values of components of skew symmetric properties. The components of a skew symmetric $\underline{\underline{A}}$ (two indices) are invariably found in the literature as the components of the dual condensed $\underline{u}$ (one index).

Retrieving $\underline{\underline{A}}$ from $\underline{u}$ is the operation inverse to (9.5):

$$A_{jk} = \epsilon_{ijk} u_i \qquad \left(A_{jk} = \epsilon_{ijk} u_i = \epsilon_{jki} u_i \Rightarrow \underline{\underline{A}} = \underline{\underline{\epsilon}} \cdot \underline{u} \right) \tag{9.7}$$

so that the components of $\underline{\underline{A}}$ in terms of those of $\underline{u}$ are:

$$A_{11} = \epsilon_{111} u_1 + \epsilon_{211} u_2 + \epsilon_{311} u_3 = 0$$

$$A_{22} = \epsilon_{122} u_1 + \epsilon_{222} u_2 + \epsilon_{322} u_3 = 0$$

---

[5] Almost; see (9.10) below.

[6] There is a good reason to choose this name, but it is of no relevance for the purposes of this book.

$$A_{33} = \epsilon_{133}u_1 + \epsilon_{233}u_2 + \epsilon_{333}u_3 = 0$$
$$A_{23} = \epsilon_{123}u_1 + \epsilon_{223}u_2 + \epsilon_{323}u_3 = u_1 = -A_{32}$$
$$A_{13} = \epsilon_{113}u_1 + \epsilon_{213}u_2 + \epsilon_{313}u_3 = -u_2 = -A_{31}$$
$$A_{12} = \epsilon_{112}u_1 + \epsilon_{212}u_2 + \epsilon_{312}u_3 = u_3 = -A_{21}$$

If we represent $\underline{u}$ and $\underline{\underline{A}}$ as vector/matrix:

$$\llbracket \underline{u} \rrbracket = \begin{bmatrix} u_1 \\ u_2 \\ u_3 \end{bmatrix} \quad \Rightarrow \quad \llbracket \underline{\underline{A}} \rrbracket = \begin{bmatrix} 0 & u_3 & -u_2 \\ -u_3 & 0 & u_1 \\ u_2 & -u_1 & 0 \end{bmatrix} \tag{9.8}$$

Dual condensation $\underline{\underline{A}} \to \underline{u}$ (9.6) is, for skew symmetric tensors, the analog of Voigt index condensation for symmetric tensors, i.e. $\underline{\underline{\tau}} \to \vec{\tau}$.

The operation $\underline{u} \to \underline{\underline{A}}$ (9.8) undoes dual index condensation for skew symmetric tensors; it is the analog of $\vec{\tau} \to \underline{\underline{\tau}}$ for symmetric tensors.

An important difference between Voigt and dual index condensation is that the dual condensed $\underline{u}$ of a tensor $\underline{\underline{A}}$ is a tensor, while the Voigt condensed of a tensor is *not* a tensor. Thus, the dual condensed can be directly transformed to another reference frame (Sect. 3.3), without having to first undo the condensation.

Let us now see how a skew symmetric second rank tensor acts on a vector. We expect the contraction $\underline{\underline{A}} \cdot \underline{n}$ to result in a vector which is "deviated" from the original $\underline{n}$. As a check of the amount of deviation $\underline{\underline{A}}$ produces we perform the scalar product of $\underline{\underline{A}} \cdot \underline{n}$ and $\underline{n}$:

$$(\underline{\underline{A}} \cdot \underline{n}) \cdot \underline{n} = A_{ij}n_j n_i$$

In this expression, three of the nine terms are zero because the diagonal terms of $\underline{\underline{A}}$ are $A_{ii} = 0, \forall i$. The remaining six terms can be gathered in three pairs, each of which cancels because of the opposite signs of A_{ij} and A_{ji}:

$$(\underline{\underline{A}} \cdot \underline{n}) \cdot \underline{n} = A_{ij}n_j n_i = \underbrace{A_{11}}_{=0} n_1 n_1 + A_{12}n_1 n_2 + \underbrace{A_{21}}_{=-A_{12}} n_2 n_1 + \ldots =$$

$$A_{12}n_1 n_2 - A_{12}n_2 n_1 + A_{13}n_1 n_3 - A_{13}n_3 n_1 + A_{23}n_2 n_3 - A_{23}n_3 n_2 = 0$$

which implies orthogonality of $\underline{\underline{A}} \cdot \underline{n}$ and $\underline{n}$. If $\underline{\underline{A}}$ is a skew symmetric second rank tensor, its contraction $\underline{\underline{A}} \cdot \underline{n}$ with any rank one tensor $\underline{n}$ is orthogonal to $\underline{n}$.

The exact same result is obtained if instead of contracting $\underline{\underline{A}}$ with $\underline{n}$, we carry out the vector product of $\underline{n}$ with the dual of $\underline{\underline{A}}$:

$$\underline{\underline{A}} \cdot \underline{n} = \underline{n} \times \underline{u}$$

which is a more familiar way of understanding the effect of $\underline{\underline{A}}$ on $\underline{n}$.

The dual vector $\underline{u}$ of $\underline{\underline{A}}$ thus has a clear geometrical effect which is also evident from the eigenvalues and eigenvectors of $\underline{\underline{A}}$. In terms of the A_{ij} and of

$$\lambda \equiv \sqrt{A_{12}^2 + A_{13}^2 + A_{23}^2} = |\underline{u}| \text{ the eigenvalues are:}$$

$$\lambda_1 = 0 \quad \lambda_2, \lambda_3 = \pm i\lambda$$

and the (unnormalized) eigenvector associated with the 0 eigenvalue is:

$$\begin{bmatrix} A_{23} \\ -A_{13} \\ A_{12} \end{bmatrix}$$

which is nothing but the dual $\underline{u}$. The other two eigenvectors have one real and two complex conjugate components without especially simple expressions. The skew symmetric tensor $\underline{\underline{A}}$ can be brought to the form:

$$[\![\underline{\underline{A}}]\!] = \begin{bmatrix} 0 & \lambda & 0 \\ -\lambda & 0 & 0 \\ 0 & 0 & 0 \end{bmatrix} \tag{9.9}$$

by a special orthogonal transformation.[7] In this reference frame, the direction of the axis ③ coincides with the dual $\underline{u}$.[8]

There is something a bit unusual about the dual $\underline{u}$. As a consequence of its definition, the same result should be obtained by $\underline{\underline{A}} \cdot \underline{n}$ or by $\underline{n} \times \underline{u}$ in all cases. Let's check this with very simple $\underline{\underline{A}}$ and $\underline{n}$, which expressed in a given reference frame ①②③ are:

$$[\![\underline{\underline{A}}]\!] = \begin{bmatrix} 0 & 1 & 0 \\ -1 & 0 & 0 \\ 0 & 0 & 0 \end{bmatrix} \Rightarrow [\![\underline{u}]\!] = \begin{bmatrix} 0 \\ 0 \\ 1 \end{bmatrix}$$

so that the effect on $[\![\underline{n}]\!] = \begin{bmatrix} 1 \\ 0 \\ 0 \end{bmatrix}$ is:

$$[\![\underline{n}]\!] = \begin{bmatrix} 1 \\ 0 \\ 0 \end{bmatrix} \Rightarrow \begin{cases} [\![\underline{\underline{A}} \cdot \underline{n}]\!] = \begin{bmatrix} 0 \\ -1 \\ 0 \end{bmatrix} \\[4mm] [\![\underline{n} \times \underline{u}]\!] = \begin{bmatrix} 0 \\ -1 \\ 0 \end{bmatrix} \end{cases}$$

As expected, both $\underline{\underline{A}}$ and its dual $\underline{u}$ rotate $\underline{n}$ in the same way, e.g. according to the right-hand rule of the vector product (the example can also be written as

---

[7]Can you write the $\underline{L}$ for the transformation that brings the general $\underline{\underline{A}}$ to the form (9.9)?

[8]It can also be brought to any of the two other variants in which λ and $-\lambda$ occupy the elements 13 or 23. In the first case, it is the axis ② that points along $\underline{u}$, in the second, it is ①.

$\delta_1 \times \delta_3 = -\delta_2$). Let's now carry out the same operation, but changing the reference frame by means of this L:

$$[\![L]\!] = \begin{bmatrix} 1 & 0 & 0 \\ 0 & 1 & 0 \\ 0 & 0 & -1 \end{bmatrix}$$

which is a reflection on the ①② plane.

This new reference frame has the same first and second base vectors: $\delta'_1 = \delta_1$, $\delta'_2 = \delta_2$ (or ①' = ①, ②' = ②), and a reversed third base vector: $\delta'_3 = -\delta_3$ (or ③' = −③). Since the result of $\underline{\underline{A}} \cdot \underline{n}$ or $\underline{n} \times \underline{u}$ lies in the plane ①②, it should not be altered by this reflection L.

Carrying out the same operations as above with the transformed factors yields:

$$\begin{cases} [\![\underline{\underline{A}'}]\!] = L[\![\underline{\underline{A}}]\!]L^T = \begin{bmatrix} 0 & 1 & 0 \\ -1 & 0 & 0 \\ 0 & 0 & 0 \end{bmatrix} & \qquad [\![\underline{\underline{A}'} \cdot \underline{n}']\!] = \begin{bmatrix} 0 \\ -1 \\ 0 \end{bmatrix} \\ & \text{so that} \\ [\![\underline{u}']\!] = L[\![\underline{u}]\!] = \begin{bmatrix} 0 \\ 0 \\ -1 \end{bmatrix} \quad [\![\underline{n}']\!] = L[\![\underline{n}]\!] = \begin{bmatrix} 1 \\ 0 \\ 0 \end{bmatrix} & \qquad [\![\underline{n}' \times \underline{u}']\!] = \begin{bmatrix} 0 \\ 1 \\ 0 \end{bmatrix} \end{cases}$$

As expected, $\underline{\underline{A}}$ and $\underline{u}$ are both unchanged, and the third component of $\underline{u}$ changes sign. But now $\underline{\underline{A}} \cdot \underline{n}$ and $\underline{n} \times \underline{u}$ do not agree.

The immediate reason is quite obvious: since the dual, but not $\underline{n}$, has changed by the transformation and it now points the other way, it is not surprising that the result of the cross product also changes sign.

It seems that the dual $\underline{u}$ is a slightly special type of vector, that should transform in a different way if we expect $\underline{\underline{A}} \cdot \underline{n}$ and $\underline{n} \times \underline{u}$ to yield the same result.

This is indeed the case: most tensors we have met up to now are so-called *polar*. Their transformation rule is the one (3.31) we met in Sect. 3.3:

$$T'_{ijkmn} \ldots = l_{ip}l_{jq}l_{kr}l_{ms}l_{nt} \ldots T_{pqrst} \ldots$$

But there is a second type of tensors, called *axial*, that transform in a slightly different fashion:

$$T'_{ijkmn} \ldots = \det(L)l_{ip}l_{jq}l_{kr}l_{ms}l_{nt} \ldots T_{pqrst} \ldots \tag{9.10}$$

the only difference being the factor $\det(L)$ which is either 1 or −1:

- $\det(L) = 1$ for proper rotations, i.e. those that do not change the handedness of the reference frame,
- $\det(L) = -1$ for improper rotations, i.e. those that reverse the handedness of the reference frame.

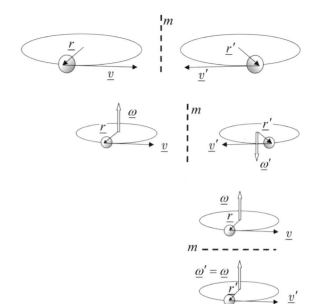

Fig. 9.9 Reflection of polar vectors on a mirror plane

Fig. 9.10 (Wrong) reflection of an axial vector, drawn as if it were polar, on a mirror plane parallel to it

Fig. 9.11 (Wrong) reflection of an axial vector, drawn as if it were polar, on a mirror plane perpendicular to it

It is now clear why things didn't work out correctly in the example above: the dual $\underline{u}$ is an axial vector; since $\underset{\sim}{L}$ in the example above reverses handedness, $\det(\underset{\sim}{L}) = -1$, and the correct transformation of $\underline{u}$ should have been:

$$[\![\underline{u'}]\!] = \det(\underset{\sim}{L})\underset{\sim}{L}[\![\underline{u}]\!] = \begin{bmatrix} 0 \\ 0 \\ 1 \end{bmatrix}$$

Now $\underline{\underline{A}} \cdot \underline{n}$ and $\underline{n} \times \underline{u}$ do agree as expected. The origin of the apparent problem is the appearance of ϵ_{ijk} in the definition of the dual (9.5).[9]

Axial vectors are also found in other contexts. Consider the movement of a particle about an axis. The position $\underline{r}$ and the velocity $\underline{u}$ of the particle are polar vectors. Under reflection in a mirror perpendicular to the plane of motion they reflect in the intuitively correct way (Fig. 9.9). If we now introduce the angular velocity vector as the vector that satisfies $\underline{\omega} \times \underline{r} = \underline{u}$, and construct it for the original and for the reflected $\underline{r'}$ and $\underline{u'}$, we see that $\underline{\omega}$ and $\underline{\omega'}$ must be different, which is *not* the way an arrow would reflect in the mirror (Fig. 9.10).

If the mirror is now placed perpendicular to $\underline{\omega}$, then $\underline{r}$ and $\underline{u}$ are reflected into $\underline{r'}$ and $\underline{u'}$, and finally $\underline{\omega'} \times \underline{r'} = \underline{u'}$ constructed, it turns out that $\underline{\omega}$ and $\underline{\omega'}$ now point in the same direction, again contradicting the behavior of polar vectors (Fig. 9.11).

---

[9]The sign of the vector product depends on the sign of the volume 3-form, in this case $\underset{=}{\epsilon}$, while the sign of the two-form onto which the vector product is mapped by duality does not.

Fig. 9.12 Correct reflection
of an axial vector on a mirror
plane parallel to it

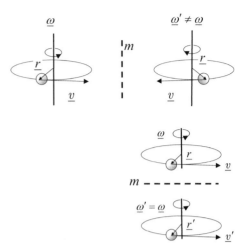

Fig. 9.13 Correct reflection
of an axial vector on a mirror
plane perpendicular to it

This apparently anomalous behavior is typical of axial tensors under transformations that imply a change of handedness, such as caused by reflection in a mirror. Representing an axial vector by an arrow is clearly not correct.

The simple and traditional way out of this is to represent axial vectors by an "arrowless" vector, i.e. just a line, plus a schematic rotation. This type of representation (a loop with a circulation around a segment) reflects correctly on a plane parallel or perpendicular to it (Figs. 9.12 and 9.13).

As a general rule, magnitudes defined through cross products, or which depend on the handedness of the reference frame, will behave axially. Because of the extra term $\det(\underset{\sim}{L})$ in their transformation law, they are often called pseudotensors (or if of rank zero, pseudoscalars).

Returning now to the way a second rank skew symmetric property $\underline{\underline{A}}$ operates, it will often couple two rank one tensors in a constitutive equation of the form $\underline{e} = \underline{\underline{A}} \cdot \underline{c}$, with $\underline{c}$ the cause and $\underline{e}$ the effect, by the usual cross product of the cause vector $\underline{c}$ and the dual $\underline{u}$ of $\underline{\underline{A}}$. The effect $\underline{e}$ will lie in the plane defined by $\underline{c}$ and $\underline{u}$ and its magnitude will be $|\underline{e}| = |\underline{u}||\underline{c}| \sin \alpha$, where α is the angle formed by $\underline{c}$ and $\underline{u}$.

The effect of $\underline{\underline{A}}$ is "as transversal" as it can possibly be: orthogonal to both the cause and the effect. Not surprisingly, the longitudinal RS of a skew symmetric tensor:

$$\underline{\underline{A}} : \underline{n}\,\underline{n} = (\underline{\underline{A}} \cdot \underline{n}) \cdot \underline{n} = 0$$

collapses to a point at the origin; not exactly a very useful RS.

It would be possible to visualize a skew symmetric tensor $\underline{\underline{A}}$ by means of the RS of its dual $\underline{u}$, which, if we ignored that it is an axial vector, would be the oriented sphere characteristic of rank one properties. But this would not be a useful representation, because it is not the scalar product $r = \underline{u} \cdot \underline{n}$ what we wish to visualize, but the way $\underline{u}$ or $\underline{\underline{A}}$ acts on $\underline{n}$.

A more practical way to visualize $\underline{\underline{A}}$ is by constructing the RS of the modulus of the effect. For the usual unit $\underline{n}$ that samples all orientations in 3D, the RS of the modulus squared is:

$$r = (\underline{\underline{A}} \cdot \underline{n}) \cdot (\underline{\underline{A}} \cdot \underline{n}) = (\underline{n} \cdot \underline{\underline{A}})^T \cdot (\underline{\underline{A}} \cdot \underline{n}) = \underline{n} \cdot \underline{\underline{A}}^T \cdot \underline{\underline{A}} \cdot \underline{n} = \underline{\underline{A}}^T \cdot \underline{\underline{A}} : \underline{n} \, \underline{n}$$

which is the usual RS of the product $\underline{\underline{P}} \equiv \underline{\underline{A}}^T \cdot \underline{\underline{A}} = -\underline{\underline{A}} \cdot \underline{\underline{A}}$. Since the dual $\underline{u}$ is an eigenvector of $\underline{\underline{A}}$ (and automatically of $\underline{\underline{A}}^T$) with eigenvalue $\lambda_1 = 0$, then:

$$\underline{\underline{P}} \cdot \underline{u} = \underline{\underline{A}}^T \cdot \underline{\underline{A}} \cdot \underline{u} = \underline{\underline{A}}^T \cdot \underline{0} = \underline{0}$$

so the dual $\underline{u}$ is also an eigenvector of $\underline{\underline{P}}$ with the same eigenvalue $\lambda_1 = 0$. The product tensor $\underline{\underline{P}}$ is then semidefinite positive with eigenvalues 0 and λ^2 (double). The RS of $\underline{\underline{P}}$:

$$r = \underline{\underline{A}}^T \cdot \underline{\underline{A}} : \underline{n} \, \underline{n} = \underline{\underline{P}} : \underline{n} \, \underline{n} = P_{ij} n_j n_i = A_{ki} A_{kj} n_j n_i \qquad (9.11)$$

is represented in Fig. 9.14. It is a surface of revolution with cross section $\sin^2 \alpha$, the revolution axis parallel to $\underline{u}$, with α the angle between the dual $\underline{u}$ and the unit vector $\underline{n}$.

Since $\underline{\underline{P}}$ is second rank symmetric, we can draw its representation quadric RQ, which for the eigenvalues 0, λ^2, λ^2 is a circular cylinder of radius $\dfrac{1}{\sqrt{\lambda}} = \dfrac{1}{|\underline{u}|}$, with axis parallel to $\underline{u}$.

As the RQ has the property that the length of the radius vector in direction $\underline{n}$ is proportional to the reciprocal of the square root of the property in that direction (6.16), the radius vector of the RQ for $\underline{\underline{P}} \equiv \underline{\underline{A}}^T \cdot \underline{\underline{A}}$ directly gives the value of $\dfrac{1}{|\underline{\underline{A}} \cdot \underline{n}|}$. As usual, in directions where the radius vector from the origin to the RS is small, the radius vector to the RQ is large, and vice versa.

To sum up: for a given $\underline{\underline{A}}$, the result (vector) of $\underline{\underline{A}} \cdot \underline{n}$ has a modulus we can obtain either from the RS or from the RQ:

- the value of $|\underline{\underline{A}} \cdot \underline{n}|$ is the square root of the radius vector to the RS, because the radius vector to the RS is $|\underline{\underline{A}} \cdot \underline{n}|^2 = |\underline{u}|^2 \sin^2 \alpha$,
- the value of $|\underline{\underline{A}} \cdot \underline{n}|$ is the reciprocal of the radius vector to the RQ, because the radius vector to the RQ is $\dfrac{1}{|\underline{\underline{A}} \cdot \underline{n}|} = \dfrac{1}{|\underline{u}| \sin \alpha}$.

As to the direction, the resulting $\underline{\underline{A}} \cdot \underline{n}$ is perpendicular to the plane defined by the dual $\underline{u}$ and by $\underline{n}$, which is the plane of the paper in Fig. 9.15. The product $\underline{\underline{A}} \cdot \underline{n}$ points in the direction given by the right-hand rule of the cross product $\underline{n} \times \underline{u}$.

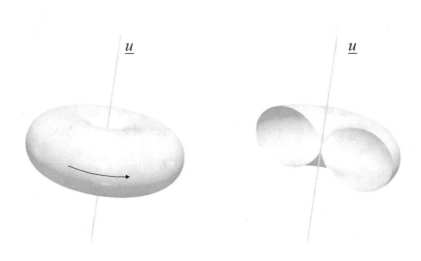

Fig. 9.14 Representation surface RS and representation quadric RQ for $\underline{\underline{P}} \equiv \underline{\underline{A}}^T \cdot \underline{\underline{A}}$. The axis is parallel to the dual $\underline{u}$ of $\underline{\underline{A}}$. The RS is axisymmetric, its cross section (right panel) $|\underline{u}|^2 \sin^2 \alpha$. The semi-transparent (infinitely long) cylinder is the RQ

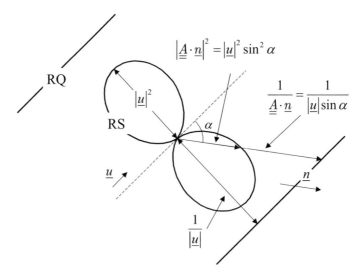

Fig. 9.15 Section of the RS and the RQ for $\underline{\underline{P}} \equiv \underline{\underline{A}}^T \cdot \underline{\underline{A}}$ of Fig. 9.14 through the plane defined by the dual $\underline{u}$ and $\underline{n}$. The thick parallel lines are the section of the RQ cylinder

9.3.1 Asymmetric Second Rank Properties

Some second rank properties (generically $\underline{\underline{T}}$) like the Seebeck $\underline{\underline{\Sigma}}$, Peltier $\underline{\underline{\Pi}}$ and Thomson $\underline{\underline{\gamma}}$ coefficients for some point groups (all triclinic and all monoclinic classes) have a $\text{str}(\underline{\underline{T}})$ that is neither symmetric nor skew symmetric (Table 9.1).

Visualizing an asymmetric $\underline{\underline{T}}$ by its RS is best done by separating it in its symmetric and its skew symmetric parts. Adding and subtracting half the transpose (3.11) produces the desired result:

$$\underline{\underline{T}} = \underline{\underline{S}} + \underline{\underline{A}} \quad \text{with} \quad \begin{cases} \underline{\underline{S}} \equiv \underbrace{\frac{1}{2}\left(\underline{\underline{T}} + \underline{\underline{T}}^T\right)}_{\text{symmetric}} \\ \underline{\underline{A}} \equiv \underbrace{\frac{1}{2}\left(\underline{\underline{T}} - \underline{\underline{T}}^T\right)}_{\text{skew symmetric}} \end{cases} \tag{9.12}$$

The symmetric and skew symmetric parts are orthogonal to each other in the sense that their mutual projection vanishes:

$$\left(\underline{\underline{T}} + \underline{\underline{T}}^T\right) : \left(\underline{\underline{T}} - \underline{\underline{T}}^T\right) = 0$$

and they can be handled separately when constructing the RS. However, the (otherwise formally correct) RS defined by:

$$r = \underline{\underline{T}} : \underline{n}\,\underline{n} = (\underline{\underline{S}} + \underline{\underline{A}}) : \underline{n}\,\underline{n} \tag{9.13}$$

is of little use, because of the transversality of $\underline{\underline{A}}$. The term $\underline{\underline{A}} : \underline{n}\,\underline{n}$ vanishes identically, so that the RS (9.13) would contain information on the symmetric part of $\underline{\underline{T}}$ only.

Since we now know how to represent symmetric and skew-symmetric second rank tensors, treating them separately is the simplest way to visualize them. The symmetric part is visualized by its RS or its RQ as defined in Sect. 6.8; the skew symmetric part, as a second RS or RQ as explained in the previous section.

$$\begin{aligned} \text{symmetric part:} \quad & r = \underline{\underline{S}} : \underline{n}\,\underline{n} \\ \text{skew symmetric part:} \quad & r = (\underline{\underline{A}}^T \cdot \underline{\underline{A}}) : \underline{n}\,\underline{n} \end{aligned}$$

The splitting (9.12) also helps clarify the question of the minimal set of geometric objects required to represent a skew-symmetric second rank property. The RS of the symmetric part $\underline{\underline{S}}$ is an ovaloid with point group symmetry equal to or greater than *mmm* for all classes.

In the general case (triclinic) in which the conventional axes do not coincide with the principal axes of $\underline{\underline{S}}$ nor with the dual $\underline{u}$ of the skew symmetric part, the contributions of the symmetric and skew symmetric parts fill $\underline{\underline{T}}$ in this way:

$$[[\underline{\underline{T}}]] = \begin{bmatrix} S_{11} & S_{12} + A_{12} & S_{13} + A_{13} \\ S_{12} - A_{12} & S_{22} & S_{23} + A_{23} \\ S_{13} - A_{13} & S_{23} - A_{23} & S_{33} \end{bmatrix} \text{ so that } \text{str}(\underline{\underline{T}}) = \begin{bmatrix} \bullet & \bullet & \bullet \\ \bullet & \bullet & \bullet \\ \bullet & \bullet & \bullet \end{bmatrix}$$

$$(9.14)$$

with nine independent coefficients, three of which describe the arbitrary orientation of the eigenvalues of $\underline{\underline{S}}$ with respect to the conventional axes (or viceversa).

Higher symmetry due to equality among principal values of $\underline{\underline{A}}$, and coincidence of eigenvectors with conventional axes result in simpler $\text{str}(\underline{\underline{S}})$. If the eigenvectors of $\underline{\underline{S}}$ are taken as conventional axes ①②③, the contribution of $\underline{\underline{S}}$ to $\text{str}(\underline{\underline{T}})$ is of the form:

$$[[\underline{\underline{S}}]] = \begin{bmatrix} S_{11} & 0 & 0 \\ 0 & S_{22} & 0 \\ 0 & 0 & S_{33} \end{bmatrix}$$

If, in the general case, the dual $\underline{u}$ of the skew symmetric part $\underline{\underline{A}}$ does not coincide with any of the eigenvectors of the symmetric part, the contribution of $\underline{\underline{A}}$ to $\text{str}(\underline{\underline{T}})$ in the conventional axes ①②③, will be of the form:

$$[[\underline{\underline{A}}]] = \begin{bmatrix} 0 & A_{12} & A_{13} \\ -A_{12} & 0 & A_{23} \\ -A_{13} & -A_{23} & 0 \end{bmatrix} \text{ with } [[\underline{u}]] = \begin{bmatrix} A_{23} \\ -A_{13} \\ A_{12} \end{bmatrix}$$

for a total:

$$[[\underline{\underline{T}}]] = \begin{bmatrix} S_{11} & A_{12} & A_{13} \\ -A_{12} & S_{22} & A_{23} \\ -A_{13} & -A_{23} & S_{33} \end{bmatrix} \text{ so that } \text{str}(\underline{\underline{T}}) = \begin{bmatrix} \bullet & \diagup & \diagup \\ \circ & \bullet & \diagup \\ \circ & \circ & \bullet \end{bmatrix}$$

$$(9.15)$$

which only has six independent components: nine of the general triclinic case minus three that are not needed, since the conventional axes and the principal axes of $\underline{\underline{S}}$ now coincide. Or seen from another point of view, three for the three eigenvalues of the symmetric part (diagonal elements), plus three for the three components of the dual $\underline{u}$.[10]

---

[10]Three and not two are needed, because $\underline{u}$ is not of unit modulus, so three components are necessary, or two to specify its arbitrary orientation, one to specify its magnitude.

Thus, in the most general case, an ovaloid for the symmetric part plus the torus of the skew symmetric part, and the six numerical parameters just described are required to visualize an asymmetric second rank tensor.[11]

If we recall that the general ovaloid is constructed from the three isomers of Fig. 6.7 of the dyadic $\underline{u}\,\underline{u}$ of Fig. 6.1, and the torus of the skew symmetric part is nothing but Fig. 6.4 and therefore also built from two of them, we conclude that only the three isomers of the object of Fig. 6.1 are necessary to construct the general RS of an asymmetric second rank property.[12]

In all the following figures, the ovaloid and the torus have been shown separated by an arbitrary distance for clarity. For the determination of the point group of the property, they have to be considered as having the same origin.

None of the structures in Table 9.1 are like (9.15), because the latter does not match the standard orientation of the conventional axes for any of the crystal systems.

On the other hand, only triclinic and monoclinic classes have truly asymmetric str($\underline{\underline{T}}$):

1. triclinic classes have the str($\underline{\underline{T}}$) of (9.14), the RS of the symmetric part is oriented arbitrarily with respect to its eigenvectors; the same is true for the dual and the RS of the skew symmetric part:

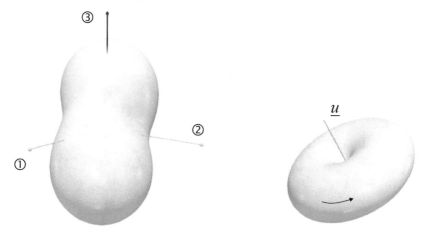

2. for monoclinic classes, the dual $\underline{u}$ points in the direction of the conventional ② axis,[13] which at the same time is the only axis that coincides with an eigenvector of the symmetric part:

---

[11] Or only five if absolute size is ignored.
[12] Why is this not surprising?
[13] Why?

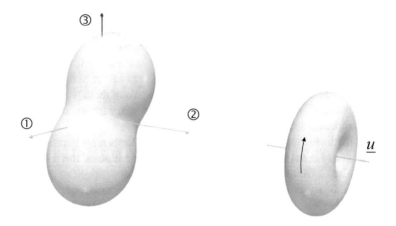

3. for (in second rank) axisymmetric classes 4, $\bar{4}$, $4/m$, 3, $\bar{3}$, 6, $\bar{6}$, $6/m$, ∞, ∞/m, both the infinite order axis of the symmetric part and the dual $\underline{u}$ point in the direction of the conventional ③ axis:

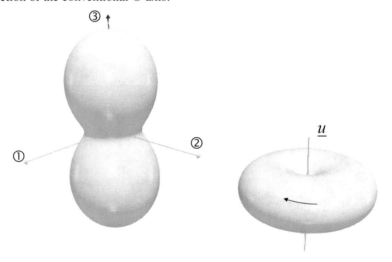

4. for all remaining classes, the property has no skew symmetric contribution. There is only the usual ovaloidal RS for the symmetric part (not represented).

Going now briefly back to Table 9.1, the tetragonal, trigonal, hexagonal, and non-spherical Curie classes split in two sets: one having a purely diagonal str( ), the second one being purely skew symmetric. Having learned what an axial vector is and using its RS as visual aid, can you explain the reason of this splitting and predict which classes belong in each set?

9.3.2 The Hall Effect Coefficient

The same condensation explained above for second rank skew symmetric tensors can also be applied to higher rank tensors which are skew symmetric in a pair of subindices. Perhaps the most familiar one is the Hall effect coefficient $\underset{\equiv}{R}$, which is third rank and skew symmetric in the first two indices.

In the case of $\underset{\equiv}{R}$, its components are reported in the literature as the components of a partially dual condensed second rank tensor, which, somewhat perversely, is often given the same name $\underline{R}$. In spite of having the same symbol, there is no confusion between $\underline{R}$ and $\underset{\equiv}{R}$ because of the different tensorial ranks.

For the Hall effect coefficient, the first two indices of $\underset{\equiv}{R}$ are dual condensed into the first index of $\underline{R}$, while the third index becomes the second index of $\underline{R}$. Thus:

- the first index, m, of the dual condensed R_{mk} results from condensing the first two indices of R_{ijk} (i, j in which $\underset{\equiv}{R}$ is skew symmetric).
- the second index, k, of R_{mk} is the third index of R_{ijk} and remains intact. Thus:

$$R_{mk} \equiv \frac{1}{2}\epsilon_{ijm} R_{ijk} \qquad \frac{1}{2}\epsilon_{mij} R_{ijk} = -\frac{1}{2}\epsilon_{mji} R_{ijk} \Rightarrow \underline{R} = -\frac{1}{2}\underset{\equiv}{\epsilon} : \underset{\equiv}{R} \qquad (9.16)$$

The structure of the property, $\text{str}(\underline{R})$, which in general is asymmetric, is reported in Table 9.1 for the different crystallographic and limit classes. Taking an orthorhombic material as illustrative example, let's explicitly write the components of $\underset{\equiv}{R}$ in terms of the components of $\underline{R}$.

From Table 9.1 the structure of $\underline{R}$ for all orthorhombic materials is:

$$\text{str}(\underline{R}) = \begin{bmatrix} \bullet & \cdot & \cdot \\ \cdot & \bullet & \cdot \\ \cdot & \cdot & \bullet \end{bmatrix}$$

which only has three independent, nonzero components: R_{11}, R_{22}, R_{33}. Undoing the condensation of the first index of R_{11}, R_{22}, R_{33} results in:

$$\text{for } k = 1 \quad \begin{aligned} R_{11} &\neq 0 \Rightarrow R_{231} = R_{11} = -R_{321} \\ R_{21} &= 0 \Rightarrow -R_{131} = 0 = R_{311} \\ R_{31} &= 0 \Rightarrow R_{121} = 0 = -R_{211} \end{aligned}$$

$$\text{for } k = 2 \quad \begin{aligned} R_{12} &= 0 \Rightarrow R_{232} = 0 = -R_{322} \\ R_{22} &\neq 0 \Rightarrow R_{132} = -R_{22} = -R_{312} \\ R_{32} &= 0 \Rightarrow R_{122} = 0 = -R_{212} \end{aligned}$$

$$\text{for } k = 3 \quad \begin{aligned} R_{13} &= 0 \Rightarrow R_{233} = 0 = -R_{323} \\ R_{23} &= 0 \Rightarrow R_{133} = 0 = -R_{313} \\ R_{33} &\neq 0 \Rightarrow R_{123} = R_{33} = -R_{213} \end{aligned}$$

In addition, due to the skew symmetry of $\underline{\underline{R}}$ in its first two indices: $R_{iik} = 0, \forall k$ (no sum). The 27 components of $\underline{\underline{R}}$ can be presented as three 3×3 tables, one for each value $k = 1, 2, 3$:

$$
\underbrace{\begin{array}{ccc} 0 & 0 & 0 \\ 0 & 0 & R_{11} \\ 0 & -R_{11} & 0 \end{array}}_{k=1} \qquad
\underbrace{\begin{array}{ccc} 0 & 0 & -R_{22} \\ 0 & 0 & 0 \\ R_{22} & 0 & 0 \end{array}}_{k=2} \qquad
\underbrace{\begin{array}{ccc} 0 & R_{33} & 0 \\ -R_{33} & 0 & 0 \\ 0 & 0 & 0 \end{array}}_{k=3}
$$

Only six of the 27 components of the full $\underline{\underline{R}}$ are non-zero, and only three of them are independent.

The constitutive equation for the Hall effect describes one of the several effects a magnetic field has on electric conductivity. An electric current density $\underline{J}$ circulating through a material in the presence of a magnetic field $\underline{H}$ gives rise to an electric field given by:

$$E = \underline{\underline{R}} : \underline{H}\,\underline{J} \tag{9.17}$$

where the Hall effect coefficient $\underline{\underline{R}}$, the field $\underline{H}$ and the current density $\underline{J}$ are all local values.

The contraction of $\underline{\underline{R}}$ with the rank one $\underline{H}$ does not affect the first two indices of $\underline{\underline{R}}$ and so produces $\underline{\underline{R}} \cdot \underline{H} = \delta_i \delta_j R_{ijk} H_k$ which is second rank and skew symmetric. For this reason, the electric field $\underline{E} = (\underline{\underline{R}} \cdot \underline{H}) \cdot \underline{J}$, is orthogonal to $\underline{J}$. We can also write the Hall effect CE as:

$$\underline{E} = \underline{\underline{R}} : \underline{H}\,\underline{J} = (\underline{\underline{R}} \cdot \underline{H}) \cdot \underline{J} = \rho^{\text{Hall}} \cdot \underline{J} \quad \text{where} \quad \rho^{\text{Hall}} \equiv \underline{\underline{R}} \cdot \underline{H}$$

is a skew symmetric resistivity. Unlike the ohmic resistivity, the Hall resistivity deviates effect ($\underline{E}$) from cause ($\underline{J}$) by exactly 90°. Hence, dissipation (per unit volume) due to the Hall effect:

$$\underline{E} \cdot \underline{J} = \underline{\underline{R}} : \underline{H}\,\underline{J} \cdot \underline{J} = \underbrace{[(\underline{\underline{R}} \cdot \underline{H}) \cdot \underline{J}]}_{\perp \text{ to } \underline{J}} \cdot \underline{J} = 0$$

vanishes. Energy dissipation is due to physical mechanisms (Ohm's law, piezoresistivity, magnetoresistivity, etc.) that contribute to the symmetric part of the resistivity.

We can build a RS to visualize Hall's effect by noting that $\underline{\underline{R}} \cdot \underline{n}$ is second rank skew symmetric, and thus the analog of $\underline{A}$ in (9.11). Its RS:

$$r = (\underline{\underline{R}} \cdot \underline{n})^T \cdot (\underline{\underline{R}} \cdot \underline{n}) : \underline{n}\,\underline{n} = R_{mik} R_{mnp} n_i n_k n_n n_p$$

which can be rewritten as:

$$r = \left({}^T\!(\underline{\underline{R}}^T) \cdot \underline{\underline{R}} \right) \vdots \underline{n}\,\underline{n}\,\underline{n}\,\underline{n}$$

i.e. as the RS of the fourth rank tensor ${}^T\!(\underline{\underline{R}}^T) \cdot \underline{\underline{R}}$, which has neither minor nor major symmetries.

As in the case of a second rank skew symmetric tensor, the radius vector to the RS represents the squared modulus of $(\underline{\underline{R}} \cdot \underline{n}) \cdot \underline{n} = \underline{\underline{R}} : \underline{n}\,\underline{n}$, i.e. the squared modulus of the electric field $\underline{E}$ that appears due to Hall's effect when a unit electric current density flows through a material in the direction of $\underline{n}$ at a point where a unit magnetic field also points in direction $\underline{n}$.

Similarly to the construction of RSs for transversal elastic properties (where shear moduli and Poisson coefficients appear) as suitable projections of $\underline{\underline{s}}$, it is also possible to define transversal RSs for the Hall effect by allowing the current density and the field to point in different directions. The RS defined by:

$$r = \left({}^T\!(\underline{\underline{R}}^T) \cdot \underline{\underline{R}} \right) \vdots \underline{n}^B\,\underline{n}^A\,\underline{n}^B\,\underline{n}^A$$

corresponds to the Hall effect when the magnetic field points in the direction of $\underline{n}^B$ and the current density along $\underline{n}^A$. Obviously, the same caveat holds regarding orientation sampling for $\underline{n}^A$ and $\underline{n}^B$ as in the elastic case.

Index

© Springer Nature Switzerland AG 2020
M. Laso and N. Jimeno, *Representation Surfaces for Physical Properties*
of Materials, Engineering Materials, https://doi.org/10.1007/978-3-030-40870-1

Printed in the United States
by Baker & Taylor Publisher Services